Birkhäuser

Operator Theory: Advances and Applications

Volume 215

Founded in 1979 by Israel Gohberg

David V. Cruz-Uribe
José Maria Martell
Carlos Pérez

Weights, Extrapolation and the Theory of Rubio de Francia

Birkhäuser

Authors
David V. Cruz-Uribe
Department of Mathematics
Trinity College
300 Summit Street
Hartford, CT 06106-3100, USA
david.cruzuribe@trincoll.edu

José María Martell
Instituto de Ciencias Matemáticas
Consejo Superior de Investigaciones Científicas
c/Serrano 121
28006 Madrid, Spain
chema.martell@icmat.es

Carlos Pérez
Departamento de Análisis Matemático
Facultad de Matemáticas
Universidad de Sevilla
41080 Sevilla, Spain
carlosperez@us.es

2010 Mathematics Subject Classification: 42, 35, 46, 47

ISBN 978-3-0348-0071-6 e-ISBN 978-3-0348-0072-3
DOI 10.1007/978-3-0348-0072-3

Library of Congress Control Number: 2011926300

Cover design: deblik, Berlin

Printed on acid-free paper

Springer Basel AG is part of Springer Science+Business Media

www.birkhauser-science.com

Contents

Preface

This monograph has been a labor of many years, and it reflects many of the themes that have dominated our joint research: weighted norm inequalities, Rubio de Francia extrapolation and its generalizations, the Calderón-Zygmund decomposition, etc. Though there does not exist a single starting point for our work, there was one seminal event. In 1998, the first author, with the generous support of the Ford Foundation and the Dean of Trinity College, was able to spend a month in Madrid. The last day of his visit found him and the third author double-parked at Barajas airport while he explained an idea he had had the night before. Twelve hours later, when he arrived home and checked his email he found three long messages waiting. These led to a new way of thinking about weak type inequalities and contained the germ of the extrapolation results proved in Chapter 8.

Over the years we have benefited from conversations with and ideas from many other mathematicians. We particularly want to thank: Jose García-Cuerva, advisor of the second author; Javier Duoandikoetxea, for his many insights on extrapolation and especially for sharing with us his new proof of extrapolation with sharp constants; Janine Wittwer and Michael Wilson, whose work on the dyadic square function informed our own; C.J. Neugebauer, for his groundbreaking work on two-weight extrapolation; and Javier Parcet, for his work in collaboration with the second author on the generalized Calderón-Zygmund decomposition contained in the Appendix. And finally, we must acknowledge our debt to the late José Luis Rubio de Francia, who created the theory of extrapolation.

We gratefully acknowledge the financial support which has made writing this monograph possible. As it is being completed, the first author is supported by the Stewart-Dorwart Faculty Development Fund at Trinity College; the first and third authors are supported by grant MTM2009-08934 from the Spanish Ministry of Science and Innovation; the second author is supported by grant MTM2007-60952 from the Spanish Ministry of Science and Innovation and by CSIC PIE 200850I015.

Personal Dedications

I want to thank my children, Nicolás, Antonio and Francisco, for accepting my extended absences while this book was written. I also want to thank Donald Sarason and Christoph Neugebauer for the faith they showed in me as a mathematician.
DCU

I want to thank my wife Isabel, and my sons Chema and Javier for filling my life with happiness.
JMM

I would like to thank my family, Cristina, Sergio and Adriana for their support.
CP

Preliminaries

We assume that the reader knows real analysis as presented in standard works such as Rudin [197] or Royden [192]. The reader should also be familiar with the basics of harmonic analysis as contained in the first few chapters of Duoandikoetxea [68] or Grafakos [91]. Here we review some notation and basic results that will be used throughout this monograph.

We will work primarily in $\mathbb{R}^n$. The norm in $\mathbb{R}^n$ will be denoted by $|\cdot|$ and Lebesgue measure by dx. Given a measurable set E, $|E|$ will also denote the Lebesgue measure of E, and χ_E will denote the characteristic function of the set E. Given a cube Q, $\ell(Q)$ will denote the side-length of Q, so that $\ell(Q)^n = |Q|$. Given Q and $\lambda > 0$, λQ will denote the cube with the same center as Q and such that $\ell(\lambda Q) = \lambda \ell(Q)$.

By a weight we will mean a non-negative function u that is positive on a set of positive measure. If a condition is given on a weight involving an integral, we will implicitly assume that the integral is finite. Given a weight u and a measurable set E, let

$$u(E) = \int_E u(x)\,dx.$$

A weight u is called a doubling measure if there exists a constant $C > 0$ such that for all cubes Q, $u(2Q) \leq Cu(Q)$. If $|E| > 0$, define

$$\fint_E u(x)\,dx = \frac{1}{|E|}\int_E u(x)\,dx;$$

if $|E| = 0$, set it equal to 0.

The collection of smooth functions of compact support will be denoted by C_c^∞.

For $1 \leq p < \infty$, L^p will denote the Banach function space with norm

$$\|f\|_{L^p} = \left(\int_{\mathbb{R}^n} |f(x)|^p\,dx\right)^{1/p}.$$

Given a measurable set E we define the localized L^p norm on E by

$$\|f\|_{p,E} = \left(\fint_E |f(x)|^p\,dx\right)^{1/p}.$$

For $1 < p < \infty$, p' denotes the conjugate exponent of p:

$$\frac{1}{p} + \frac{1}{p'} = 1.$$

The space $L^{p'}$ is the dual space of L^p. Given a locally integrable weight u, $L^p(u)$ will denote the Banach function space with norm

$$\|f\|_{L^p(u)} = \left(\int_{\mathbb{R}^n} |f(x)|^p u(x)\,dx\right)^{1/p}.$$

The dual space of $L^p(u)$ is $L^{p'}(u)$. Given a vector-valued function $f = \{f_i\}$ and q, $1 \le q < \infty$, let

$$\|f\|_{\ell^q} = \left(\sum_{i=1}^{\infty} |f_i|^q\right)^{1/q}.$$

We say that $f = \{f_i\} \in L^p(u)$ if $\|f(\cdot)\|_{\ell^q} \in L^p(u)$.

Given p, q, $1 \le p, q < \infty$, and a pair of weights (u, v), an operator T is of strong type (p, q) if there exists a constant C such that for all $f \in L^p(v)$,

$$\|Tf\|_{L^q(u)} \le C\|f\|_{L^p(v)}.$$

We also denote this by $T : L^p(u) \to L^q(u)$. If $u = v$ and $p = q$, then we say that T is bounded on $L^p(u)$, and we denote the infimum of the constant C by $\|T\|_{L^p(u)}$.

Given a locally integrable weight u, $L^{p,\infty}(u)$, $1 \le p < \infty$, will denote the Lorentz space with quasi-norm

$$\|f\|_{L^{p,\infty}(u)} = \sup_{\lambda>0} \lambda u(\{x \in \mathbb{R}^n : |f(x)| > \lambda\})^{1/p}.$$

An operator T is of weak type (p, q) if there exists a constant C such that, for all $f \in L^p(v)$,

$$\|Tf\|_{L^{q,\infty}(u)} \le C\|f\|_{L^p(v)},$$

or equivalently, for all $\lambda > 0$,

$$u(\{x \in \mathbb{R}^n : |Tf(x)| > \lambda\}) \le C\left(\frac{1}{\lambda^p}\int_{\mathbb{R}^n} |f(x)|^p v(x)\,dx\right)^{q/p}.$$

We denote this by $T : L^p(v) \to L^{q,\infty}(u)$.

Given a locally integrable function f, the Hardy-Littlewood maximal function, Mf, is defined by

$$Mf(x) = \sup_{Q\ni x} \fint_Q |f(y)|\,dy,$$

where the supremum is taken over all cubes Q in $\mathbb{R}^n$ with sides parallel to the coordinate axes. An operator that is pointwise equivalent is gotten if the supremum is taken over all cubes in $\mathbb{R}^n$, cubes centered at x, or balls. The maximal operator M is a bounded operator on L^p, $1 < p < \infty$, and satisfies the weak $(1,1)$ inequality $M : L^1 \to L^{1,\infty}$.

For each $j \in \mathbb{Z}$, define the set

$$\mathcal{D}_j = \{[0, 2^{-j})^n + k : k \in \mathbb{Z}^n\};$$

the set of dyadic cubes $\mathcal{D}$ is the union $\cup_j \mathcal{D}_j$.

Finally, throughout this monograph, C, c, etc. will denote positive constants whose values may change even in a chain of inequalities. Specific values that the constants depend on will be noted as necessary.

Part I

One-Weight Extrapolation

Chapter 1

Introduction to Norm Inequalities and Extrapolation

The extrapolation theorem of Rubio de Francia is one of the deepest results in the study of weighted norm inequalities in harmonic analysis: it is simple to state but has profound and diverse applications. The goal of this book is to give a systematic development of the theory of extrapolation, one which unifies known results and expands them in new directions. In addition, we want to show how extrapolation theory, broadly defined, can be applied to the theory of weighted norm inequalities. We describe new and simpler proofs of known results, and then prove new results and show how these lead to additional open questions.

The primary audience for our work is researchers and graduate students who are working on weighted norm inequalities and related topics. However, we believe that many of our results will be useful to mathematicians working in other areas of harmonic analysis and partial differential equations. While the more technical results and proofs will require specialized knowledge to be fully understood, we have striven to make the broad outline of the theory and the statement of our main results accessible to a broader audience. The minimum we have assumed and the basic notation we use is given in the Preliminaries at the beginning of the book.

In this chapter, to put our results in context, we first review the history of the theory of weighted norm inequalities. Unfortunately, no recent survey of the field exists, but beyond the specific articles we cite below, we refer the reader to the books by García-Cuerva and Rubio de Francia [88], Duoandikoetxea [68] and Grafakos [92], and the early survey articles by Muckenhoupt [150] and Dynkin and Osilenker [72]. We then describe the theory of extrapolation and summarize the subsequent chapters. A more detailed overview of the contents of Part I is given in the second half of Chapter 2.

1.1 Weighted norm inequalities

By a weight we mean a non-negative, locally integrable function that is positive on a set of positive measure. The integrability condition can be relaxed, but for simplicity we consider here this important special case. The basic problems in the study of weighted norm inequalities are to prove estimates of the form

$$\int_{\mathbb{R}^n} |Tf(x)|^p u(x)\,dx \leq C \int_{\mathbb{R}^n} |f(x)|^p v(x)\,dx$$

or

$$u(\{x \in \mathbb{R}^n : |Tf(x)| > \lambda\}) \leq \frac{C}{\lambda^p} \int_{\mathbb{R}^n} |f(x)|^p v(x)\,dx,$$

where $1 \leq p < \infty$, and T is an operator, usually one of the classical operators of harmonic analysis: i.e., a maximal function, singular integral, square function, etc. These problems divides naturally into two classes: when we have a single weight function w (i.e., $u = v = w$) and when we have a pair of weights (u, v). These are referred to as one-weight and two-weight inequalities.

The theory of one-weight inequalities began with the study of power weights of the form $u(x) = |x|^a$. See for example, Stein [213, 214] (see also Soria and Weiss [212]). Shortly thereafter came the celebrated Helson-Szegö theorem [101], which characterized one-weight inequalities for the conjugate function (i.e., the periodic Hilbert transform on the unit circle) using complex analysis.

A period of sustained research in this area began in the 1970s with the work of Muckenhoupt and others. In [148] Muckenhoupt introduced the A_p weights (now often referred to as Muckenhoupt weights): for $1 < p < \infty$, $w \in A_p$ if there exists a constant K such that for every cube $Q \subset \mathbb{R}^n$,

$$\fint_Q w(x)\,dx \left(\fint_Q w(x)^{1-p'}\,dx\right)^{p-1} \leq K < \infty.$$

(A variant of this condition was introduced earlier by Rosenblum [191].) The weight w is in A_1 if there exists a constant K such that for almost every $x \in \mathbb{R}^n$, $Mw(x) \leq Kw(x)$, where M is the Hardy-Littlewood maximal operator. In each case the infimum of all such K is denoted by $[w]_{A_p}$. The prototypical A_p weights are the power weights: for all $a \in \mathbb{R}$, $|x|^a \in A_1$ if and only if $-n < a \leq 0$, and for $p > 1$, $|x|^a \in A_p$ if and only if $-n < a < (p-1)n$.

The centrality of the A_p condition is shown by the following result.

Theorem 1.1. *Given p, $1 \leq p < \infty$, and $w \in A_p$, then*

$$w(\{x \in \mathbb{R}^n : |Tf(x)| > \lambda\}) \leq \frac{C}{\lambda^p} \int_{\mathbb{R}^n} |f(x)|^p w(x)\,dx,$$

where T is the Hardy-Littlewood maximal operator, the Hilbert transform, or a Riesz transform. If $p > 1$, then the corresponding strong type inequality holds:

$$\int_{\mathbb{R}^n} |Tf(x)|^p w(x)\,dx \leq C \int_{\mathbb{R}^n} |f(x)|^p w(x)\,dx.$$

Furthermore, the A_p condition is necessary: if the strong or weak (p,p) inequality holds for a weight w and one of these operators, then $w \in A_p$.

The sufficiency of the A_p weights for the Hardy-Littlewood maximal operator to be bounded on $L^p(w)$ was proved by Muckenhoupt [148]; for the Hilbert transform by Hunt, Muckenhoupt and Wheeden [104]; and for Riesz transforms (and indeed for any singular integral with a sufficiently smooth kernel) by Coifman and Fefferman [25]. Their proof involved proving an intermediate inequality: for $0 < p < \infty$,

$$\int_{\mathbb{R}^n} |Tf(x)|^p w(x)\,dx \le C \int_{\mathbb{R}^n} Mf(x)^p w(x)\,dx, \tag{1.1}$$

where w satisfies the so-called A_∞ condition. (This is defined in Theorem 1.3 below.) The necessity of the A_p condition for the maximal operator is also due to Muckenhoupt, and a similar argument works for the Hilbert transform [104]. This argument can be extended to show that if all the Riesz transforms are bounded on $L^p(w)$, then $w \in A_p$. (See [88].) The fact that if a single Riesz transform is bounded, then $w \in A_p$ is due to Stein [216].

Similar results were soon proved for a variety of other operators, and there now exists an extensive literature on one-weight norm inequalities. For a partial list, we refer the reader to [68, 72, 88, 92] and the references they contain. A common approach has been to prove inequalities that are similar to (1.1), and we will collectively refer to these as Coifman-Fefferman inequalities.

One important variation that emerged was the class of "dyadic" A_p weights, A_p^d. This condition is defined as the general A_p conditions but with the cubes restricted to dyadic cubes. This class is the appropriate one to consider for a variety of dyadic operators. For more on this subject, we refer the reader to the lecture notes by Pereyra [169] and the references they contain.

Central to the original proofs of Theorem 1.1 is the rich structure of the A_p weights. Several properties are immediate consequences of the definition.

Proposition 1.2. *The Muckenhoupt weights have the following properties:*

(a) *for $1 < p < \infty$, $w \in A_p$ if and only if $w^{1-p'} \in A_{p'}$;*

(b) *if $1 \le p < q < \infty$, then $A_p \subset A_q$;*

(c) *given $w_1, w_2 \in A_1$, for $1 < p < \infty$, $w_1 w_2^{1-p} \in A_p$.*

Property (a) follows at once from the definition; property (b) from Hölder's inequality; and property (c) from the fact that if $w \in A_1$, then for almost every $x \in Q$,

$$\fint_Q w(y)\,dy \le Mw(x) \le [w]_{A_1} w(x).$$

Beyond the elementary results of Proposition 1.2, Muckenhoupt weights have many deeper properties. We begin with a definition: define the class of weights A_∞

by

$$A_\infty = \bigcup_{p\geq 1} A_p.$$

Theorem 1.3. *The Muckenhoupt weights can be characterized by the following properties:*

(a) $w \in A_\infty$ *if and only if there exist constants* C, $\delta > 0$ *such that given any cube* Q *and any measurable set* $E \subset Q$,

$$\frac{w(E)}{w(Q)} \leq C\left(\frac{|E|}{|Q|}\right)^\delta;$$

(b) $w \in A_\infty$ *if and only if for some* $s > 1$, $w \in RH_s$: *there exists a constant* K *such that for every cube* Q,

$$\left(\fint_Q w(x)^s\,dx\right)^{1/s} \leq K\fint_Q w(x)\,dx;$$

(c) *If* $w \in A_p$, $p > 1$, *there exists* ϵ, $0 < \epsilon < p-1$, *such that* $w \in A_{p-\epsilon}$;

(d) *If* $w \in A_p$, $p > 1$, *there exist* w_1, $w_2 \in A_1$ *such that* $w = w_1 w_2^{1-p}$.

The A_∞ condition was discovered independently by Coifman and Fefferman [25] and Muckenhoupt [149]. The RH_s condition is referred to as the reverse Hölder inequality and was also proved by Coifman and Fefferman [25]. (The RH_s classes were considered independently by Gehring [89].) The A_p condition itself is a kind of reverse Hölder inequality, since the opposite inequality,

$$1 \leq \fint_Q w(x)\,dx \left(\fint_Q w(x)^{1-p'}\,dx\right)^{p-1},$$

is a consequence of Hölder's inequality. Property (c) follows immediately from the reverse Hölder inequality for the weight $w^{1-p'}$; it was first proved directly by Muckenhoupt [148]. Property (d) is referred to as the Jones factorization theorem; it was first conjectured by Muckenhoupt at the Williamstown conference in 1979 (see [150]) and proved by P. Jones at the same conference [111]. A much simpler proof was later given by Coifman, Jones and Rubio de Francia [26]. (We will say more about this proof below.) The expression "the factorization theorem" usually refers to both property (d) and its much simpler converse, property (c) in Proposition 1.2 above, but we will reserve this name for property (d). No standard terminology exists for property (c) in Proposition 1.2, but we will refer to it as "reverse factorization."

In the 1970s the rapid progress in the study of one-weight norm inequalities initially fed hopes that the corresponding problems for two-weight inequalities would soon be solved as well. The immediate candidate for a condition on a pair of weights (u, v) was the two-weight A_p condition: for $p > 1$, $(u, v) \in A_p$ if

$$\fint_Q u(x)\,dx \left(\fint_Q v(x)^{1-p'}\,dx\right)^{p-1} \le K < \infty,$$

and $(u, v) \in A_1$ if $Mu(x) \le Kv(x)$. (In particular, given any weight u, $(u, Mu) \in A_1$.) Muckenhoupt [148] noted that the same proof as in the one-weight case immediately shows that for all p, $1 \le p < \infty$, $(u, v) \in A_p$ if and only if the maximal operator satisfies the weak (p, p) inequality. However, it was soon discovered that while the two-weight A_p condition is necessary for the strong (p, p) inequality for the maximal operator and the strong and weak type inequalities for the Hilbert transform, it is not sufficient. (See Muckenhoupt and Wheeden [155].)

This led Muckenhoupt and Wheeden [147] to focus not on the structural or geometric properties of A_p weights but on their relationship to the maximal operator, in particular, the fact that $w \in A_p$ was necessary and sufficient for the maximal operator to be bounded on $L^p(w)$ and $L^{p'}(w^{1-p'})$. This led them to make the following conjecture which is still open: given a pair of weights (u, v), a sufficient condition for the Hilbert transform to satisfy the strong (p, p) inequality $H : L^p(v) \to L^p(u)$, $1 < p < \infty$, is that the maximal operator satisfy the pair of inequalities

$$M : L^p(v) \to L^p(u), \tag{1.2}$$

$$M : L^{p'}(u^{1-p'}) \to L^{p'}(v^{1-p'}). \tag{1.3}$$

(Even though M is not a linear operator, inequality (1.3) is referred to as the dual of (1.2).) Additionally, Muckenhoupt and Wheeden conjectured that if the dual inequality (1.3) holds for a pair (u, v), then the weak (p, p) inequality $H : L^p(v) \to L^{p,\infty}(u)$ also holds. For the weak $(1, 1)$ inequality, they conjectured that

$$u(\{x \in \mathbb{R}^n : |Hf(x)| > \lambda\}) \le \frac{C}{\lambda} \int_{\mathbb{R}^n} |f(x)| Mu(x)\,dx.$$

Each of these conjectures can be generalized naturally to other singular integrals. All of them are in the spirit of Calderón and Zygmund, whose philosophy was that to control a singular integral one should control the maximal operator.

The study of two-weight norm inequalities has proved to be considerably more difficult than it is in the one-weight case. As B. Muckenhoupt recently noted [152], fundamental problems in the two-weight case, including the conjectures just discussed, remain open. Progress has been made, but slowly, and seemingly small improvements in results have required the development of sophisticated techniques. In many cases interesting results have been proved but they have remained isolated and could not be developed further. Cotlar and Sadosky [29, 30], working in the spirit of the Helson-Szegö theorem, gave a necessary and sufficient condition on a pair of weights for the conjugate function to satisfy a two-weight strong (p, p)

inequality. Leckband [122] and Fujii [82] found two-weight conditions that generalized the A_p condition by incorporating measure-theoretic properties similar to the A_∞ condition. Rakotondratsimba [186, 187, 188] gave A_p type conditions for weights that are radial and monotone.

Currently there are two major approaches to two-weight norm inequalities, which we will refer to as "testing conditions" and "A_p bump conditions." The latter are central to our understanding of norm inequalities and extrapolation theory, and so determine the point of view we have adopted in this book. However, though testing conditions do not play a direct role in our work, we want to describe them before discussing our own. We do so for two reasons: first, they are very important in the study of weighted norm inequalities and an area of active research today. Second, despite its importance, we believe that this approach has some shortcomings and we want to highlight these to suggest to the reader the advantages of our approach. We do not claim that the A_p bump conditions are "better" in any normative sense: we just want to illustrate the reasons why we prefer one over the other.

Testing conditions were originally introduced by Sawyer [201]. He proved that a necessary and sufficient condition on a pair of weights (u, v) for the strong (p, p) inequality, $1 < p < \infty$,

$$\int_{\mathbb{R}^n} Mf(x)^p u(x)\,dx \le C \int_{\mathbb{R}^n} |f(x)|^p v(x)\,dx,$$

is that for every cube Q,

$$\int_Q M(v^{1-p'}\chi_Q)(x)^p u(x)\,dx \le C \int_Q v(x)^{1-p'}\,dx. \tag{1.4}$$

The necessity of this condition is immediate: simply apply the norm inequality to the family of test functions $v^{1-p'}\chi_Q$. We denote the fact that a pair of weights satisfies (1.4) by writing $(u, v) \in S_p$.

Sawyer [207, 208] later extended this approach to linear operators with positive kernels, for instance, the fractional integral operator I_α, $0 < \alpha < n$. Because of linearity, strong (p, p) inequalities with respect to the weights (u, v) are equivalent to strong (p', p') inequalities for the weights $(v^{1-p'}, u^{1-p'})$. This led naturally to two testing conditions: one from the L^p inequality and one from the dual $L^{p'}$ inequality. More precisely, I_α satisfies the (p, p) inequality $I_\alpha : L^p(v) \to L^p(u)$ if and only if the weights (u, v) satisfy

$$\int_Q I_\alpha(v^{1-p'}\chi_Q)(x)^p u(x)\,dx \le C \int_Q v(x)^{1-p'}\,dx, \tag{1.5}$$

$$\int_Q I_\alpha(u\chi_Q)(x)^{p'} v(x)^{1-p'} \le C \int_Q u(x)\,dx. \tag{1.6}$$

Sawyer [204] also proved that the dual testing condition (1.6) was necessary and sufficient for the fractional integral operator to satisfy the weak (p, p) inequality $I_\alpha : L^p(v) \to L^{p,\infty}(u)$.

These results for fractional integrals led to the following conjectures: if T is a singular integral operator (e.g., the Hilbert transform), then $T : L^p(v) \to L^p(u)$ if and only if

$$\int_Q |T(v^{1-p'}\chi_Q)(x)|^p u(x)\,dx \le C \int_Q v(x)^{1-p'}\,dx, \tag{1.7}$$

$$\int_Q |T(u\chi_Q)(x)|^{p'} v(x)^{1-p'} \le C \int_Q u(x)\,dx. \tag{1.8}$$

Testing conditions such as these are referred to generically as Sawyer-type conditions.

After the original work of Sawyer, no progress was made on these conjectures until the groundbreaking work of Nazarov, Treil and Volberg. They realized that there is a close connection between Sawyer type conditions and the testing conditions that are part of the $T1$ theorem of David and Journé [58] (also see [92]). As part of their work on the Vitushkin conjecture, they developed a theory of singular integrals on non-homogeneous spaces (e.g., $\mathbb{R}^n$ with a non-doubling measure) including a Tb theorem. (See [157, 159, 160].) Building on these ideas they have been able to prove L^2 Sawyer-type conditions for several operators. In [158] they proved that Sawyer-type conditions were necessary and sufficient for families of Haar multipliers H_a to satisfy $H_a : L^2(v) \to L^2(u)$ with uniform bounds. (Haar multipliers are dyadic singular integral operators that are "localized" and so easier to deal with. They provide a good model for singular integrals, and more general dyadic operators can be used to approximate Hilbert and Riesz transforms—see [105, 180, 181, 182, 183, 184].) They also proved that the single testing condition (1.7) was necessary and sufficient for the dyadic square function to satisfy $S_d : L^2(v) \to L^2(u)$. In [227] they proved that (1.7), (1.8) and a stronger version of the two-weight A_2 condition—the so-called invariant A_2 condition—are necessary and sufficient for the Hilbert transform to satisfy $H : L^2(v) \to L^2(u)$, provided that u and v satisfy doubling conditions. In [161] they proved L^2 Sawyer-type conditions for individual Haar multipliers and other dyadic operators (without doubling conditions).

Despite the elegance of these results, we believe that they have some drawbacks. First, it is not clear if they are the correct conditions when $p \neq 2$. In [158], Nazarov, Treil and Volberg noted (without proof) that their results for families of Haar multipliers are not true when $p \neq 2$. In addition, for operators other than singular integrals we have reason to believe that Sawyer-type conditions may not be the correct ones to consider. As we discuss in detail Chapter 10 below, in the two weight case the dyadic square function behaves very differently for $p \le 2$ and for $p > 2$; this in turn suggests that the Sawyer-type condition (which is in some sense the same for all p) may not be sufficient when $p > 2$.

Second, we believe that Sawyer-type conditions are of limited utility in practice. Given a pair of weights (u, v), it seems almost as difficult to check whether the Sawyer-type conditions hold for a specific operator as it does to prove a strong type norm inequality. Conversely, it seems equally difficult to construct examples of weights that satisfy them. Moreover, unlike the A_p weights, the Sawyer-type conditions and the weights that satisfy them are bound to individual operators: if the operator is changed, the work of finding or checking pairs of weights must be started over.

Our approach to two-weight norm inequalities is quite different: our goal has been to find two-weight, A_p-type conditions that are sufficient for large classes of operators. Our work has close connections with the deep conjecture of Muckenhoupt and Wheeden discussed above, which we will explain below. Its proximate origins are in our work to generalize an often overlooked paper by Neugebauer [163]. To best understand his result, we first restate the A_p condition, $1 < p < \infty$, in terms of localized L^p norms: $(u, v) \in A_p$ if for every cube Q,

$$\|u^{1/p}\|_{p,Q}\|v^{-1/p}\|_{p',Q} \leq K < \infty.$$

Neugebauer showed that given a pair of weights (u, v), there exist $w \in A_p$ and positive constants c_1, c_2 such that $c_1u(x) \leq w(x) \leq c_2v(x)$ if and only if there exists $r > 1$ such that for every cube Q,

$$\|u^{1/p}\|_{rp,Q}\|v^{-1/p}\|_{rp',Q} \leq K < \infty. \tag{1.9}$$

From this condition we immediately get a large number of two-weight norm inequalities as corollaries to the analogous one-weight results. In particular, we have that the two inequalities (1.2) and (1.3) hold for the maximal operator. We refer to (1.9) as an A_p bump condition.

An immediate question was whether this condition could be weakened and still get that the maximal operator satisfies $M : L^p(v) \to L^p(u)$. This was answered in [174], where it was shown that a sufficient condition was that the pair of weights satisfies

$$\|u^{1/p}\|_{p,Q}\|v^{-1/p}\|_{B,Q} \leq K < \infty,$$

where the norm on the right-hand term is a normalized Orlicz space norm and B is a Young function that satisfies an easily checked growth condition. We defer the statement of this condition to Chapter 5, Theorem 5.14, as it requires some additional definitions; intuitively, it says that $B(t)$ is "infinitesimally" larger than $t^{p'}$. For example, we could take $B(t) = t^{rp'}$, $r > 1$; in this, case we see that we can eliminate one "bump" from Neugebauer's condition. But we can also take $B(t) = t^{p'} \log(e+t)^{p'-1+\delta}$, where $\delta > 0$. The centrality of this growth condition is shown by the fact that it is necessary for the maximal operator to be bounded—see Remark 5.15 below.

As an immediate consequence of this result, we have that the maximal operator satisfies inequalities (1.2) and (1.3) provided that the pair of weights (u, v) satisfy the A_p bump condition

$$\|u^{1/p}\|_{A,Q}\|v^{-1/p}\|_{B,Q} \leq K < \infty, \tag{1.10}$$

where the Young functions A, B satisfy the appropriate growth conditions. This led naturally to the following version of the conjecture of Muckenhoupt and Wheeden: a sufficient condition on the pair of weights (u, v) for any singular integral to satisfy $T : L^p(v) \to L^p(u)$ is that (1.10) holds. Moreover, our version of their conjecture for weak (p, p) inequalities is that $T : L^p(v) \to L^{p,\infty}(u)$ if the pair satisfies

$$\|u^{1/p}\|_{A,Q}\|v^{-1/p}\|_{p',Q} \leq K < \infty.$$

In a series of papers over the past fifteen years [33, 47, 53, 55, 56, 171] we have made considerable progress on these conjectures, and in the final two chapters of this book we expand upon our earlier work.

In contrast to the testing conditions discussed above, it is usually straightforward to determine if a pair of weights satisfies (1.10), though we must admit that it can be computationally tedious depending on the Young functions A and B. Moreover, it is very easy to construct examples of pairs (u, v) that satisfy (1.10)—see (1.12) below.

1.2 The theory of extrapolation

With the theory of weighted norm inequalities as a foundation, we can now discuss the extrapolation theorem of Rubio de Francia. Here we state the essential version of the theorem, though we defer the proof to Chapter 2 below.

Theorem 1.4. *Given an operator T, suppose that for some p_0, $1 \leq p_0 < \infty$, and every $w \in A_{p_0}$, there exists a constant C depending on $[w]_{A_{p_0}}$ such that*

$$\int_{\mathbb{R}^n} |Tf(x)|^{p_0} w(x)\, dx \leq C \int_{\mathbb{R}^n} |f(x)|^{p_0} w(x)\, dx.$$

Then for every p, $1 < p < \infty$, and every $w \in A_p$ there exists a constant depending on $[w]_{A_p}$ such that

$$\int_{\mathbb{R}^n} |Tf(x)|^{p} w(x)\, dx \leq C \int_{\mathbb{R}^n} |f(x)|^{p} w(x)\, dx.$$

The extrapolation theorem was an unexpected and very surprising result. It was discovered by Rubio de Francia (see [193, 194, 195]) whose background in functional analysis gave him a very different perspective on the theory of weighted norm inequalities. (For this background, see the survey articles by Torrea *et al.* [223].) The philosophy underlying this result was pithily summarized by Rubio de Francia's colleague Antonio Cordoba [84]:

There are no L^p spaces, only weighted L^2.

Beyond the original work of Rubio de Francia, there are a number of proofs of Theorem 1.4 and we will discuss these in more detail in Chapter 2. A key feature of many of these proofs is the iteration algorithm of Rubio de Francia: given a positive, sublinear operator T that is bounded on $L^p(w)$, define a new operator $\mathcal{R}$ by

$$\mathcal{R}h = \sum_{k=0}^{\infty} \frac{T^k h}{2^k \|T\|_{L^p(w)}^k}.$$

(This was first referred to as the Rubio de Francia algorithm in [13].) The crucial, but deceptively simple property of the iteration algorithm is that it is almost "invariant" under the operator T: more precisely,

$$T(\mathcal{R}h) \le 2\|T\|_{L^p(w)} \mathcal{R}h.$$

The iteration algorithm is central to our own work and we discuss it in more detail in subsequent chapters. Here, we want to note that it is also used in a central way in the simplest proofs of the Jones factorization theorem (Theorem 1.3 above; see [88, 92]), and this shows that there is a very deep connection between extrapolation and factorization.

Given the ongoing work of creating a theory of two-weight norm inequalities parallel to the one-weight theory, it is not surprising that a number of authors considered two-weight extrapolation and factorization theorems: see Neugebauer [163, 164], Bloom [13], Hernández [102], Ruiz and Torrea [198], and Segovia and Torrea [209, 210]. In every case these authors worked with pairs of weights (u, v) such that the maximal operator satisfied inequality (1.2) and the dual inequality (1.3); in other words, there is a close connection between their results and the Muckenhoupt-Wheeden conjecture for singular integrals discussed in the previous section. This points to one drawback of these results: given the current state of knowledge they cannot be applied, since it is not possible to prove the "base case" (e.g., weighted L^2 inequalities) needed to use extrapolation. A different approach to two-weight extrapolation using pairs $(u, v) \in A_1$ was developed in [54] and was implicit in [53].

The importance of the Rubio de Francia extrapolation theorem lies not only in its intrinsic beauty, but also in its powerful applications. There have been many; here we describe four in detail. Additional applications that arise from our generalizations of one-weight extrapolation are given in Chapters 3 and 4. The first important application was to rough singular integrals. Let S^{n-1} denote the unit sphere in $\mathbb{R}^n$, let $\Omega \in L^\infty(S^{n-1})$ be such that $\int_{S^{n-1}} \Omega(x)\,dx = 0$ and let $h \in L^\infty(\mathbb{R}_+)$. We consider the singular integral T with kernel

$$K(x) = h(|x|)\frac{\Omega(x/|x|)}{|x|^n}.$$

T is bounded on L^p, $1 < p < \infty$; when $h \equiv 1$ this follows from the method of rotations (see [68]); for general h this was proved by R. Fefferman [77]. Duoandikoetxea and Rubio de Francia [71] proved that T is bounded on $L^p(w)$, $1 < p < \infty$, for

$w \in A_p$. Key to the proof was the extrapolation theorem, since this reduced the problem to proving that T is bounded on $L^2(w)$, $w \in A_2$. For this case they used the Fourier transform, square function estimates and interpolation with change of measure to deduce the weighted inequality from the unweighted L^2 estimate.

Another important application of extrapolation was given by R. Fefferman and Pipher [78]. They were considering singular integral operators $T_{\mathcal{Z}}$ on $\mathbb{R}^3$ that commute with the family of multiparameter dilations $\phi_{s,t}(x,y,z) = (sx, ty, stz)$, $s,\, t > 0$. The closely related maximal operators $M_{\mathcal{Z}}$ are defined as the supremum of averages over rectangles in the Zygmund basis $\mathcal{Z}$ whose side-lengths are of the form (s,t,st). Unweighted estimates for such operators were proved by Ricci and Stein [190]. Fefferman and Pipher proved one-weight estimates on $L^p(w)$, where $w \in A_{p,\mathcal{Z}}$, the A_p class defined with respect to rectangles in $\mathcal{Z}$. Again central to the proof was Rubio de Francia extrapolation (which they noted could be extended to weights in $A_{p,\mathcal{Z}}$): they showed that in $L^2(w)$ the proof reduced to a certain square function estimate, but this approach does not work when $p \neq 2$. They also proved sharp embedding theorems for these operators in Orlicz spaces close to L^1. They showed that these followed from sharp L^p estimates for the Hilbert transform; implicit in their proof of these sharp estimates is a duality argument that is reminiscent of our approach to both one and two-weight extrapolation.

The third application is to elliptic differential equations. The Beltrami equation in the plane is $f_z - \mu f_{\bar{z}} = 0$, where μ is a bounded function such that $\|\mu\|_\infty = k < 1$. Astala, Iwaniec and Saksman [6] showed that solutions of this equation are continuous if $f \in W^{1,q}_{\mathrm{loc}}$ for $q > k+1$ and that there were discontinuous solutions if $q < k+1$. Further, they showed that solutions in $W^{1,1+k}_{loc}$ were continuous if the Beurling-Ahlfors operator (a complex-valued analog of the Hilbert transform) satisfied a certain sharp weighted A_p estimate. This estimate was proved by Petermichl and Volberg [184]; via extrapolation they reduced the problem to weighted L^2 estimates, which they proved using Bellman function techniques.

As a final application we consider our work on a conjecture by Sawyer. In [205] he proved that if $u,\, v \in A_1$, then

$$uv(\{x \in \mathbb{R} : M(fv)(x) > \lambda v(x)\}) \leq \frac{C}{\lambda} \int_{\mathbb{R}} |f(x)|u(x)v(x)\, dx; \tag{1.11}$$

this inequality arose naturally when trying to prove weighted norm inequalities by combining interpolation with change of measure (see Stein and Weiss [217]) and the Jones factorization theorem. Sawyer conjectured that this inequality is true if the maximal operator is replaced by the Hilbert transform. This conjecture was proved in [45] and was shown to hold in higher dimensions for Calderón-Zygmund singular integral operators. The proof consisted of several steps. First, inequality (1.11) was extended to higher dimensions for the dyadic maximal operator. Then, using a version of Rubio de Francia extrapolation adapted to this kind of inequality, it was shown that it holds in $\mathbb{R}^n$ for both the Hardy-Littlewood maximal operator and singular integrals.

1.3 The organization of this book

The starting point for this book is a new and much simpler proof of the Rubio de Francia extrapolation theorem. It does not require cases depending on the size of p, and it only uses very elementary structural properties of weights and the fact that the maximal operator is bounded on $L^p(w)$. In some sense, we are able to make clear "what is really going on" in the proof, and this provides a springboard for a number of generalizations of the extrapolation theorem. We had already begun to do so in earlier work (see, for instance, [40, 44, 46, 57]); here we develop these extensions systematically.

Our material divides naturally into two parts. In Part I we consider the one-weight theory. In Chapter 2 we briefly describe earlier proofs of Theorem 1.4 and then give our own proof. We analyze the proof to highlight the key features—the boundedness of the maximal operator, duality, and reverse factorization. We then describe the numerous generalizations which our approach makes possible. These generalizations are developed and proved in Chapters 3 and 4. In Chapter 3 we focus on weighted L^p results, and in Chapter 4 we show that Rubio de Francia extrapolation can be generalized to prove norm inequalities for operators in large families of Banach function spaces. These generalizations may be succinctly captured by expanding upon Cordoba's remark given above:

There are no Banach function spaces, only weighted L^2.

At the end of both of these chapters we sketch a number of applications of our extrapolation theory. These include a new approach to Coifman-Fefferman inequalities that avoids the so-called good-λ inequalities, vector-valued inequalities, modular inequalities for singular integrals, and norm inequalities for operators on the variable Lebesgue spaces.

In Part II we treat two-weight extrapolation and factorization theory. To a certain extent this half of the book is independent of Part I, though the reader should consult Chapter 2 to get a better sense of our overall philosophy. Our approach to two-weight extrapolation grew naturally out of our approach to weighted norm inequalities; in fact, some special cases of our extrapolation results were implicit in earlier work [33, 53, 54]. We show that one can extrapolate in the scale of weights that satisfy (1.10): given an operator T, suppose that for some p_0 and Young functions A, B in a certain class, $\|Tf\|_{L^{p_0}(u_0)} \leq C\|f\|_{L^{p_0}(v_0)}$ whenever (u_0, v_0) satisfy (1.10) with p replaced by p_0. Then given any p we give sufficient conditions on Young functions A and B so that $\|Tf\|_{L^p(u)} \leq C\|f\|_{L^p(v)}$ whenever (u, v) satisfy (1.10).

The presence of the Orlicz space norms in (1.10) causes our proofs to be more technical than the proofs in the one-weight case in Part I. Further, to get the sharpest possible results we diverge considerably from the specific proofs given in Part I. Nevertheless, the proofs in the two-weight case rely on the same essential ingredients: boundedness of the maximal operator, duality, and reverse factorization.

The material in Part II is organized as follows: in Chapter 5 we gather preliminary information about Young functions, Orlicz spaces, and Orlicz maximal operators that is needed in subsequent chapters. In particular, we characterize the Young functions such that (1.10) implies that the maximal operator satisfies (1.2) and the dual inequality (1.3).

In Chapter 6 we discuss factorization in the two-weight setting. We first define the appropriate A_1-type weights and prove a reverse factorization theorem for weights that satisfy (1.10). Since there is a close connection between reverse factorization, factorization and extrapolation, we also develop a two-weight factorization theory for weights that satisfy (1.10). We introduce an important new class of weights—the so-called factored weights,

$$(\tilde{u}, \tilde{v}) = \big(w_1(M_\Psi w_2)^{1-p}, (M_\Phi w_1)w_2^{1-p}\big), \tag{1.12}$$

where M_Φ and M_Ψ are Orlicz maximal operators—that are gotten from the reverse factorization theorem and which satisfy (1.10). These weights are of particular interest in applications since we can prove a number of results for this special class that generalize known results in surprising ways, and these lead to new conjectures for two-weight inequalities in general.

Chapters 7 and 8 are the theoretical heart of Part II. In Chapter 7 we prove the main two-weight extrapolation theorems. This chapter is unavoidably technical, both because of the nature of the conditions on the weights and because we wanted to develop our results in a fairly general setting. To clarify the situation we give a number of examples and special cases. In Chapter 8 we further develop the theory of two-weight extrapolation, focusing particularly on endpoint results and rescaling such as we used to develop the so-called A_∞ extrapolation in Chapter 3.

Throughout Chapters 5–8 we will primarily consider A_p-type conditions and maximal operators defined with respect to arbitrary cubes. However, unless we specifically say otherwise, all of our results hold when we restrict ourselves to operators and conditions defined in terms of dyadic cubes. At certain points we will point out when other, stronger, results hold, but in Chapter 10 we will often apply results from these chapters to the dyadic case without comment.

In the last two chapters we give applications of extrapolation to the study of two-weight norm inequalities. In Chapter 9 we consider three kinds of operators: the sharp maximal operator, Calderón-Zygmund singular integrals, and fractional integral operators. For the sharp maximal operator, we give a two-weight inequality which is a generalization of the Fefferman-Stein inequality [76] and compare our result to another two-weight version due to Fujii [81]. We then use these results to develop a two-weight theory of Coifman-Fefferman inequalities. For singular and fractional integrals we give conjectures for sharp conditions for two-weight weak and strong type inequalities that are based on the original conjectures of Muckenhoupt and Wheeden. We then review the known results, showing in some cases that they are easy consequences of extrapolation, and then prove new results, including new results for pairs of factored weights.

In Chapter 10 we consider applications to the dyadic square function and the vector-valued maximal operator. In the one-weight case these operators are often treated as vector-valued singular integrals, following the approach of Benedek, Calderón and Panzone [11]. (See, for example, [88, 196].) This leads naturally to conjectures for two-weight norm inequalities for both operators. Extrapolation, however, suggests another approach, one which exploits the critical index p_0 of each operator ($p_0 = 2$ for the square function, and $p_0 = q$ for the vector-valued maximal operator defined using the ℓ^q norm, $1 < q < \infty$). Roughly, we show that for p below the critical index these operators behave like maximal functions, and for p above the critical index more like singular integrals, but with the degree of singularity depending on the ratio p/p_0. This leads to completely different conjectures for these operators. As in the previous chapter, for both sets of conjectures we review the known results and prove new ones. In particular, for these operators we give a number of new results for factored weights.

Finally, in Appendix A we give the details of the Calderón-Zygmund decompositions that we use in Chapters 9 and 10. Most of these results are not new, but their proofs are scattered through the literature and we gather them here for the convenience of the reader.

As this book was going to press, we were able to prove several of the conjectures made in Chapters 9 and 10 via completely different techniques from the ones we use here. These new results can be found in [48, 49]. Though now partly superseded, the applications in these chapters are still an important illustration of our techniques and the foundation on which our later work is built.

Chapter 2

The Essential Theorem

In this chapter we give our new proof of the Rubio de Francia extrapolation theorem, Theorem 1.4, and discuss how our proof allows a number of powerful generalizations. For the convenience of the reader we restate it here.

Theorem 1.4. *Given an operator T, suppose that for some p_0, $1 \leq p_0 < \infty$, and every $w \in A_{p_0}$, there exists a constant C depending on $[w]_{A_{p_0}}$ such that*

$$\int_{\mathbb{R}^n} |Tf(x)|^{p_0} w(x)\, dx \leq C \int_{\mathbb{R}^n} |f(x)|^{p_0} w(x)\, dx.$$

Then for every p, $1 < p < \infty$, and every $w \in A_p$ there exists a constant depending on $[w]_{A_p}$ such that

$$\int_{\mathbb{R}^n} |Tf(x)|^{p} w(x)\, dx \leq C \int_{\mathbb{R}^n} |f(x)|^{p} w(x)\, dx.$$

Before giving our proof of Theorem 1.4 we want to describe briefly earlier proofs. The original proof of Rubio de Francia [193, 194, 195] is quite complex and depends on a connection between vector-valued estimates and weighted norm inequalities. A more direct proof that depends only on weighted norm inequalities was given by García-Cuerva [83] (see also [88]). However, this approach requires two complicated lemmas on the structure of A_p weights, and the proof itself is divided into two cases, depending on whether $p > p_0$ or $p < p_0$. A more refined version of this proof appears in Grafakos [92] and in Dragičević, Grafakos, Pereyra and Petermichl [65]. (We will consider this proof again below.)

As we noted in Section 1.2 above, Rubio de Francia and García-Cuerva used the iteration algorithm of Rubio de Francia: given a positive, sublinear operator T that is bounded on $L^p(w)$, define a new operator $\mathcal{R}$ by

$$\mathcal{R}h = \sum_{k=0}^{\infty} \frac{T^k h}{2^k \|T\|^k_{L^p(w)}}.$$

A simpler proof of Theorem 1.4 that avoided the iteration algorithm was given by Duoandikoetxea [68] when $p_0 > 1$. This proof has two steps: first prove that the desired inequality holds for $1 < p < p_0$ and $w \in A_1$, and then use this to prove the full result. The proof requires the very deep property of A_p weights that if $w \in A_p$, then there exists $\epsilon > 0$ such that $w \in A_{p-\epsilon}$ (Theorem 1.3).

2.1 The new proof

Our proof of the Rubio de Francia extrapolation theorem is simpler and more direct than any previous proof, since it yields the desired inequality directly without cases or intermediate steps, and uses only the iteration algorithm and basic properties of A_p weights.

Proof of Theorem 1.4. Fix p, $1 < p < \infty$, and $w \in A_p$. We first introduce two versions of the iteration algorithm. Since $w \in A_p$, M is bounded on $L^p(w)$, so given $h \in L^p(w)$ we can define

$$\mathcal{R}h(x) = \sum_{k=0}^{\infty} \frac{M^k h(x)}{2^k \|M\|_{L^p(w)}^k},$$

where for $k \geq 1$, $M^k = M \circ \cdots \circ M$ is k iterations of the maximal operator, and $M^0 h = |h|$. The operator $\mathcal{R}$ has the following properties:

- for all x, $|h(x)| \leq \mathcal{R}h(x)$;
- $\|\mathcal{R}h\|_{L^p(w)} \leq 2\|h\|_{L^p(w)}$;
- $\mathcal{R}h \in A_1$ with $[\mathcal{R}h]_{A_1} \leq 2\|M\|_{L^p(w)}$.

The first two follow immediately from the definition; to see the third, note that since M is sublinear we have that

$$M(\mathcal{R}h)(x) \leq \sum_{k=0}^{\infty} \frac{M^{k+1} h(x)}{2^k \|M\|_{L^p(w)}^k} \leq 2\|M\|_{L^p(w)} \mathcal{R}h(x).$$

Now define the operator $M'f = M(fw)/w$. Since $w^{1-p'} \in A_{p'}$, M is bounded on $L^{p'}(w^{1-p'})$ and so M' is bounded on $L^{p'}(w)$. Therefore, we can define another iteration algorithm:

$$\mathcal{R}'h(x) = \sum_{k=0}^{\infty} \frac{(M')^k h(x)}{2^k \|M'\|_{L^{p'}(w)}^k}.$$

(Again, $(M')^0 h = |h|$.) Arguing exactly as before we have that:

- for all x, $|h(x)| \leq \mathcal{R}'h(x)$;
- $\|\mathcal{R}'h\|_{L^{p'}(w)} \leq 2\|h\|_{L^{p'}(w)}$;

- $M'(\mathcal{R}'h)(x) \le 2\|M'\|_{L^{p'}(w)}\mathcal{R}'h(x)$, and so $\mathcal{R}'h\,w \in A_1$ with $[\mathcal{R}'h\,w]_{A_1} \le 2\|M'\|_{L^{p'}(w)}$.

Given the two iteration algorithms, the proof is now straightforward. Fix $f \in L^p(w)$. By duality there exists a non-negative function $h \in L^{p'}(w)$, $\|h\|_{L^{p'}(w)} = 1$, such that

$$\begin{aligned}\|Tf\|_{L^p(w)} &= \int_{\mathbb{R}^n} |Tf(x)|h(x)w(x)\,dx \\ &\le \int_{\mathbb{R}^n} |Tf(x)|\mathcal{R}f(x)^{-1/p_0'}\mathcal{R}f(x)^{1/p_0'}\mathcal{R}'h(x)w(x)\,dx,\end{aligned}$$

where we have used that $h \le \mathcal{R}'h$, and if $p_0 = 1$ we let $1/p_0' = 0$. Since $\mathcal{R}f$, $\mathcal{R}'h\,w \in A_1$, by the reverse factorization property, Proposition 1.2 (c), $(\mathcal{R}f)^{1-p_0}\mathcal{R}'h\,w \in A_{p_0}$. Therefore, by Hölder's inequality with respect to the measure $\mathcal{R}'h\,w$ (if $p_0 > 1$), by our hypothesis, and since $|f| \le \mathcal{R}f$,

$$\begin{aligned}\|Tf\|_{L^p(w)} &\le \left(\int_{\mathbb{R}^n} |Tf(x)|^{p_0}\mathcal{R}f(x)^{1-p_0}\mathcal{R}'h(x)w(x)\,dx\right)^{1/p_0} \\ &\qquad\times\left(\int_{\mathbb{R}^n} \mathcal{R}f(x)\mathcal{R}'h(x)w(x)\,dx\right)^{1/p_0'} \\ &\le C\left(\int_{\mathbb{R}^n} |f(x)|^{p_0}\mathcal{R}f(x)^{1-p_0}\mathcal{R}'h(x)w(x)\,dx\right)^{1/p_0} \\ &\qquad\times\left(\int_{\mathbb{R}^n} \mathcal{R}f(x)\mathcal{R}'h(x)w(x)\,dx\right)^{1/p_0'} \\ &\le C\int_{\mathbb{R}^n} \mathcal{R}f(x)\mathcal{R}'h(x)w(x)\,dx.\end{aligned}$$

Again by Hölder's inequality and since $\mathcal{R}$ is bounded on $L^p(w)$ and $\mathcal{R}'$ is bounded on $L^{p'}(w)$,

$$\begin{aligned}\|Tf\|_{L^p(w)} &\le C\left(\int_{\mathbb{R}^n} \mathcal{R}f(x)^p w(x)\,dx\right)^{1/p}\left(\int_{\mathbb{R}^n} \mathcal{R}'h(x)^{p'} w(x)\,dx\right)^{1/p'} \\ &\le C\left(\int_{\mathbb{R}^n} |f(x)|^p w(x)\,dx\right)^{1/p}\left(\int_{\mathbb{R}^n} h(x)^{p'} w(x)\,dx\right)^{1/p'} \\ &= C\left(\int_{\mathbb{R}^n} |f(x)|^p w(x)\,dx\right)^{1/p}.\end{aligned}$$

□

Beyond its simplicity, an important feature of our proof of the Rubio de Francia extrapolation theorem is that it makes clear exactly what the essential ingredients are. They are three-fold: norm inequalities for the Hardy-Littlewood maximal operator, duality, and the reverse factorization property of A_p weights. More precisely, we need the following:

(a) M is sublinear, positive and bounded on $L^p(w)$ if $w \in A_p$;

(b) M' is sublinear, positive and bounded on $L^{p'}(w)$ if $w \in A_p$;

(c) $L^{p'}(w)$ is the dual space of $L^p(w)$;

(d) Hölder's inequality;

(e) if $w_1, w_2 \in A_1$, then $w_1 w_2^{1-p} \in A_p$.

Properties (a) and (b) let us define the iteration algorithm, and Properties (c), (d), and (e) are all that we use in the second part of the proof.

This list of essential properties can be simplified further. Property (b) follows from (a) and another structural property of A_p weights:

(f) $w \in A_p$ if and only if $w^{1-p'} \in A_{p'}$.

Furthermore, in the proof we do not use that $L^{p'}(w)$ is the dual space of $L^p(w)$; it suffices to assume that it is the associate space, thereby giving us the reverse of Hölder's inequality. We can avoid explicitly using duality if we simply define the function $h = |Tf|^{p-1}/\|Tf\|_{L^p(w)}^{p-1}$. With some minor modifications to the proof we can actually take $h = |Tf|^{p-1}$.

As we mentioned in Chapter 1, Property (e) is usually subsumed into the Jones factorization theorem, but we emphasize that we only need the reverse factorization property, and not the converse, which is the heart of this result and more difficult to prove.

Conspicuously missing from this list of properties is any mention of the operator T: we do not assume that T is linear or even sublinear. In the original proofs of Rubio de Francia and García-Cuerva, T was assumed to be sublinear; it was later noted that this hypothesis is superfluous provided that T is well defined on the union of $L^p(w)$ for all $1 < p < \infty$ and $w \in A_p$.

2.2 Extensions of the extrapolation theorem

A very important feature of our proof is that we can adapt it to prove a number of non-trivial extensions of the Rubio de Francia extrapolation theorem. The following are the principal generalizations which we will consider in Chapters 3 and 4.

Generalized maximal operators

The Hardy-Littlewood maximal operator is defined in terms of averages over cubes, as are the Muckenhoupt A_p classes. However, the maximal operator can be generalized to averages over other families of sets: dyadic cubes, rectangles with sides parallel to the coordinate axes, etc. For each such maximal operator there is a corresponding A_p class, and in many important examples the maximal operator satisfies one-weight norm inequalities with respect to this class. In each of these cases there is an extrapolation theorem: the original proofs of Rubio de Francia and García-Cuerva both go through, as each author noted.

Our approach makes this extension immediate: since the structural properties (e) and (f) automatically hold for these generalized A_p classes, our proof extends at once if we assume that the maximal operator satisfies property (a). We will develop these ideas carefully in Chapter 3 and use this approach throughout Part I. (In Part II we will restrict ourselves to the Hardy-Littlewood maximal operator.)

Elimination of the operator

Since we make no assumptions on the operator T, we can reinterpret Theorem 1.4 as follows: if an $L^{p_0}(w)$ inequality holds for pairs of the form $(|Tf|, |f|)$, then an $L^p(w)$ inequality also holds for such pairs. In fact, we can eliminate the operator T entirely, and restate the extrapolation theorem for pairs of non-negative functions (f, g): given a suitably chosen family of pairs of functions (f, g), if for some p_0 and all $w \in A_{p_0}$, $\|f\|_{L^{p_0}(w)} \leq \|g\|_{L^{p_0}(w)}$, then for all p and $w \in A_p$, $\|f\|_{L^p(w)} \leq \|g\|_{L^p(w)}$.

This perspective was first described in passing in [54], but it was not fully exploited until later in a series of papers by the authors and their collaborators (see [40, 44, 45, 57, 93]). The advantage of our approach is that a number of different results become special cases of a single extrapolation theorem. We consider three important examples. For clarity we state them in terms of operators, but below we will treat them in full generality in terms of pairs of functions.

Weak type inequalities

Given an operator T, suppose that for some p_0 and all $w \in A_{p_0}$, $T : L^{p_0}(w) \to L^{p_0,\infty}(w)$. Let

$$E_\lambda = \{x \in \mathbb{R}^n : |Tf(x)| > \lambda\};$$

then we can rewrite the weak type (p_0, p_0) inequality as

$$\|\lambda \chi_{E_\lambda}\|_{L^{p_0}(w)} \leq C\|f\|_{L^{p_0}(w)}.$$

Hence, if we apply extrapolation to the family of pairs $(\lambda\chi_{E_\lambda}, |f|)$, we get that for all p and $w \in A_p$ that $\|\lambda\chi_{E_\lambda}\|_{L^p(w)} \leq C\|f\|_{L^p(w)}$, or equivalently, that T is of weak type (p, p).

This idea first appeared in [93]. Extrapolation for weak type inequalities was proved by both Rubio de Francia and García-Cuerva [83, 195]. However, each gave a separate proof by adapting the proof in the case of strong type inequalities.

Vector-valued inequalities

Given an operator T, suppose that it is bounded on $L^{p_0}(w)$ for all $w \in A_{p_0}$. Then by extrapolation it is bounded on $L^q(w)$ whenever $w \in A_q$. If we let $f = \{f_i\}$, we can extend T to a vector-valued operator by defining $Tf = \{Tf_i\}$. Then we immediately have that

$$\int_{\mathbb{R}^n} \|Tf(x)\|_{\ell^q}^q w(x)\, dx \le C \int_{\mathbb{R}^n} \|f(x)\|_{\ell^q}^q w(x)\, dx.$$

We can, therefore, apply extrapolation again, this time to the pairs $(\|Tf\|_{\ell^q}, \|f\|_{\ell^q})$, and conclude that for all p and $w \in A_p$,

$$\int_{\mathbb{R}^n} \|Tf(x)\|_{\ell^q}^p w(x)\, dx \le C \int_{\mathbb{R}^n} \|f(x)\|_{\ell^q}^p w(x)\, dx.$$

Extrapolation to vector-valued inequalities was proved by Rubio de Francia [195] as part of the original extrapolation theorem. He was led to this extension because his proof relied on the connection between vector-valued inequalities and weighted norm inequalities. In [88] it was noted in passing that extrapolation can be used to prove vector-valued inequalities, but no details were given. Later authors only considered the scalar case.

Rescaling

There are two versions of the extrapolation theorem that yield $L^p(w)$ inequalities for weights w that are not in A_p. Rubio de Francia [195] observed that his proof of the extrapolation theorem could be modified to prove the following: given $r > 1$, suppose that for some $p_0 \ge r$ the operator T is bounded on $L^{p_0}(w)$ whenever $w \in A_{p_0/r}$. Then for all $p > r$, T is bounded on $L^p(w)$ whenever $w \in A_{p/r}$. (See also Duoandikoetxea [68].) As an application, Rubio de Francia used this result to proved weighted Littlewood-Paley inequalities. Other operators that satisfy such inequalities include the square function g_λ^* (see [153, 219]) and singular integrals with rough kernels (see [67, 119, 196, 228]).

This version of the extrapolation theorem is an immediate consequence of the general result for pairs of functions. We can restate the L^{p_0} inequality as an $L^{p_0/r}$ inequality: $\||Tf|^r\|_{L^{p_0/r}(w)} \le C\||f|^r\|_{L^{p_0/r}(w)}$. Hence, we can apply extrapolation to the pairs $(|Tf|^r, |f|^r)$ to get the desired inequality for all p and $w \in A_{p/r}$.

An extrapolation theorem for A_∞ weights was introduced in [44]: given a pair of operators S and T, suppose that for some p_0, $0 < p_0 < \infty$, and for all $w \in A_\infty$, $\|Tf\|_{L^{p_0}(w)} \le C\|Sf\|_{L^{p_0}(w)}$. Then for all p, $0 < p < \infty$, $\|Tf\|_{L^p(w)} \le C\|Sf\|_{L^p(w)}$ whenever $w \in A_\infty$. The original proof of this result did not use Rubio de Francia extrapolation; the proof was direct and had two steps, similar to the proof of Theorem 1.4 due to Duoandikoetxea [68]. However, A_∞ extrapolation is an immediate corollary of the general extrapolation theorem for pairs of functions.

Since the A_p classes are nested, $w \in A_\infty$ is equivalent to $w \in A_{p_0/r}$ for some r, $0 < r < p_0$. Therefore, this result follows by the same rescaling argument as before.

As we noted in Chapter 1, inequalities of this type were introduced by Coifman and Fefferman [25], who showed that if T is a Calderón-Zygmund singular integral, then for all p and $w \in A_\infty$, $\|Tf\|_{L^p(w)} \leq C\|Mf\|_{L^p(w)}$. This and related estimates were originally proved using good-λ inequalities, but in [44] we showed that they can also be proved using extrapolation. As there is no standard terminology, we refer to all such inequalities involving pairs of operators as Coifman-Fefferman inequalities.

In Chapter 3 we will prove the extrapolation theorem for pairs of functions, and we will also provide the details on the above applications. Throughout this monograph we will state and prove all the extrapolation theorems in this generality.

Sharp constants

Initially, little attention was paid to the exact constant obtained via extrapolation: the primary concern was to establish weighted L^p estimates. However, beginning with the work of Buckley [15], there has been increasing interest in the best constants in weighted norm inequalities (in terms of the A_p constant of the weight). In particular, the results of Astala, Iwaniec and Saksman [6] on the Beltrami equation (discussed in Section 1.2) showed that sharp constants had important consequences. Sharp constants for singular integrals and other operators have been considered by a number of authors: see [48, 49, 65, 121, 127, 129, 130, 179, 181, 182, 184]. For every operator except the Hardy-Littlewood maximal operator, sharp constants were proved for a specific value of p (usually but not universally $p = 2$) and then extrapolation was used to find the best constant for all other values of p.

If the constant in the initial L^{p_0} inequality is $N_{p_0}([w]_{A_{p_0}})$, where N_{p_0} is an increasing function with values in $[1, \infty)$, then it can be shown that the constants gotten for L^p inequalities are

$$\begin{cases} 2^{1/p_0} N_{p_0}(C_{n,p,p_0}[w]_{A_p}) & p > p_0, \\ 2^{1/p_0'} N_{p_0}(C_{n,p,p_0}[w]_{A_p}^{\frac{p_0-1}{p-1}}) & p < p_0. \end{cases}$$

These bounds are sharp in the sense that for many operators (e.g. the Hilbert transform) the resulting constants are the best possible. These bounds were first obtained by Petermichl and Volberg [184] for $p > p_0 = 2$, and then for all p and p_0 by Dragičević, *et al.* [65]. These proofs required a careful adaptation of the two case proof of García-Cuerva. A simpler proof was given by Grafakos [92].

The singular weakness of our proof is that it does not yield these sharp constants. A close examination of the proof shows that the constant is

$$N_{p_0}(C_{n,p_0,p}[w]_{A_p}^{1+\frac{p_0-1}{p-1}}).$$

This estimate depends on the best constant for $\|M\|_{L^p(w)}$ due to Buckley [15] (see also the recent proof by Lerner [127]) and details are left to the reader. This appears to be intrinsic to our proof: since we treat all values of p simultaneously we must use both iteration algorithms, and this yields a larger constant. However, by treating the cases separately we can refine our approach to get the sharp constants, and we do so in the context of Muckenhoupt bases. Our proof uses some ideas from recent work of Duoandikoetxea [69].

Off-diagonal extrapolation

We can extend our proof of Theorem 1.4 to prove extrapolation for "off-diagonal" inequalities. More precisely, given p, q, $1 < p < q < \infty$, we say that $w \in A_{p,q}$ if for every cube Q,

$$\left(\fint_Q w(x)^q\,dx\right)^{1/q}\left(\fint_Q w(x)^{-p'}\,dx\right)^{1/p'} \le K < \infty.$$

Suppose an operator T is such that for some p_0, q_0, and every $w \in A_{p_0,q_0}$, $T : L^{p_0}(w^{p_0}) \to L^{q_0}(w^{q_0})$. Then for all pairs (p,q) such that $1/p - 1/q = 1/p_0 - 1/q_0$, and all $w \in A_{p,q}$, $T : L^p(w^p) \to L^q(w^q)$. In addition, though we do not explore it in detail, our approach yields many of the same extensions and generalizations described above (such as vector-valued inequalities) in the off-diagonal case.

Off-diagonal extrapolation was first proved by Harboure, Macías and Segovia [96] by adapting the proof of García-Cuerva. Our proof simplifies and extends theirs. It is applicable to the study of the so-called fractional operators, for instance the fractional integral operators, also known as the Riesz potentials.

Extrapolation for arbitrary pairs of operators

In the proof of Theorem 1.4 we can replace M (and so M') not just with a more general maximal operator as we discussed above, but with an arbitrary pair of positive, sublinear operators, T, T'. If we do so, however, we need to replace A_p weights with weight classes associated to these operators. This generalization was implicit in Coifman, Jones and Rubio de Francia [26], and was made explicit by Jawerth [108], who replaced the maximal operator with a positive sublinear operator T. (Also see Bloom [13].) Later, Hernández [102] and Ruiz and Torrea [198] extended the argument to two arbitrary operators. A version of this technique was used by Watson [229] to prove norm inequalities for a family of rough operators.

Using our approach we can easily deduce what we need to assume about the operators T and T', though the situation is complicated by the fact that property (e), reverse factorization, does not necessarily hold in this context. We will give a precise statement and proof in Chapter 3. We consider the special case of extrapolation for the one-sided A_p weights associated with the one-sided maximal operators. These weights were introduced by Sawyer [206], and extensively explored by Martín-Reyes, *et al.* [139, 140, 141, 142]. Extrapolation results were proved in [134, 141, 198].

Limited range extrapolation

The conclusion of Theorem 1.4 yields that the operator T is bounded on $L^p(w)$, $1 < p < \infty$. Therefore, extrapolation cannot be applied to operators T that are only bounded, for instance, on L^p if $1 < p_- < p < p_+ < \infty$. Operators of this type include the Riesz transforms and square functions associated with divergence form elliptic operators; see [7, 8] for precise definitions and results.

A restricted range extrapolation theorem can be gotten, however, by restricting the class of weights to $A_p \cap RH_s$ for some $s > 1$ depending on p. Results of this kind have been obtained by Johnson and Neugebauer [109] and by Duoandikoetxea *et al.* [70]. Here we use our techniques to prove a limited range extrapolation theorem that generalizes the extrapolation result in [9] and includes the above results as special cases.

We note in passing that a different kind of limited range extrapolation theorem was proved by Passarelli di Napoli [168].

Extrapolation to Banach function spaces

Since our proof of Theorem 1.4 only uses a basic property of $L^p(w)$—the existence of an associate space—we can replace $L^p(w)$ by more general Banach function spaces. Given a Banach function space $\mathbb{X}$, with modest assumptions we have that M is bounded on $\mathbb{X}$ and M' is bounded on its associate space $\mathbb{X}'$. Thus we can show that if T is bounded on $L^p(w)$ whenever $w \in A_p$, then T is bounded on $\mathbb{X}$. If $\mathbb{X}$ is rearrangement invariant, then we can also get estimates for T on the weighted spaces $\mathbb{X}(w)$.

Further, as we noted above, by a clever choice of the "dual function" h we do not have to use duality explicitly. This point of view can be extended to let us extrapolate to so-called modular spaces (see [156]) and so obtain weighted modular inequalities,

$$\int_{\mathbb{R}^n} \Phi(|Tf(x)|)w(x)\,dx \le C \int_{\mathbb{R}^n} \Phi(|f(x)|)w(x)\,dx,$$

where Φ is a Young function, as a consequence of weighted L^p inequalities. Such inequalities have been considered extensively by many authors: see [90, 114, 115] for details and further references. We will consider all of these results and their applications in Chapter 4.

This extension of extrapolation is new: the idea of extending extrapolation to modular inequalities and Banach function spaces first appeared in [57] where the authors and their collaborators used it to prove A_∞ extrapolation theorems for rearrangement invariant Banach function spaces and modular spaces. In [40] extrapolation results were obtained for variable L^p spaces, Banach function spaces that are not rearrangement invariant. Here we unite and extend both results.

Chapter 3

Extrapolation for Muckenhoupt Bases

In this chapter we prove our fundamental generalization of the Rubio de Francia extrapolation theorem. Our result combines the two most important generalizations we discussed in Chapter 2: generalized maximal operators and the elimination of the operator. We then consider the results we get by rescaling, particularly A_∞ extrapolation. Next, we prove four variants of our main result: extrapolation with sharp constants, off-diagonal extrapolation, extrapolation with pairs of positive operators in place of the maximal operator (which we apply to one-sided A_p weights), and limited range extrapolation. Finally, we survey some of the many possible applications of our results. Extrapolation in the context of Banach function spaces and modular spaces will be discussed in Chapter 4.

3.1 Preliminaries

In this section we give some basic results necessary to state and prove our extrapolation theorems.

Muckenhoupt bases

We first give a generalization of the Hardy-Littlewood maximal operator and the Muckenhoupt A_p classes. We are following the development in [170], which in turn is based on the ideas in [108]. The importance of this generalization is two-fold. First, it allows us to simultaneously consider the Hardy-Littlewood maximal operator and its common generalizations—e.g., the dyadic maximal operator and the strong maximal operator. Second, we are able to avoid any direct appeal to the underlying geometry: all the relevant properties are derived from the weighted norm inequality. Our proofs do not use any covering lemmas.

Hereafter we adopt the following (standard) conventions: $0\cdot\infty = 0$, $t\cdot\infty = \infty$ for $t > 0$, and $1/\infty = 0$.

Define a basis $\mathcal{B}$ to be a collection of open sets $B \subset \mathbb{R}^n$, and let $\Omega_{\mathcal{B}} = \cup_{B\in\mathcal{B}} B$. Given a basis $\mathcal{B}$, the maximal operator associated with $\mathcal{B}$ is defined by

$$M_{\mathcal{B}}f(x) = \sup_{B\ni x} \fint_B |f(y)|\,dy$$

if $x \in \Omega_{\mathcal{B}}$ and $M_{\mathcal{B}}f(x) = 0$ otherwise.

A weight w is a non-negative measurable function. We will say that a weight is trivial on a measurable E if either $w = 0$ or $w = \infty$ almost everywhere on E.

Given a basis $\mathcal{B}$ and a weight w, we will always assume that w is non-trivial on $\Omega_{\mathcal{B}}$. We say that w belongs to the Muckenhoupt class associated to $\mathcal{B}$, $A_{p,\mathcal{B}}$, $1 < p < \infty$, if there exists a constant K such that for every $B \in \mathcal{B}$,

$$\left(\fint_B w(x)\,dx\right)\left(\fint_B w(x)^{1-p'}\,dx\right)^{p-1} \leq K < \infty.$$

When $p = 1$, we say that $w \in A_{1,\mathcal{B}}$ if $M_{\mathcal{B}}w(x) \leq K\,w(x)$ for almost every $x \in \mathbb{R}^n$. The infimum of all such K, denoted by $[w]_{A_{p,\mathcal{B}}}$, is called the $A_{p,\mathcal{B}}$ constant of w. Let

$$A_{\infty,\mathcal{B}} = \bigcup_{p\geq 1} A_{p,\mathcal{B}}.$$

It is immediate from the definition that $w \in A_{p,\mathcal{B}}$ if and only if $w^{1-p'} \in A_{p',\mathcal{B}}$. Further, by Hölder's inequality, if $q > p$, then $A_{p,\mathcal{B}} \subset A_{q,\mathcal{B}}$.

While the definition of basis is quite general, we are going to restrict our attention to a particular class of bases.

Definition 3.1. A basis $\mathcal{B}$ is a Muckenhoupt basis if for each p, $1 < p < \infty$, and for every $w \in A_{p,\mathcal{B}}$, the maximal operator $M_{\mathcal{B}}$ is bounded on $L^p(w)$: for every $f \in L^p(w)$,

$$\int_{\mathbb{R}^n} M_{\mathcal{B}}f(x)^p\,w(x)\,dx \leq C \int_{\mathbb{R}^n} |f(x)|^p\,w(x)\,dx,$$

where the constant C is independent of f and depends only on the $A_{p,\mathcal{B}}$ constant of w.

Remark 3.2. If $\mathcal{B}$ is a Muckenhoupt basis, then $M_{\mathcal{B}}$ is bounded on $L^p(\mathbb{R}^n)$, $1 < p < \infty$. However, we make no assumption about the behavior of $M_{\mathcal{B}}$ at the endpoint. In fact $M_{\mathcal{B}}$ may not be weak type $(1,1)$.

Given a Muckenhoupt basis $\mathcal{B}$, we may assume without loss of generality that $|B| < \infty$ for every $B \in \mathcal{B}$. For if $|B| = \infty$, then the average of any function on B is 0. Therefore, if we let $\mathcal{B}'$ be the sets in $\mathcal{B}$ with finite measure, $A_{p,\mathcal{B}} = A_{p,\mathcal{B}'}$, and $M_{\mathcal{B}} = M_{\mathcal{B}'}$. Since the support of $M_{\mathcal{B}}$ is contained in $\Omega_{\mathcal{B}}$, and since the $A_{p,\mathcal{B}}$

condition only depends on the behavior of the weight on $\Omega_\mathcal{B}$, it is convenient to assume that w is identically zero on $\mathbb{R}^n \setminus \Omega_\mathcal{B}$, and we shall do so hereafter.

Three immediate examples of Muckenhoupt bases are $\mathcal{Q}$, the set of all cubes in $\mathbb{R}^n$, $\mathcal{D}$, the set of all dyadic cubes in $\mathbb{R}^n$, and $\mathcal{S}$, the set of all rectangles (i.e., parallelepipeds) in $\mathbb{R}^n$ whose sides are parallel to the coordinate axes. (See [68].) Technically, $\mathcal{D}$ is not a basis since dyadic cubes are not open, but without loss of generality we can replace them by their interiors. Another interesting example is the basis $\mathcal{Z}$ of Zygmund cubes in $\mathbb{R}^3$: rectangles in $\mathbb{R}^3$ whose sides are parallel to the coordinate axes and have lengths s, t and st with $s, t > 0$. (See [78].)

Given a Muckenhoupt basis $\mathcal{B}$, it is possible to have weights $w \in A_{p,\mathcal{B}}$ that are trivial on sets of positive measure. For example, there exist weights in $A_{p,\mathcal{D}}$ that are 0 on one quadrant in $\mathbb{R}^n$—e.g., $w = 1$ on the first quadrant and $w = 0$ elsewhere is a dyadic A_1 weight. This causes minor technical difficulties in the proofs, and we will discuss these at the end of this section.

Definition 3.3. A basis $\mathcal{B}$ has the $A_{p,\mathcal{B}}$ openness condition, or more simply, is $A_{p,\mathcal{B}}$ open, if for every p, $1 < p < \infty$, given any $w \in A_{p,\mathcal{B}}$, there exists $\epsilon > 0$ depending on p and $[w]_{A_{p,\mathcal{B}}}$ such that $w \in A_{p-\epsilon,\mathcal{B}}$.

The four examples of Muckenhoupt bases given above all have this property; in each case it is a consequence of a reverse Hölder inequality. However, not every basis has this property. For example, in $\mathbb{R}^2$, define the basis $\mathcal{B}$ to consist of the single set B defined in terms of polar coordinates:

$$B = \{(r,\theta) : 0 < r < 1/2, |\theta| < \log(r)^{-2}\}.$$

Since $\mathcal{B}$ contains a single set, it is a Muckenhoupt basis: given p, $1 < p < \infty$, and $w \in A_{p,\mathcal{B}}$, by Hölder's inequality,

$$\int_{\mathbb{R}^n} M_\mathcal{B} f(x)^p w(x)\,dx = \left(\fint_B |f(x)|\,dx \right)^p w(B) \le C \int_B |f(x)|^p w(x)\,dx.$$

Now let $w(x) = |x|^2$. Then $0 < w(B) < \infty$, and

$$\int_B w(x)^{-1}\,dx = 2 \int_0^{1/2} \frac{dr}{r \log(r)^2} < \infty.$$

Hence, $w \in A_{2,\mathcal{B}}$. But for any $p < 2$, since $p' > 2$,

$$\int_B w(x)^{1-p'}\,dx = 2 \int_0^{1/2} \frac{r^{3-2p'}}{\log(r)^2}\,dr = \infty.$$

Hence $w \notin A_{p,\mathcal{B}}$.

In this chapter we will not use the $A_{p,\mathcal{B}}$ openness property except in Proposition 3.21. However, it will be necessary to assume it in parts of Chapter 4.

Remark 3.4. In a recent paper, Lerner and Ombrosi [128], working with a slightly different definition of a basis, gave a sufficient condition for a basis to be $A_{p,\mathcal{B}}$ open.

Finally, we give some basic properties of $M_{\mathcal{B}}$ and the $A_{p,\mathcal{B}}$ classes needed in the next section. These should be compared to the list of properties in Section 2.1 used in the proof of Theorem 1.4. Given $w \in A_{\infty,\mathcal{B}}$, we first define the dual operator $M'_{\mathcal{B}}$ by

$$M'_{\mathcal{B}}f(x) = \frac{M_{\mathcal{B}}(f\,w)(x)}{w(x)}$$

if $x \in \Omega_{\mathcal{B}}$, and $M'_{\mathcal{B}}f(x) = 0$ otherwise. (Although the definition of $M'_{\mathcal{B}}$ depends on w, we will not make this explicit unless it is not clear from the context.)

Proposition 3.5. *If $\mathcal{B}$ is a Muckenhoupt basis and $1 < p < \infty$:*

(a) *$M_{\mathcal{B}}$ is sublinear, positive and bounded on $L^p(w)$ if $w \in A_{p,\mathcal{B}}$;*

(b) *$M'_{\mathcal{B}}$ is sublinear, positive and bounded on $L^{p'}(w)$ if $w \in A_{p,\mathcal{B}}$;*

(c) *if $w_1, w_2 \in A_{1,\mathcal{B}}$, then $w_1 w_2^{1-p} \in A_{p,\mathcal{B}}$.*

Proof. Property (a) is simply the definition of a Muckenhoupt basis. Property (b) also follows from the definition and the fact that $w \in A_{p,\mathcal{B}}$ implies that $w^{1-p'} \in A_{p',\mathcal{B}}$. Finally, the reverse factorization property follows from the definition of $A_{p,\mathcal{B}}$ and the fact that if $w \in A_{1,\mathcal{B}}$, then for $B \in \mathcal{B}$ and almost every $x \in B$,

$$\fint_B w(y)\,dy \le M_{\mathcal{B}}w(x) \le [w]_{A_{1,\mathcal{B}}} w(x).$$

Note that in the proof of (b) and (c) we have been using the conventions $0 \cdot \infty = 0$, etc. □

Pairs of functions

As we discussed in Chapter 2, operators do not play a role in the proof of Theorem 1.4. Therefore, all of our extrapolation theorems will be stated in terms of pairs of functions (f, g). Hereafter, by $\mathcal{F}$ we will mean a family of pairs (f, g) of non-negative, measurable functions that are not identically zero. Given such a family $\mathcal{F}$, $p > 0$ and a weight $w \in A_{q,\mathcal{B}}$, if we say that

$$\int_{\mathbb{R}^n} f(x)^p\, w(x)\,dx \le C \int_{\mathbb{R}^n} g(x)^p\, w(x)\,dx, \qquad (f, g) \in \mathcal{F},$$

we mean that this inequality holds for all pairs $(f, g) \in \mathcal{F}$ such that the left-hand side is finite, and that the constant C depends only upon p and the $A_{q,\mathcal{B}}$ constant of w. In Chapter 4 similar estimates will appear with the $L^p(w)$ norm replaced by a Banach function space norm or a modular space estimate; all of these should be interpreted in the same way.

In applications, the family $\mathcal{F}$ often consists of pairs of the form $(|Tf|,|f|)$, where T is the operator under consideration, and f belongs to some "nice" family of functions (e.g., bounded functions of compact support, C_c^∞, $\cup_{p>1}L^p$) that makes the left-hand side finite. We will discuss this in more detail in Section 3.8.

A technical reduction

As we noted above, there are important examples of bases $\mathcal{B}$ and weights $w \in A_{p,\mathcal{B}}$ such that w is trivial on a subset of $\Omega_{\mathcal{B}}$ with positive measure. This leads to a number of technical problems in our proofs; all of them are minor, but to avoid dealing with them as they arise we would like to make a single reduction to an important special case. Throughout this chapter, given a weight $w \in A_{p,\mathcal{B}}$, we will implicitly replace w with $w^* = w\chi_{\Omega^*_{\mathcal{B}}}$, where $\Omega^*_{\mathcal{B}} = \{x \in \Omega_{\mathcal{B}} : 0 < w(x) < \infty\}$. Having done so, whenever we make an assertion of integrability, or use duality or Hölder's inequality, we will assume that the underlying measure space is $\Omega^*_{\mathcal{B}}$ and not $\mathbb{R}^n$. In the classical case—that is, for the basis $\mathcal{Q}$ of cubes—$\Omega^*_{\mathcal{B}} = \mathbb{R}^n$ up to a set of measure 0, so this reduction is not necessary. But in the case of the basis $\mathcal{D}$ of dyadic cubes, this reduction means that we restrict our argument to those quadrants where the weight w is non-trivial.

We can do this because, as we will show below, if $w \in A_{p,\mathcal{B}}$, then $w^* \in A_{p,\mathcal{B}}$, and given any function $h \in L^p(w)$, then $h \in L^p(w^*)$ and $\|h\|_{L^p(w)} = \|h\|_{L^p(w^*)}$. This means that in all of our theorems, we can pass seamlessly in both our hypotheses and our conclusions between w and w^* and $\mathbb{R}^n$ and $\Omega^*_{\mathcal{B}}$.

Remark 3.6. This issue will arise in Chapter 4 as well, except for unweighted estimates on Banach function spaces. For rearrangement invariant Banach function spaces we will have to use a similar reduction to define weighted Banach function spaces. For modular spaces we will use exactly the same technical reduction without comment.

To make this reduction, we proceed as follows. Fix a basis $\mathcal{B}$; we define an equivalence relationship on it. Given two sets B, $B' \in \mathcal{B}$, we say $B \sim B'$ if there exists a finite collection $\{B_k\}_{k=0}^N \subset \mathcal{B}$, such that $B_0 = B$, $B_N = B'$, and for all k, $0 \leq k \leq N-1$, $B_k \cap B_{k+1} \neq \emptyset$. Clearly this is an equivalence relationship. Let $\{\mathcal{B}_j\}$ be the family of equivalence classes. Let $\Omega_j = \cup_{B\in\mathcal{B}_j} B$; since these are disjoint open sets in $\mathbb{R}^n$, it follows that there are at most a countable number of equivalence classes.

In our examples, the bases $\mathcal{Q}$, $\mathcal{R}$ and $\mathcal{Z}$ each have a single equivalence class; the basis $\mathcal{D}$ has 2^n equivalence classes, corresponding to the 2^n quadrants in $\mathbb{R}^n$. The number of equivalence classes could be infinite; for example, take the basis consisting of all dyadic cubes of side-length less than or equal to 1.

Proposition 3.7. *Given a Muckenhoupt basis $\mathcal{B}$, let $\{\mathcal{B}_j\}$ and $\{\Omega_j\}$ be as above. Then for every j and every weight $w \in A_{\infty,\mathcal{B}}$, either w is trivial on Ω_j, or $0 < w(B) < \infty$ for $B \in \mathcal{B}_j$, and $0 < w(x) < \infty$ for almost every $x \in \Omega_j$.*

Proof. We prove this result by adapting an argument in [88]. Fix $w \in A_{\infty,\mathcal{B}}$ and p, $1 < p < \infty$, such that $w \in A_{p,\mathcal{B}}$. First, we prove that there exists $C > 0$ such that if $B \in \mathcal{B}$ and $S \subset B$ is measurable, then

$$\left(\frac{|S|}{|B|}\right)^p w(B) \le Cw(S). \tag{3.1}$$

Let $f = \chi_S$; then for all $x \in B$, $M_{\mathcal{B}}f(x) \ge |S|/|B|$. Then inequality (3.1) follows from the boundedness of $M_{\mathcal{B}}$ on $L^p(w)$.

Now, given $B \in \mathcal{B}$, if $w = 0$ or $w = \infty$ on a subset of B of positive measure, then w is trivial on B. For if $S = \{x \in B : w(x) = 0\}$ has positive measure, then by (3.1), $w(B) = 0$. On the other hand, if $\{x \in B : w(x) = \infty\}$ has positive measure, then $w(B) = \infty$; therefore, by (3.1), $w(S) = \infty$ for any measurable set $S \subset B$ with positive measure, which implies $w = \infty$ almost everywhere.

For any pair B, $B' \in \mathcal{B}$, if $B \cap B' \neq \emptyset$, then w is trivial on B if and only it is trivial on B'. This follows at once from the above: for if w is trivial on B, then it is either equal to 0 or ∞ on $B \cap B'$, which is a subset of positive measure of B' since B and B' are open. Therefore, by the definition of our equivalence classes, given $B \in \mathcal{B}_j$, w is trivial on B if and only if it is trivial on Ω_j.

Now suppose w is non-trivial on Ω_j for some j. Then for all $B \in \mathcal{B}_j$, $0 < w(B) < \infty$. For if $w(B) = 0$, then $w = 0$ almost everywhere on B, and so w would be trivial on Ω_j. On the other hand, if $w(B) = \infty$, by (3.1) and arguing as before, we get that $w = \infty$ almost everywhere on B, and again w would be trivial on Ω_j. Furthermore, $0 < w(x) < \infty$ for almost every $x \in \Omega_j$. For if it were trivial on a set of positive measure, it would be trivial on a set of positive measure contained in some $B \in \mathcal{B}_j$, which in turn would imply it was trivial on B and so on Ω_j. □

Remark 3.8. For the basis of cubes $\mathcal{Q}$, Proposition 3.7 implies that if $w \in A_{\infty,\mathcal{Q}}$, then $0 < w(x) < \infty$ a.e. On the other hand, for the basis of dyadic cubes $\mathcal{D}$, if $w \in A_{\infty,\mathcal{D}}$, then in each quadrant w is trivial or satisfies this inequality.

If $\mathcal{B}$ is a Muckenhoupt basis and $w \in A_{p,\mathcal{B}}$, then by Proposition 3.7, $\Omega_{\mathcal{B}}^*$ is the union of the components Ω_j where w is non-trivial. Therefore, if we let $w^* = w\chi_{\Omega_{\mathcal{B}}^*}$, then $w^* \in A_{p,\mathcal{B}}$ with $[w^*]_{A_{p,\mathcal{B}}} = [w]_{A_{p,\mathcal{B}}}$. For if $B \subset \Omega_{\mathcal{B}}^*$, then $w = w^*$ on B; if $B \subset \Omega_{\mathcal{B}} \setminus \Omega_{\mathcal{B}}^*$, then both w and w^* are trivial on B and there is nothing to show.

If $h \in L^p(w)$, h non-negative, then

$$\int_{\mathbb{R}^n} h(x)^p w(x)\, dx < \infty,$$

which in turn means that h must be equal to zero almost everywhere on the set where $w = \infty$. Hence, for almost every $x \in \mathbb{R}^n$, $h(x)^p w(x) = h(x)^p w^*(x)$, and so $\|h\|_{L^p(w)} = \|h\|_{L^p(w^*)}$. Further, since in the right-hand term the integration is over $\Omega_{\mathcal{B}}^*$, we may assume that $\Omega_{\mathcal{B}}^*$ is the underlying measure space.

3.2 A_p extrapolation

In this section we state and prove our main extrapolation theorem, and give two immediate consequences of working with pairs of functions.

Theorem 3.9. *Let $\mathcal{B}$ be a Muckenhoupt basis. Suppose that for some p_0, $1 \le p_0 < \infty$, and every $w_0 \in A_{p_0,\mathcal{B}}$,*

$$\int_{\mathbb{R}^n} f(x)^{p_0}\, w_0(x)\, dx \le C \int_{\mathbb{R}^n} g(x)^{p_0}\, w_0(x)\, dx, \qquad (f,g) \in \mathcal{F}. \tag{3.2}$$

Then for all p, $1 < p < \infty$, and for all $w \in A_{p,\mathcal{B}}$,

$$\int_{\mathbb{R}^n} f(x)^{p}\, w(x)\, dx \le C \int_{\mathbb{R}^n} g(x)^{p}\, w(x)\, dx, \qquad (f,g) \in \mathcal{F}. \tag{3.3}$$

Before proving Theorem 3.9 we show that by a careful choice of the family $\mathcal{F}$ we can restate it as an extrapolation theorem for weak type inequalities, and deduce vector-valued inequalities.

Corollary 3.10. *Let $\mathcal{B}$ be a Muckenhoupt basis. Suppose that for some p_0, $1 \le p_0 < \infty$, and every $w_0 \in A_{p_0,\mathcal{B}}$,*

$$\|f\|_{L^{p_0,\infty}(w_0)} \le C\|g\|_{L^{p_0}(w_0)}, \qquad (f,g) \in \mathcal{F}. \tag{3.4}$$

Then for all p, $1 < p < \infty$, and for all $w \in A_{p,\mathcal{B}}$,

$$\|f\|_{L^{p,\infty}(w)} \le C\|g\|_{L^{p}(w)}, \qquad (f,g) \in \mathcal{F}. \tag{3.5}$$

Remark 3.11. Corollary 3.12 was proved in the classical setting (that is for the basis of cubes or balls) by both Rubio de Francia and García-Cuerva [195, 83]. Their arguments adapt the proof used for strong type inequalities; here we see that a single proof covers both cases.

Proof. We use the argument in [93]. (Also see [44].) Define a new family $\mathcal{F}_{\text{weak}}$ consisting of the pairs

$$(f_\lambda, g) = (\lambda\chi_{\{x: f(x)>\lambda\}}, g),$$

where $(f,g) \in \mathcal{F}$ and $\lambda > 0$. Then for every $w \in A_{p_0,\mathcal{B}}$ and pair $(f_\lambda, g) \in \mathcal{F}_{\text{weak}}$, (3.4) implies that

$$\|f_\lambda\|_{L^{p_0}(w)} = \lambda\, w(\{x \in \mathbb{R}^n : f(x) > \lambda\})^{1/p_0} \le \|f\|_{L^{p_0,\infty}(w)} \le C\|g\|_{L^{p_0}(w)}.$$

Therefore, we can apply Theorem 3.9 with the family $\mathcal{F}_{\text{weak}}$. Thus (3.3) holds for any pair $(f_\lambda, g) \in \mathcal{F}_{\text{weak}}$, which in turn immediately implies (3.5). □

Corollary 3.12. *Given a Muckenhoupt basis $\mathcal{B}$, assume that (3.2) holds. Then for all p and q, $1 < p, q < \infty$, $w \in A_{p,\mathcal{B}}$, and sequences $\{(f_i, g_i)\}_i \subset \mathcal{F}$,*

$$\Big\|\Big(\sum_i f_i^q\Big)^{1/q}\Big\|_{L^p(w)} \leq C\Big\|\Big(\sum_i g_i^q\Big)^{1/q}\Big\|_{L^p(w)}. \tag{3.6}$$

Proof. For each q, $1 < q < \infty$, define the new family $\mathcal{F}_{\ell^q}$ consisting of the pairs (F, G), where

$$F(x) = \Big(\sum_i f_i(x)^q\Big)^{1/q}, \qquad G(x) = \Big(\sum_i g_i(x)^q\Big)^{1/q},$$

and $\{(f_i, g_i)\}_i \subset \mathcal{F}$. By (3.3) with $p = q$, for every $(F, G) \in \mathcal{F}_{\ell^q}$ and $w \in A_q$,

$$\|F\|_{L^q(w)}^q = \sum_i \int_{\mathbb{R}^n} f_i(x)^q\, w(x)\, dx \leq C \sum_i \int_{\mathbb{R}^n} g_i(x)^q\, w(x)\, dx = C\|G\|_{L^q(w)}^q.$$

In other words, inequality (3.2) holds with $p_0 = q$ for the family $\mathcal{F}_{\ell^q}$. Therefore, we can apply Theorem 3.9 to obtain (3.3) for the family $\mathcal{F}_{\ell^q}$; but this is the desired vector-valued estimate (3.6). □

Proof of Theorem 3.9. The proof is very similar to the proof of Theorem 1.4. However, the use of pairs of functions introduces some key differences and so we give all the details. After the proof we will comment on the changes.

Fix p, $1 < p < \infty$, and $w \in A_{p,\mathcal{B}}$. We introduce two versions of the iteration algorithm. Since $w \in A_{p,\mathcal{B}}$, by Proposition 3.5, $M_\mathcal{B}$ is bounded on $L^p(w)$ and $M'_\mathcal{B}$ is bounded on $L^{p'}(w)$. Therefore, given non-negative functions $h_1 \in L^p(w)$ and $h_2 \in L^{p'}(w)$ we define

$$\mathcal{R}h_1(x) = \sum_{k=0}^{\infty} \frac{M_\mathcal{B}^k h_1(x)}{2^k \|M_\mathcal{B}\|_{L^p(w)}^k}, \qquad \mathcal{R}'h_2(x) = \sum_{k=0}^{\infty} \frac{(M'_\mathcal{B})^k h_2(x)}{2^k \|M'_\mathcal{B}\|_{L^{p'}(w)}^k},$$

where for $k \geq 1$, $M_\mathcal{B}^k = M_\mathcal{B} \circ \cdots \circ M_\mathcal{B}$ is k iterations of $M_\mathcal{B}$ and $M_\mathcal{B}^0$ is the identity operator; $(M'_\mathcal{B})^k$ is defined similarly. Then the following are true:

(A) $h_1(x) \leq \mathcal{R}h_1(x)$ (A′) $h_2(x) \leq \mathcal{R}'h_2(x)$

(B) $\|\mathcal{R}h_1\|_{L^p(w)} \leq 2\|h_1\|_{L^p(w)}$ (B′) $\|\mathcal{R}'h_2\|_{L^{p'}(w)} \leq 2\|h_2\|_{L^{p'}(w)}$

(C) $[\mathcal{R}h_1]_{A_{1,\mathcal{B}}} \leq 2\|M_\mathcal{B}\|_{L^p(w)}$ (C′) $[\mathcal{R}'h_2\, w]_{A_{1,\mathcal{B}}} \leq 2\|M'_\mathcal{B}\|_{L^{p'}(w)}$.

Properties (A), (A′), (B), (B′) are immediate consequences of the definitions. To get property (C), observe that

$$M_\mathcal{B}(\mathcal{R}h_1)(x) \leq \sum_{k=0}^{\infty} \frac{M_\mathcal{B}^{k+1} h_1(x)}{2^k \|M_\mathcal{B}\|_{L^p(w)}^k} \leq 2\|M_\mathcal{B}\|_{L^p(w)} \mathcal{R}h_1(x).$$

That same argument shows that $M'_{\mathcal{B}}(\mathcal{R}'h_2)(x) \le 2\|M'_{\mathcal{B}}\|_{L^{p'}(w)}\mathcal{R}'h_2(x)$, and so $M_{\mathcal{B}}(\mathcal{R}'h_2 \cdot w)(x) \le 2\|M'_{\mathcal{B}}\|_{L^{p'}(w)}(\mathcal{R}'h_2(x) \cdot w(x))$; this gives (C′).

Given these properties of the iteration algorithms the proof is now straightforward. Fix $(f, g) \in \mathcal{F}$; without loss of generality f and g are non-trivial and in $L^p(w)$. Define

$$h_1(x) = \frac{f(x)}{\|f\|_{L^p(w)}} + \frac{g(x)}{\|g\|_{L^p(w)}}.$$

Then $h_1 \in L^p(w)$ and $\|h_1\|_{L^p(w)} \le 2$. Since $f \in L^p(w)$, by duality there exists a non-negative function $h_2 \in L^{p'}(w)$, $\|h_2\|_{L^{p'}(w)} = 1$, such that

$$\|f\|_{L^p(w)} = \int_{\mathbb{R}^n} f(x)\, h_2(x)\, w(x)\, dx.$$

By property (A′) and Hölder's inequality with respect to the measure $\mathcal{R}'h_2\, w$ (if $p_0 = 1$, we let $1/p_0' = 0$) we have that

$$\begin{aligned}
\|f\|_{L^p(w)} &\le \int_{\mathbb{R}^n} f(x)\, \mathcal{R}'h_2(x)\, w(x)\, dx \\
&= \int_{\mathbb{R}^n} f(x)\, \mathcal{R}h_1(x)^{-1/p_0'}\, \mathcal{R}h_1(x)^{1/p_0'}\, \mathcal{R}'h_2(x)\, w(x)\, dx \\
&\le \left(\int_{\mathbb{R}^n} f(x)^{p_0}\, \mathcal{R}h_1(x)^{1-p_0}\, \mathcal{R}'h_2(x)\, w(x)\, dx\right)^{1/p_0} \\
&\qquad \times \left(\int_{\mathbb{R}^n} \mathcal{R}h_1(x)\, \mathcal{R}'h_2(x) w(x)\, dx\right)^{1/p_0'}.
\end{aligned}$$

We estimate the last two terms separately. By Hölder's inequality and properties (B) and (B′),

$$\begin{aligned}
\left(\int_{\mathbb{R}^n} \mathcal{R}h_1(x)\, \mathcal{R}'h_2(x) w(x)\, dx\right)^{1/p_0'} &\le \|\mathcal{R}h_1\|_{L^p(w)}^{1/p_0'} \|\mathcal{R}'h_2\|_{L^{p'}(w)}^{1/p_0'} \\
&\le 4^{1/p_0'} \|h_1\|_{L^p(w)}^{1/p_0'} \|h_2\|_{L^{p'}(w)}^{1/p_0'} \le 8^{1/p_0'}.
\end{aligned}$$

We want to estimate the first term by applying (3.2); to do so we must first check that

$$\left(\int_{\mathbb{R}^n} f(x)^{p_0}\, \mathcal{R}h_1(x)^{1-p_0}\, \mathcal{R}'h_2(x)\, w(x)\, dx\right)^{1/p_0} < \infty.$$

But by (A) we have that $f/\|f\|_{L^p(w)} \le h_1 \le \mathcal{R}h_1$, and so by the previous estimate

$$\left(\int_{\mathbb{R}^n} f(x)^{p_0}\,\mathcal{R}h_1(x)^{1-p_0}\,\mathcal{R}'h_2(x)\,w(x)\,dx\right)^{1/p_0}$$
$$\leq \|f\|_{L^p(w)}\left(\int_{\mathbb{R}^n} \mathcal{R}h_1(x)\,\mathcal{R}'h_2(x)\,w(x)\,dx\right)^{1/p_0} \leq 8^{1/p_0}\|f\|_{L^p(w)} < \infty.$$

By (C), (C′) and the reverse factorization property (Proposition 3.5), $(\mathcal{R}h_1)^{1-p_0}\,\mathcal{R}'h_2\,w \in A_{p_0,\mathcal{B}}$. Hence, by (3.2),

$$\left(\int_{\mathbb{R}^n} f(x)^{p_0}\,\mathcal{R}h_1(x)^{1-p_0}\,\mathcal{R}'h_2(x)\,w(x)\,dx\right)^{1/p_0}$$
$$\leq C\left(\int_{\mathbb{R}^n} g(x)^{p_0}\,\mathcal{R}h_1(x)^{1-p_0}\,\mathcal{R}'h_2(x)\,w(x)\,dx\right)^{1/p_0}.$$

By property (A) we have that $g/\|g\|_{L^p(w)} \leq h_1 \leq \mathcal{R}h_1$. Therefore, if we combine the above estimates we get that

$$\begin{aligned}\|f\|_{L^p(w)} &\leq C\left(\int_{\mathbb{R}^n} g(x)^{p_0}\,\mathcal{R}h_1(x)^{1-p_0}\,\mathcal{R}'h_2(x)\,w(x)\,dx\right)^{1/p_0}\\ &\leq C\|g\|_{L^p(w)}\left(\int_{\mathbb{R}^n} \mathcal{R}h_1(x)\,\mathcal{R}'h_2(x)\,w(x)\,dx\right)^{1/p_0}\\ &\leq 8^{1/p_0}\,C\|g\|_{L^p(w)}.\end{aligned}$$

□

Remark 3.13. As we noted above, this proof is very similar to the proof of Theorem 1.4, since all of the properties of A_p weights and the iteration algorithms that we used there hold for Muckenhoupt bases. The major difference is that we introduced the function h_1 and inserted $\mathcal{R}h_1$ instead of $\mathcal{R}g$. We did this since, when working with pairs of functions, it is necessary to assume that the left-hand side of (3.2) is finite. Our definition of h_1 is a technicality that insures that this is the case. If we omitted that step we could replace h_1 by g (or by $g/\|g\|_{L^p(w)}$) and the proof would be essentially the same.

3.3 Rescaling and extrapolation

In this section we prove two corollaries of Theorem 3.9 that are gotten by rescaling. These results further illustrate the value of proving extrapolation for families of pairs of functions; in Section 3.8 we will discuss their applications to, for example, Coifman-Fefferman inequalities.

Our first theorem yields weighted L^p inequalities with weights in $A_{p/r,\mathcal{B}}$, $p \geq r > 0$.

Corollary 3.14. *Let $\mathcal{B}$ be a Muckenhoupt basis. Suppose that for some $r > 0$ and some p_0, $r \le p_0 < \infty$, and for every $w_0 \in A_{p_0/r,\mathcal{B}}$,*

$$\int_{\mathbb{R}^n} f(x)^{p_0} w_0(x)\,dx \le C \int_{\mathbb{R}^n} g(x)^{p_0} w_0(x)\,dx, \qquad (f,g) \in \mathcal{F}. \tag{3.7}$$

Then for all p, $r < p < \infty$ and all $w \in A_{p/r,\mathcal{B}}$,

$$\int_{\mathbb{R}^n} f(x)^p w(x)\,dx \le C \int_{\mathbb{R}^n} g(x)^p w(x)\,dx, \qquad (f,g) \in \mathcal{F}. \tag{3.8}$$

Proof. Define a new family $\mathcal{F}_r$ consisting of the pairs (f^r, g^r), $(f,g) \in \mathcal{F}$. Then for $w_0 \in A_{p_0/r,\mathcal{B}}$, by (3.7),

$$\begin{aligned}\int_{\mathbb{R}^n} (f(x)^r)^{p_0/r} w_0(x)\,dx &= \int_{\mathbb{R}^n} f(x)^{p_0} w_0(x)\,dx \\ &\le C \int_{\mathbb{R}^n} g(x)^{p_0} w_0(x)\,dx = C \int_{\mathbb{R}^n} (g(x)^r)^{p_0/r} w_0(x)\,dx.\end{aligned}$$

Therefore, we can apply Theorem 3.9 with this as our initial hypothesis and conclude that for all $q > 1$ and all $w \in A_{q,\mathcal{B}}$,

$$\int_{\mathbb{R}^n} (f(x)^r)^q w(x)\,dx \le C \int_{\mathbb{R}^n} (g(x)^r)^q w(x)\,dx, \qquad (f^r, g^r) \in \mathcal{F}_r.$$

Since $q = p/r$ for some $p > r$, this is equivalent to (3.8). □

Our second theorem yields norm inequalities on $L^p(w)$ for $w \in A_{\infty,\mathcal{B}}$ and for all $p > 0$. We refer to this result as A_∞ extrapolation, to distinguish it from Theorem 3.9, which gives A_p extrapolation.

Corollary 3.15. *Let $\mathcal{B}$ be a Muckenhoupt basis. Suppose that for some p_0, $0 < p_0 < \infty$, and every $w_0 \in A_{\infty,\mathcal{B}}$,*

$$\int_{\mathbb{R}^n} f(x)^{p_0}\, w_0(x)\,dx \le C \int_{\mathbb{R}^n} g(x)^{p_0}\, w_0(x)\,dx, \qquad (f,g) \in \mathcal{F}. \tag{3.9}$$

Then for all p, $0 < p < \infty$, and for all $w \in A_{\infty,\mathcal{B}}$,

$$\int_{\mathbb{R}^n} f(x)^p\, w(x)\,dx \le C \int_{\mathbb{R}^n} g(x)^p\, w(x)\,dx, \qquad (f,g) \in \mathcal{F}. \tag{3.10}$$

Proof. The proof is very similar to the proof of Corollary 3.14. Fix r, $1 < r < \infty$, and define a new family $\mathcal{F}_r$ consisting of the pairs $(f^{p_0/r}, g^{p_0/r})$ with $(f,g) \in \mathcal{F}$. Then by inequality (3.9), for every $w \in A_{r,\mathcal{B}} \subset A_{\infty,\mathcal{B}}$ and $(f^{p_0/r}, g^{p_0/r}) \in \mathcal{F}_r$,

$$\int_{\mathbb{R}^n} (f(x)^{p_0/r})^r w(x)\,dx = \int_{\mathbb{R}^n} f(x)^{p_0}\, w(x)\,dx$$
$$\le C \int_{\mathbb{R}^n} g(x)^{p_0}\, w(x)\,dx = C \int_{\mathbb{R}^n} (g(x)^{p_0/r})^r\, w(x)\,dx.$$

This gives us the initial estimate (3.2) for $\mathcal{F}_r$ and with r in place of p_0. Therefore, by Theorem 3.9 we have that for every s, $1 < s < \infty$, and $w \in A_{s,\mathcal{B}}$,

$$\int_{\mathbb{R}^n} (f(x)^{p_0/r})^s\, w(x) \le C \int_{\mathbb{R}^n} (g(x)^{p_0/r})^s\, w(x)\,dx, \quad (f^{p_0/r}, g^{p_0/r}) \in \mathcal{F}_r.$$

To get the desired conclusion, fix $p > 0$ and $w \in A_{\infty,\mathcal{B}}$. By Proposition 3.5 the $A_{p,\mathcal{B}}$ classes are nested, so we may assume without loss of generality that $w \in A_{s,\mathcal{B}}$ with $s > p/p_0$. Therefore, we can choose $r > 1$ such that $p_0 s/r = p$, and the above inequality yields (3.10). □

Remark 3.16. It is possible to adapt the proof of Theorem 3.9 to give a direct proof of Corollary 3.15. This proof is an improvement over the original in [44] since it does not require two steps. The only substantive change in the proof is in the definition and use of the auxiliary function h_1. Given $w \in A_{\infty,\mathcal{B}}$ and p, $0 < p < \infty$, fix q, $0 < q < \min(p_0, p)$, such that $w \in A_{r,\mathcal{B}}$ with $r = p/q > 1$. Let

$$h_1(x) = \left(\frac{f(x)}{\|f\|_{L^p(w)}}\right)^q + \left(\frac{g(x)}{\|g\|_{L^p(w)}}\right)^q \in L^r(w),$$

multiply and divide by $\mathcal{R}h_1(x)^{1/(p_0/q)'}$, and then apply duality with respect to $L^r(w)$. The remainder of the proof is left to the reader.

From both corollaries we can deduce weak type extrapolation and vector-valued inequalities, as we did above in Corollaries 3.10 and 3.12. The proofs are exactly the same. Here we state them for A_∞ extrapolation; the analogous results for $A_{p/r}$ extrapolation are the same.

Corollary 3.17. *Let $\mathcal{B}$ be a Muckenhoupt basis. Suppose that for some p_0, $0 < p_0 < \infty$, and every $w_0 \in A_{\infty,\mathcal{B}}$,*

$$\|f\|_{L^{p_0,\infty}(w_0)} \le C\|g\|_{L^{p_0}(w_0)}, \qquad (f,g) \in \mathcal{F}. \tag{3.11}$$

Then for all p, $0 < p < \infty$, and for all $w \in A_{\infty,\mathcal{B}}$,

$$\|f\|_{L^{p,\infty}(w)} \le C\|g\|_{L^p(w)}, \qquad (f,g) \in \mathcal{F}. \tag{3.12}$$

Corollary 3.18. *Let $\mathcal{B}$ be a Muckenhoupt basis $\mathcal{B}$ and suppose that (3.9) holds. Then for all p and q, $0 < p, q < \infty$, $w \in A_{\infty,\mathcal{B}}$, and sequences $\{(f_i, g_i)\}_i \subset \mathcal{F}$,*

$$\Big\|\Big(\sum_i f_i^q\Big)^{1/q}\Big\|_{L^p(w)} \le C\Big\|\Big(\sum_i g_i^q\Big)^{1/q}\Big\|_{L^p(w)}. \tag{3.13}$$

Remark 3.19. Using these ideas, it is possible to extrapolate from inequalities of the form $\|f\|_{L^{p,\infty}(w)} \le C\|g\|_{L^{p,\infty}(w)}$, $w \in A_\infty$. This was proved in [44] and we refer the reader there for details. It is an open question as to whether this extrapolation argument works for other Lorentz spaces $L^{p,q}$, $q < \infty$.

A_1 extrapolation

There is an interesting connection between A_∞ extrapolation and A_1 extrapolation —that is, extrapolation for norm inequalities involving A_1 weights. In [44] the following equivalence was proved—in fact, this result was the basis of the proof of A_∞ extrapolation given there.

Proposition 3.20. *Given a Muckenhoupt basis $\mathcal{B}$, the following are equivalent:*

(a) *For all p, $0 < p < \infty$, and all $w \in A_{\infty,\mathcal{B}}$,*

$$\int_{\mathbb{R}^n} f(x)^p w(x)\,dx \le C \int_{\mathbb{R}^n} g(x)^p w(x)\,dx, \qquad (f,g) \in \mathcal{F}.$$

(b) *There exists $p_0 > 0$ such that for all p, $0 < p < p_0$, and all $w \in A_{1,\mathcal{B}}$,*

$$\int_{\mathbb{R}^n} f(x)^p w(x)\,dx \le C \int_{\mathbb{R}^n} g(x)^p w(x)\,dx, \qquad (f,g) \in \mathcal{F}.$$

A similar equivalence holds for A_p extrapolation but with an extra assumption on the basis.

Proposition 3.21. *Given a Muckenhoupt basis $\mathcal{B}$ that is $A_{p,\mathcal{B}}$ open, the following are equivalent:*

(a) *For all p, $1 < p < \infty$, and all $w \in A_{p,\mathcal{B}}$,*

$$\int_{\mathbb{R}^n} f(x)^p w(x)\,dx \le C \int_{\mathbb{R}^n} g(x)^p w(x)\,dx, \qquad (f,g) \in \mathcal{F}.$$

(b) *There exists $p_0 > 1$ such that for all p, $1 < p < p_0$ and all $w \in A_{1,\mathcal{B}}$,*

$$\int_{\mathbb{R}^n} f(x)^p w(x)\,dx \le C \int_{\mathbb{R}^n} g(x)^p w(x)\,dx, \qquad (f,g) \in \mathcal{F}.$$

This equivalence was implicit in the proof of Rubio de Francia extrapolation in the classical case given by Duoandikoetxea [68]. He used the maximal function instead of the iteration algorithm, but the proof can be readily adapted to the more general setting. The proofs of both results require rescaling arguments similar to the one in Corollary 3.15 and we leave the details to the reader. It is an open question whether Proposition 3.21 is true if we do not assume the $A_{p,\mathcal{B}}$ openness condition.

3.4 Sharp extrapolation constants

In this section we give a different proof of Theorem 3.9, one which gives a sharp estimate for the constant. As before we use duality, reverse factorization and weighted estimates for the maximal operator, but in order to get the sharp constant we must divide the proof into two cases depending on whether $p < p_0$ or $p > p_0$. Our proof uses some ideas from a new proof in the special case of a basis of cubes shown to us by Duoandikoetxea [69]; a similar proof is in Grafakos [92].

Theorem 3.22. *Let $\mathcal{B}$ be a Muckenhoupt basis. Suppose that for some p_0, $1 \le p_0 < \infty$, there exists a positive increasing function N_{p_0} on $[1,\infty)$ such that for every $w_0 \in A_{p_0,\mathcal{B}}$,*

$$\int_{\mathbb{R}^n} f(x)^{p_0}\, w_0(x)\, dx \\ \le N_{p_0}([w_0]_{A_{p_0,\mathcal{B}}}) \int_{\mathbb{R}^n} g(x)^{p_0}\, w_0(x)\, dx, \qquad (f,g) \in \mathcal{F}. \tag{3.14}$$

Then for all p, $1 < p < \infty$, and for all $w \in A_{p,\mathcal{B}}$,

$$\int_{\mathbb{R}^n} f(x)^{p}\, w(x)\, dx \\ \le N_p(\mathcal{B}, p_0, p, [w]_{A_{p,\mathcal{B}}}) \int_{\mathbb{R}^n} g(x)^{p}\, w(x)\, dx, \qquad (f,g) \in \mathcal{F}, \tag{3.15}$$

where if $p < p_0$,

$$N_p(\mathcal{B}, p_0, p, [w]_{A_{p,\mathcal{B}}}) = 2^{\frac{p_0-p}{p_0}}\, N_{p_0}\big(2^{p_0-p}\, \|M_{\mathcal{B}}\|_{L^p(w)}^{p_0-p}\, [w]_{A_{p,\mathcal{B}}}\big),$$

and if $p > p_0$,

$$N_p(\mathcal{B}, p_0, p, [w]_{A_{p,\mathcal{B}}}) = 2^{\frac{p-p_0}{p_0\,(p-1)}}\, N_{p_0}\big(2^{\frac{p-p_0}{p-1}}\, \|M_{\mathcal{B}}\|_{L^{p'}(w^{1-p'})}^{\frac{p-p_0}{p-1}}\, [w]_{A_{p,\mathcal{B}}}^{\frac{p_0-1}{p-1}}\big).$$

When $\mathcal{B}$ is the basis of cubes $\mathcal{Q}$ and $M_{\mathcal{B}}$ is the Hardy-Littlewood maximal operator, Buckley [15] (see also [127]) proved that

$$\|M\|_{L^p(w)} \le C_n\, (p')^{1/p}\, (p)^{1/p'}\, [w]_{A_p}^{\frac{1}{p-1}}.$$

It follows immediately from this that

$$\|M\|_{L^{p'}(w^{1-p'})} \le C_{n,p} [w^{1-p'}]_{A'_p}^{\frac{1}{p'-1}} = C_{n,p}\, [w]_{A_p},$$

and so

$$N_p(\mathcal{Q}, p_0, p, [w]_{A_{p,\mathcal{Q}}}) \le C_{p,p_0}\, N_{p_0}\big(C_{n,p,p_0}\, [w]_{A_p}^{\max\left(1, \frac{p_0-1}{p-1}\right)}\big).$$

This constant was first proved by Petermichl and Volberg [184] for $p > p_0 = 2$, and in general by Dragičević, *et al.* [65]. (Also see [92].) We say that this constant is sharp since for many operators—e.g., singular integrals—the results gotten by extrapolation with this constant are the best possible. See [15, 48, 49, 65].

Proof. We consider two cases: $1 < p < p_0$ and $1 \le p_0 < p < \infty$.

Case 1: $1 < p < p_0$. Fix p, $1 < p < p_0$, and $w \in A_{p,\mathcal{B}}$. Let $\mathcal{R}$ be the iteration algorithm defined in the proof of Theorem 3.9; we will use properties (A), (B) and (C) given there. Fix $(f,g) \in \mathcal{F}$; without loss of generality f and g are non-trivial and in $L^p(w)$. Let $\varepsilon > 0$ and define

$$h(x) = \varepsilon \frac{f(x)}{\|f\|_{L^p(w)}} + \frac{g(x)}{\|g\|_{L^p(w)}}.$$

Then $h \in L^p(w)$ and $\|h\|_{L^p(w)} \le 1+\varepsilon$. By Hölder's inequality with respect to the measure w and with exponents p_0/p and $(p_0/p)' = p_0/(p_0 - p)$, and by property (B) we have that

$$\begin{aligned}
\|f\|_{L^p(w)} &= \left(\int_{\mathbb{R}^n} f(x)^p\, \mathcal{R}h(x)^{-(p_0-p)\frac{p}{p_0}}\, \mathcal{R}h(x)^{(p_0-p)\frac{p}{p_0}}\, w(x)\, dx\right)^{1/p} \\
&\le \left(\int_{\mathbb{R}^n} f(x)^{p_0}\, \mathcal{R}h(x)^{-(p_0-p)}\, w(x)\, dx\right)^{1/p_0} \\
&\qquad \times \left(\int_{\mathbb{R}^n} \mathcal{R}h(x)^p\, w(x)\, dx\right)^{\frac{p_0-p}{p\, p_0}} \\
&\le (2\,(1+\varepsilon))^{\frac{p_0-p}{p_0}} \left(\int_{\mathbb{R}^n} f(x)^{p_0}\, \mathcal{R}h(x)^{-(p_0-p)}\, w(x)\, dx\right)^{1/p_0}.
\end{aligned}$$

By property (A), $f/\|f\|_{L^p(w)} \le \varepsilon^{-1}\, h \le \varepsilon^{-1}\, \mathcal{R}h$; hence,

$$\begin{aligned}
&\int_{\mathbb{R}^n} f(x)^{p_0}\, \mathcal{R}h(x)^{-(p_0-p)}\, w(x)\, dx \\
&\qquad \le \varepsilon^{-(p_0-p)}\, \|f\|_{L^p(w)}^{p_0-p} \int_{\mathbb{R}^n} f(x)^p\, w(x)\, dx = \varepsilon^{-(p_0-p)}\, \|f\|_{L^p(w)}^{p_0} < \infty.
\end{aligned}$$

We claim that $W = \mathcal{R}h^{-(p_0-p)}\, w \in A_{p_0,\mathcal{B}}$ and

$$[W]_{A_{p_0,\mathcal{B}}} \le 2^{p_0-p}\, \|M_{\mathcal{B}}\|_{L^p(w)}^{p_0-p}\, [w]_{A_{p,\mathcal{B}}}.$$

Assuming this for the moment, we can apply (3.14): since N_{p_0} is increasing and since by property (A), $g/\|g\|_{L^p(w)} \le h \le \mathcal{R}h$, we have that

$$\begin{aligned}\|f\|_{L^p(w)} &\leq \big(2\,(1+\varepsilon)\big)^{\frac{p_0-p}{p_0}}\, N_{p_0}([W]_{A_{p_0,\mathcal{B}}}) \\ &\qquad\times\left(\int_{\mathbb{R}^n} g(x)^{p_0}\, \mathcal{R}h(x)^{-(p_0-p)}\, w(x)\,dx\right)^{1/p_0} \\ &\leq \big(2\,(1+\varepsilon)\big)^{\frac{p_0-p}{p_0}}\, N_{p_0}\big(2^{p_0-p}\,\|M_{\mathcal{B}}\|_{L^p(w)}^{p_0-p}\,[w]_{A_{p,\mathcal{B}}}\big)\,\|g\|_{L^p(w)}.\end{aligned}$$

Since this is true for every ε, if we let $\varepsilon \to 0$, we get (3.15).

To complete the proof of this case we need to show that $W \in A_{p_0,\mathcal{B}}$ and estimate $[W]_{A_{p_0,\mathcal{B}}}$. Given $B \in \mathcal{B}$, by property (C) we have that

$$\begin{aligned}\fint_B W(x)\,dx &= \fint_B \mathcal{R}g(x)^{-(p_0-p)}\,w(x)\,dx \\ &\leq [\mathcal{R}g]_{A_{1,\mathcal{B}}}^{p_0-p}\left(\fint_B \mathcal{R}g(x)\,dx\right)^{-(p_0-p)} \fint_B w(x)\,dx.\end{aligned}$$

On the other hand, if we set $q=(p_0-1)/(p-1)>1$, then $q'=(p_0-1)/(p_0-p)$, and so by Hölder's inequality,

$$\begin{aligned}\fint_B W(x)^{1-p_0'}\,dx &= \fint_B \mathcal{R}g(x)^{\frac{p_0-p}{p_0-1}}\,w(x)^{1-p_0'}\,dx \\ &\leq \left(\fint_B \mathcal{R}g(x)\,dx\right)^{\frac{p_0-p}{p_0-1}}\left(\fint_B w(x)^{1-p'}\,dx\right)^{\frac{p-1}{p_0-1}}.\end{aligned}$$

If we combine these two estimates and use property (C), we obtain $W \in A_{p_0,\mathcal{B}}$ and

$$\begin{aligned}&\fint_B W(x)\,dx\left(\fint_B W(x)^{1-p_0'}\,dx\right)^{p_0-1} \\ &\qquad\leq [\mathcal{R}g]_{A_{1,\mathcal{B}}}^{p_0-p}\fint_B w(x)\,dx\left(\fint_B w(x)^{1-p'}\,dx\right)^{p-1} \\ &\qquad\leq [\mathcal{R}g]_{A_{1,\mathcal{B}}}^{p_0-p}\,[w]_{A_{p,\mathcal{B}}} \leq \big(2\,\|M_{\mathcal{B}}\|_{L^p(w)}\big)^{p_0-p}\,[w]_{A_{p,\mathcal{B}}}.\end{aligned}$$

Case 2: $p_0 < p < \infty$. Fix p, $p_0 < p < \infty$, and $w \in A_{p,\mathcal{B}}$. Let $\mathcal{R}'$ be as in the proof of Theorem 3.9; we will use properties (A′), (B′) and (C′) given there. Note that by the definition of $M'_{\mathcal{B}}$, $\|M'_{\mathcal{B}}\|_{L^{p'}(w)} = \|M_{\mathcal{B}}\|_{L^{p'}(w^{1-p'})}$.

Fix $(f,g)\in\mathcal{F}$; without loss of generality f and g are non-trivial and in $L^p(w)$. Since $f \in L^p(w)$, by duality there exists a non-negative function $h \in L^{p'}(w)$, $\|h\|_{L^{p'}(w)} = 1$, such that

$$\|f\|_{L^p(w)} = \int_{\mathbb{R}^n} f(x)\,h(x)\,w(x)\,dx.$$

Assume first that $p_0 > 1$. We claim that $W = \mathcal{R}'h^{\frac{p-p_0}{p-1}}\, w \in A_{p_0,\mathcal{B}}$ and

$$[W]_{A_{p_0,\mathcal{B}}} \le 2^{\frac{p-p_0}{p-1}}\, \|M_{\mathcal{B}}\|_{L^{p'}(w^{1-p'})}^{\frac{p-p_0}{p-1}}\, [w]_{A_{p,\mathcal{B}}}^{\frac{p_0-1}{p-1}}.$$

If we assume this for the moment, then by property (A′) and Hölder's inequality with exponents p_0 and p_0',

$$\begin{aligned}
&\int_{\mathbb{R}^n} f(x)\, h(x)\, w(x)\, dx \\
&\quad \le \int_{\mathbb{R}^n} f(x)\, \mathcal{R}'h(x)^{\frac{p-p_0}{p_0\,(p-1)}}\, h(x)^{\frac{(p_0-1)\,p}{p_0\,(p-1)}}\, w(x)\, dx \\
&\quad \le \left(\int_{\mathbb{R}^n} f(x)^{p_0}\, W(x)\, dx\right)^{1/p_0} \left(\int_{\mathbb{R}^n} h(x)^{p'}\, w(x)\, dx\right)^{1/p_0'} \\
&\quad = \left(\int_{\mathbb{R}^n} f(x)^{p_0}\, W(x)\, dx\right)^{1/p_0}.
\end{aligned}$$

By Hölder's inequality with exponents p/p_0 and $(p/p_0)'$, and by property (B′),

$$\begin{aligned}
&\int_{\mathbb{R}^n} f(x)^{p_0}\, \mathcal{R}'h(x)^{\frac{p-p_0}{p-1}}\, w(x)\, dx \\
&\qquad\qquad \le \|f\|_{L^p(w)}^{p_0} \|\mathcal{R}'h\|_{L^{p'}(w)}^{\frac{p-p_0}{p-1}} \le 2^{\frac{p-p_0}{p-1}}\, \|f\|_{L^p(w)}^{p_0} < \infty.
\end{aligned}$$

Therefore, we can apply (3.14): since N_{p_0} is increasing, by Hölder's inequality and property (B′),

$$\begin{aligned}
&\int_{\mathbb{R}^n} f(x)\, h(x)\, w(x)\, dx \\
&\quad \le N_{p_0}([W]_{A_{p_0,\mathcal{B}}}) \left(\int_{\mathbb{R}^n} g(x)^{p_0}\, \mathcal{R}'h^{\frac{p-p_0}{p-1}}\, w\, dx\right)^{1/p_0} \\
&\quad \le N_{p_0}\big(2^{\frac{p-p_0}{p-1}}\, \|M_{\mathcal{B}}\|_{L^{p'}(w^{1-p'})}^{\frac{p-p_0}{p-1}}\, [w]_{A_{p,\mathcal{B}}}^{\frac{p_0-1}{p-1}}\big)\, \|g\|_{L^p(w)}\, \|\mathcal{R}'h\|_{L^{p'}(w)}^{\frac{p-p_0}{p_0\,(p-1)}} \\
&\quad \le 2^{\frac{p-p_0}{p_0\,(p-1)}}\, N_{p_0}\big(2^{\frac{p-p_0}{p-1}}\, \|M_{\mathcal{B}}\|_{L^{p'}(w^{1-p'})}^{\frac{p-p_0}{p-1}}\, [w]_{A_{p,\mathcal{B}}}^{\frac{p_0-1}{p-1}}\big)\, \|g\|_{L^p(w)}.
\end{aligned}$$

To complete the proof we need to show that $W \in A_{p_0,\mathcal{B}}$ and estimate $[W]_{A_{p_0,\mathcal{B}}}$. Fix $B \in \mathcal{B}$. If we set $q = (p-1)/(p-p_0) > 1$, then $q' = (p-1)/(p_0-1)$, and so by Hölder's inequality,

$$\begin{aligned}
&\fint_B W(x)\, dx = \fint_B \mathcal{R}'h(x)^{\frac{p-p_0}{p-1}}\, w(x)\, dx \\
&\qquad\qquad \le \left(\fint_B \mathcal{R}'h(x)\, w(x)\, dx\right)^{\frac{p-p_0}{p-1}} \left(\fint_B w(x)\, dx\right)^{\frac{p_0-1}{p-1}}.
\end{aligned}$$

On the other hand, by property (C′),

$$\begin{aligned}
&\left(\fint_B W(x)^{1-p_0'}\,dx\right)^{p_0-1} \\
&\quad= \left(\fint_B \mathcal{R}'h(x)^{-\frac{p-p_0}{(p-1)(p_0-1)}}\,w(x)^{1-p_0'}\,dx\right)^{p_0-1} \\
&\quad\le [\mathcal{R}'h\,w]_{A_{1,\mathcal{B}}}^{\frac{p-p_0}{p-1}} \left(\fint_B \mathcal{R}'h(x)\,w(x)\,dx\right)^{-\frac{p-p_0}{p-1}} \\
&\qquad\quad\times \left(\fint_B w(x)^{\frac{p-p_0}{(p-1)(p_0-1)}}\,w(x)^{1-p_0'}\,dx\right)^{p_0-1} \\
&\quad= [\mathcal{R}'h\,w]_{A_{1,\mathcal{B}}}^{\frac{p-p_0}{p-1}} \left(\fint_B \mathcal{R}'h(x)\,w(x)\,dx\right)^{-\frac{p-p_0}{p-1}} \left(\fint_B w(x)^{1-p'}\,dx\right)^{p_0-1}.
\end{aligned}$$

If we combine these estimates and use property (C′), then we get that $W \in A_{p_0,\mathcal{B}}$ and

$$\begin{aligned}
&\fint_B W(x)\,dx \left(\fint_B W(x)^{1-p_0'}\,dx\right)^{p_0-1} \\
&\qquad\le [\mathcal{R}'h\,w]_{A_{1,\mathcal{B}}}^{\frac{p-p_0}{p-1}} \left(\fint_B w(x)\,dx\right)^{\frac{p_0-1}{p-1}} \left(\fint_B w(x)^{1-p'}\,dx\right)^{p_0-1} \\
&\qquad\qquad\le [\mathcal{R}'h\,w]_{A_{1,\mathcal{B}}}^{\frac{p-p_0}{p-1}}\,[w]_{A_{p,\mathcal{B}}}^{\frac{p_0-1}{p-1}} \le 2^{\frac{p-p_0}{p-1}}\,\|M_{\mathcal{B}}\|_{L^{p'}(w^{1-p'})}^{\frac{p-p_0}{p-1}}\,[w]_{A_{p,\mathcal{B}}}^{\frac{p_0-1}{p-1}}.
\end{aligned}$$

Finally, suppose $p_0 = 1$. In this case the argument is simpler. Let $W = \mathcal{R}'h\,w$; by (C′), $W \in A_{1,\mathcal{B}}$ and $[W]_{A_{1,\mathcal{B}}} \le 2\,\|M_{\mathcal{B}}\|_{L^{p'}(w^{1-p'})}$. Then we can argue as before to get

$$\begin{aligned}
\int_{\mathbb{R}^n} f(x)\,h(x)\,w(x)\,dx &\le \int_{\mathbb{R}^n} f(x)\,\mathcal{R}'h(x)\,dx \\
&\le N_1([W]_{A_{1,\mathcal{B}}}) \int_{\mathbb{R}^n} g(x)\,\mathcal{R}'h(x)\,w\,dx \\
&\le N_1\big(2\,\|M_{\mathcal{B}}\|_{L^{p'}(w^{1-p'})}\big)\,\|g\|_{L^p(w)}\,\|\mathcal{R}'h\|_{L^{p'}(w)} \\
&\le 2\,N_1\big(2\,\|M_{\mathcal{B}}\|_{L^{p'}(w^{1-p'})}\big)\,\|g\|_{L^p(w)}. \qquad \square
\end{aligned}$$

3.5 Off-diagonal extrapolation

In this section we extend Theorem 3.9 to the off-diagonal case: that is, norm inequalities from $L^p(w^p)$ to $L^q(w^q)$, $1 < p \le q < \infty$. In this case the correct weights are in the $A_{p,q,\mathcal{B}}$ classes: given $1 < p \le q < \infty$, then $w \in A_{p,q,\mathcal{B}}$ if for every $B \in \mathcal{B}$,

$$\left(\frac{1}{|B|}\int_B w(x)^q\,dx\right)^{1/q}\left(\frac{1}{|B|}\int_B w(x)^{-p'}\,dx\right)^{1/p'} \leq K < \infty.$$

For $p = 1$, we say that $w \in A_{1,q,\mathcal{B}}$ if for almost every $x \in B$,

$$\left(\frac{1}{|B|}\int_B w(x)^q\,dx\right)^{1/q} \leq Kw(x).$$

When $p = q$, this definition is equivalent to $w^p \in A_{p,\mathcal{B}}$; further it follows immediately that $w \in A_{p,q,\mathcal{B}}$ if and only if $w^q \in A_{1+q/p',\mathcal{B}}$. Thus the $A_{p,q,\mathcal{B}}$ classes can be seen as subsets of $A_{\infty,\mathcal{B}}$. In the classical case (i.e., when $\mathcal{B} = \mathcal{Q}$) this weight class is denoted by $A_{p,q}$. It was introduced by Muckenhoupt and Wheeden [154] to study weighted norm inequalities for fractional integral operators.

Theorem 3.23. *Given a Muckenhoupt basis $\mathcal{B}$, suppose that for some p_0, q_0, $1 \leq p_0 \leq q_0 < \infty$, and every $w_0 \in A_{p_0,q_0,\mathcal{B}}$,*

$$\left(\int_{\mathbb{R}^n} f(x)^{q_0} w_0(x)^{q_0}\,dx\right)^{1/q_0} \leq C\left(\int_{\mathbb{R}^n} g(x)^{p_0} w_0(x)^{p_0}\,dx\right)^{1/p_0}, \quad (f,g) \in \mathcal{F}. \tag{3.16}$$

Then for all p and q such that $1 < p \leq q < \infty$ and $1/p - 1/q = 1/p_0 - 1/q_0$, and for all $w \in A_{p,q,\mathcal{B}}$,

$$\left(\int_{\mathbb{R}^n} f(x)^q w(x)^q\,dx\right)^{1/q} \leq C\left(\int_{\mathbb{R}^n} g(x)^p w(x)^p\,dx\right)^{1/p}, \quad (f,g) \in \mathcal{F}. \tag{3.17}$$

Further, for all r, $1 < r < \infty$, and all sequences $\{(f_i, g_i)\}_i \subset \mathcal{F}$,

$$\Big\|\Big(\sum_i f_i^r\Big)^{1/r}\Big\|_{L^q(w^q)} \leq C\Big\|\Big(\sum_i g_i^r\Big)^{1/r}\Big\|_{L^p(w^p)}. \tag{3.18}$$

In the scalar case, Theorem 3.23 was originally proved by Harboure, Macías and Segovia [96] for a basis of cubes; the extension to vector-valued inequalities is new.

Remark 3.24. A version of Theorem 3.23 with sharp constants was proved by Lacey *et al.* [120]. A more general version, with sharp constants and including values of $q < 1$, is due to Duoandikoetxea [69].

The proof of Theorem 3.23 is similar to that of Theorem 3.9: the two main differences are that duality is not used to get to L^1 but to L^s, $s > 1$, and both iteration algorithms are defined in terms of $M_{\mathcal{B}}$, instead of $M_{\mathcal{B}}$ and $M'_{\mathcal{B}}$. The same variation is used in Theorem 3.31 below. (Also see the alternative proof of the two-weight extrapolation theorem in Section 7.6.)

Proof. The case $p_0 = q_0$ corresponds to Theorem 3.9, so we may assume that $p_0 < q_0$. We first consider the case $p_0 > 1$. Fix $1 < p < q < \infty$ such that $1/p - 1/q = 1/p_0 - 1/q_0$, and $w \in A_{p,q,\mathcal{B}}$. Choose s so that $1/s' = 1/p - 1/q = 1/p_0 - 1/q_0$. Then $1 < s < \min(q, q_0)$. Let $r_0 = q_0/s > 1$ and $r = q/s = 1 + q/p' > 1$.

We now define the Rubio de Francia iteration algorithms. Since $w \in A_{p,q,\mathcal{B}}$, we have that $w^q \in A_{1+q/p',\mathcal{B}} = A_{r,\mathcal{B}}$, and so $M_\mathcal{B}$ is bounded on $L^r(w^q)$. Therefore, given non-negative $h_1 \in L^r(w^q)$ let

$$\mathcal{R}h_1(x) = \sum_{k=0}^{\infty} \frac{M_\mathcal{B}^k h_1(x)}{2^k \|M_\mathcal{B}\|_{L^r(w^q)}^k}.$$

$\mathcal{R}h_1$ satisfies properties (A), (B) and (C) in the proof of Theorem 3.9 with the obvious changes.

Furthermore, since $w^{q(1-r')} \in A_{r',\mathcal{B}}$, $M_\mathcal{B}$ is bounded on $L^{r'}(w^{q(1-r')})$, so for non-negative $h_2 \in L^{r'}(w^{q(1-r')})$ we define

$$\mathcal{R}'h_2(x) = \sum_{k=0}^{\infty} \frac{M_\mathcal{B}^k h_2(x)}{2^k \|M_\mathcal{B}\|_{L^{r'}(w^{q(1-r')})}^k}.$$

(Unlike in the proof of Theorem 3.9, here $\mathcal{R}'$ is composed of iterations of $M_\mathcal{B}$ instead of $M'_\mathcal{B}$. To distinguish between the two algorithms we abuse our earlier notation and denote this iteration algorithm by $\mathcal{R}'$.) Again, $\mathcal{R}'h_2$ satisfies properties (A), (B) and (C).

Fix $(f, g) \in \mathcal{F}$; without loss of generality we may assume that both functions are non-trivial and $f \in L^q(w^q)$, $g \in L^p(w^p)$. Define

$$h_1(x) = \frac{f(x)}{\|f\|_{L^q(w^q)}} + \frac{g(x)^{p/q}\, w(x)^{p/q-1}}{\|g\|_{L^p(w^p)}^{p/q}};$$

then $\|h_1\|_{L^q(w^q)} \leq 2$. By duality, there exists non-negative $h_2 \in L^{r'}(w^q)$ with $\|h_2\|_{L^{r'}(w^q)} = 1$ such that

$$\|f\|_{L^q(w^q)} = \|f^s\|_{L^r(w^q)}^{1/s} = \left(\int_{\mathbb{R}^n} f(x)^s\, h_2(x)\, w(x)^q\, dx \right)^{1/s}.$$

For clarity, let

$$H_1 = \mathcal{R}(h_1^s)^{1/s}, \qquad H_2 = \mathcal{R}'(h_2\, w^q)\, w^{-q}.$$

Then by the properties of the iteration algorithms,

(A_1)	$h_1(x) \leq H_1(x)$	(A_2)	$h_2(x) \leq H_2(x)$
(B_1)	$\|H_1\|_{L^q(w^q)} \leq C$	(B_2)	$\|H_2\|_{L^{r'}(w^q)} \leq C$
(C_1)	$H_1^s \in A_{1,\mathcal{B}}$	(C_2)	$H_2\, w^q \in A_{1,\mathcal{B}}$.

Therefore, by property (A_2) and Hölder's inequality,

$$\begin{aligned}\|f\|_{L^q(w^q)} &\le \left(\int_{\mathbb{R}^n} f(x)^s\, H_2(x)\, w(x)^q\, dx\right)^{1/s}\\ &= \left(\int_{\mathbb{R}^n} f(x)^s\, H_1(x)^{-s/r_0'}\, H_1(x)^{s/r_0'}\, H_2(x)\, w(x)^q\, dx\right)^{1/s}\\ &\le \left(\int_{\mathbb{R}^n} f(x)^{q_0}\, H_1(x)^{-(q_0-s)}\, H_2(x)\, w(x)^q\, dx\right)^{1/q_0}\\ &\qquad\times \left(\int_{\mathbb{R}^n} H_1(x)^s\, H_2(x)\, w(x)^q\, dx\right)^{1/(sr_0')}\\ &= I_1 \times I_2^{1/(sr_0')}.\end{aligned}$$

We estimate each term separately. By Hölder's inequality, (B_1) and (B_2),

$$I_2 \le \|H_1\|^s_{L^q(w^q)}\, \|H_2\|_{L^{r'}(w^q)} \le C.$$

To estimate I_1 using (3.16), we first show that it is finite. By (A_1), $f/\|f\|_{L^q(w^q)} \le h_1 \le H_1$, so

$$I_1^{q_0} \le \|f\|^{q_0}_{L^q(w^q)} \int_{\mathbb{R}^n} H_1(x)^s\, H_2(x)\, w(x)^q\, dx = \|f\|^{q_0}_{L^q(w^q)} \cdot I_2 < \infty.$$

We claim that $W = \left(H_1^{-(q_0-s)}\, H_2\, w^q\right)^{1/q_0} \in A_{p_0,q_0,\mathcal{B}}$. Assuming this for the moment, since by (A_1), $g^{p/q}\, w^{p/q-1}/\|g\|^{p/q}_{L^p(w^p)} \le h_1 \le H_1$, it follows from (3.16) that

$$\begin{aligned}I_1 &\le C\left(\int_{\mathbb{R}^n} g(x)^{p_0}\, W(x)^{p_0}\, dx\right)^{1/p_0}\\ &\le \|g\|_{L^p(w^p)} \left(\int_{\mathbb{R}^n} H_1(x)^{\frac{p_0\, q}{p} - \frac{p_0\,(q_0-s)}{q_0}}\, H_2(x)^{p_0/q_0} w(x)^{p_0\,(\frac{q}{p}-1)+\frac{q\, p_0}{q_0}}\, dx\right)^{1/p_0}\\ &= \|g\|_{L^p(w^p)} \left(\int_{\mathbb{R}^n} H_1(x)^{\frac{p_0\, q}{p} - \frac{p_0\,(q_0-s)}{q_0}}\, H_2(x)^{p_0/q_0}\, w(x)^q\, dx\right)^{1/p_0}.\end{aligned}$$

Let $\alpha = r'\, q_0/p_0 > 1$; then

$$\alpha' \left(\frac{p_0\, q}{p} - \frac{p_0\,(q_0 - s)}{q_0}\right) = q.$$

Therefore, by Hölder's inequality with exponent α and properties (B_1) and (B_2) we get

$$I_1 \le C\|g\|_{L^p(w^p)}\, \|H_1\|^{\frac{q}{p_0\,\alpha'}}_{L^q(w^q)}\, \|H_2\|^{1/q_0}_{L^{r'}(w^q)} \le C\|g\|_{L^p(w^p)}.$$

To complete the proof we need to show that $W \in A_{p_0,q_0,\mathcal{B}}$. We will do so by generalizing the reverse factorization property (Proposition 3.5). Given $B \in \mathcal{B}$, by property (C_1) we have that

$$\left(\fint_B W(x)^{q_0}\,dx\right)^{1/q_0} = \left(\fint_B H_1(x)^{-(q_0-s)}\,H_2(x)\,w(x)^q\,dx\right)^{1/q_0}$$
$$\leq C\left(\fint_B H_1(x)^s\,dx\right)^{\frac{s-q_0}{q_0 s}}\left(\fint_B H_2(x)\,w(x)^q\,dx\right)^{1/q_0}.$$

By property (C_2),

$$\left(\fint_B W(x)^{-p_0'}\,dx\right)^{1/p_0'}$$
$$= \left(\fint_B H_1(x)^{\frac{(q_0-s)\,p_0'}{q_0}}\left(H_2(x)\,w(x)^q\right)^{-p_0'/q_0}\,dx\right)^{1/p_0'}$$
$$\leq C\left(\fint_B H_2(x)\,w(x)^q\,dx\right)^{-1/q_0}\left(\fint_B H_1(x)^{\frac{(q_0-s)\,p_0'}{q_0}}\,dx\right)^{1/p_0'}$$
$$= \left(\fint_B H_2(x)\,w(x)^q\,dx\right)^{-1/q_0}\left(\fint_B H_1(x)^s\,dx\right)^{\frac{q_0-s}{q_0 s}}.$$

Combining these estimates we get that $w \in A_{p_0,q_0,\mathcal{B}}$:

$$\left(\fint_B W(x)^{q_0}\,dx\right)^{1/q_0}\left(\fint_B W(x)^{-p_0'}\,dx\right)^{1/p_0'} \leq C.$$

We now consider the case $p_0 = 1$; the proof is very similar. Arguing as before, we let $s = q_0$, $r_0 = 1$ and $r = q/q_0 > 1$. Apply duality to get $h_2 \in L^{r'}(w^q)$ and define H_2. Let $W = (H_2\,w^q)^{1/q_0}$; then property (C_2) yields $W^{q_0} \in A_{1,\mathcal{B}}$, so $W \in A_{1,q_0,\mathcal{B}}$. Then property (B_2) implies that

$$\int_{\mathbb{R}^n} f(x)^{q_0}\,H_2(x)\,w(x)^q\,dx \leq \|f\|_{L^q(w^q)}^{q_0}\,\|H_2\|_{L^{r'}(w^q)} \leq C\|f\|_{L^q(w^q)}^{q_0} < \infty.$$

Therefore, we can use (3.16) to get the desired inequality:

$$\|f\|_{L^q(w^q)} \leq \left(\int_{\mathbb{R}^n} f(x)^{q_0}\,H_2(x)\,w(x)^q\,dx\right)^{1/q_0}$$
$$\leq C\int_{\mathbb{R}^n} g(x)\left(H_2(x)\,w(x)^q\right)^{1/q_0}\,dx$$
$$\leq \|g\|_{L^p(w^p)}\,\|H_2\|_{L^{r'}(w^q)}^{1/q_0}$$

$$\leq C\|g\|_{L^p(w^p)}.$$

Finally, we prove (3.18). Fix r, $1 < r < \infty$, and define the new family $\mathcal{F}_{\ell^r}$ consisting of the pairs (F, G) where

$$F(x) = \Big(\sum_i f_i(x)^r\Big)^{1/r}, \qquad G(x) = \Big(\sum_i g_i(x)^r\Big)^{1/r},$$

and $\{(f_i, g_i)\} \subset \mathcal{F}$. Choose the pair (p_1, q_1) so that $p_1 \leq r \leq q_1$ and $1/p_0 - 1/q_0 = 1/p_1 - 1/q_1$. This is possible for all values of r. Fix $w \in A_{p_1,q_1,\mathcal{B}}$. Then by Minkowski's inequality with exponent $q_1/r \geq 1$, by (3.17), and then again by Minkowski's inequality with exponent $r/p_1 \geq 1$,

$$\begin{aligned}
\|F\|^r_{L^{q_1}(w^{q_1})} &= \left(\int_{\mathbb{R}^n} \Big(\sum_i f_i(x)^r\Big)^{q_1/r} w(x)^{q_1}\, dx\right)^{r/q_1} \\
&\leq \sum_i \left(\int_{\mathbb{R}^n} f_i(x)^{q_1} w(x)^{q_1}\, dx\right)^{r/q_1} \\
&\leq C \sum_i \left(\int_{\mathbb{R}^n} g_i(x)^{p_1} w(x)^{p_1}\, dx\right)^{r/p_1} \\
&\leq C \left(\int_{\mathbb{R}^n} \Big(\sum_i g_i(x)^r\Big)^{p_1/r} w(x)^{p_1}\, dx\right)^{r/p_1} \\
&\leq C\|G\|^r_{L^{p_1}(w^{p_1})}.
\end{aligned}$$

This inequality gives us our initial estimate for the family $\mathcal{F}_{\ell^r}$, so by the first part of the theorem, inequality (3.17) holds for all p, q for the family $\mathcal{F}_{\ell^r}$. This is the desired estimate. □

3.6 Extrapolation for pairs of positive operators

In this section we consider a generalized extrapolation theorem in which we replace the operators M and M' by a pair of operators T_1 and T_2, and the A_p weights by the weights associated with these operators. We will first consider the special case of one-sided maximal operators and one-sided A_p weights; this example motivates the general case which we will discuss afterwards.

Extrapolation for one-sided weights

On the real line the one-sided Hardy-Littlewood maximal functions are defined by

$$M^+f(x) = \sup_{h>0} \frac{1}{h} \int_x^{x+h} |f(y)|\, dy, \qquad M^-f(x) = \sup_{h>0} \frac{1}{h} \int_{x-h}^{x} |f(y)|\, dy.$$

The original maximal operator as defined by Hardy and Littlewood [98] was equivalent to M^-. Clearly, $M^\pm f(x) \le Mf(x)$.

The weighted norm inequalities of these operators are governed by the one-sided Muckenhoupt weights. Given $1 < p < \infty$, $w \in A_p^+$ if for every $x \in \mathbb{R}$ and $h > 0$,

$$\left(\frac{1}{h}\int_{x-h}^{x} w(y)\,dy\right)\left(\frac{1}{h}\int_{x}^{x+h} w(y)^{1-p'}\,dy\right)^{p-1} \le K < \infty.$$

Similarly, $w \in A_p^-$ if

$$\left(\frac{1}{h}\int_{x}^{x+h} w(y)\,dy\right)\left(\frac{1}{h}\int_{x-h}^{x} w(y)^{1-p'}\,dy\right)^{p-1} \le K < \infty.$$

When $p = 1$, $w \in A_1^+$ if $M^-w(x) \le Cw(x)$ for a.e. $x \in \mathbb{R}$, and $w \in A_1^-$ if $M^+w(x) \le Cw(x)$ for a.e. $x \in \mathbb{R}$. The class A_∞^+ is defined as the union of the A_p^+ classes, $p \ge 1$; A_∞^- is defined analogously.

It follows from their definitions that the three classes A_p, A_p^+ and A_p^- are related as follows: $A_p = A_p^+ \cap A_p^-$, $A_p \subsetneq A_p^\pm$, and $w \in A_p^+$ if and only if $w^{1-p'} \in A_{p'}^-$. Further, we have the reverse factorization property: if $w_1 \in A_1^+$ and $w_2 \in A_1^-$, then $w_1 w_2^{1-p} \in A_p^+$. Detailed information on the properties of these weight classes is given in [31, 51, 52, 139, 140, 141, 142, 206].

Sawyer [206] showed that M^+ is bounded on $L^p(w)$, $1 < p < \infty$, if and only if $w \in A_p^+$. Martín-Reyes *et al.* [141] showed that these operators satisfy the weighted weak $(1,1)$ inequality $M^+ : L^1(w) \to L^{1,\infty}(w)$ if and only if $w \in A_1^+$. They also showed the analogous results for M^- and A_p^-.

Hereafter we will concentrate on the A_p^+ classes; the corresponding results for A_p^- are proved in exactly the same way and the details are left to the reader.

Theorem 3.25. *Suppose that for some p_0, $1 \le p_0 < \infty$, and every $w_0 \in A_{p_0}^+$,*

$$\int_{\mathbb{R}} f(x)^{p_0}\, w_0(x)\,dx \le C \int_{\mathbb{R}} g(x)^{p_0}\, w_0(x)\,dx, \qquad (f,g) \in \mathcal{F}. \tag{3.19}$$

Then for all p, $1 < p < \infty$, and for all $w \in A_p^+$,

$$\int_{\mathbb{R}} f(x)^{p}\, w(x)\,dx \le C \int_{\mathbb{R}} g(x)^{p}\, w(x)\,dx, \qquad (f,g) \in \mathcal{F}. \tag{3.20}$$

Proof. The proof is nearly identical to that of Theorem 3.9; the only substantive changes are in the definitions of the iteration algorithms.

Given $w \in A_p^+$, since M^+ is bounded on $L^p(w)$ we can define for non-negative $h_1 \in L^p(w)$ the iteration algorithm

$$\mathcal{R}^+ h_1(x) = \sum_{k=0}^{\infty} \frac{(M^+)^k h_1(x)}{2^k \|M^+\|_{L^p(w)}^k}.$$

Then $h_1(x) \le \mathcal{R}^+h_1(x)$ and $\|\mathcal{R}^+h_1\|_{L^p(w)} \le 2\|h_1\|_{L^p(w)}$. Furthermore, we have that

$$M^+(\mathcal{R}^+h_1) \le 2\,\|M^+\|_{L^p(w)}\,\mathcal{R}^+h_1,$$

so $\mathcal{R}^+h_1 \in A_1^-$. These properties are the analogs of properties (A), (B) and (C) for $\mathcal{R}$ in the proof of Theorem 3.9.

Since $w^{1-p'} \in A_p^-$, M^- is bounded on $L^{p'}(w^{1-p'})$. Therefore, the "dual" operator $(M^-)'f(x) = M^-(f\,w)(x)/w(x)$ is a bounded operator on $L^{p'}(w)$, and we can define the corresponding iteration algorithm for non-negative $h_2 \in L^{p'}(w)$:

$$\mathcal{R}^-h_2(x) = \sum_{k=0}^{\infty} \frac{((M^-)')^k h_2(x)}{2^k\|(M^-)'\|_{L^{p'}(w)}^k}.$$

We again have the analogs of the properties (A′), (B′) and (C′) satisfied by $\mathcal{R}'$ in the proof of Theorem 3.9: $h_2(x) \le \mathcal{R}^-h_2(x)$, $\|\mathcal{R}^-h_2\|_{L^{p'}(w)} \le 2\|h_2\|_{L^{p'}(w)}$, and

$$(M^-)'(\mathcal{R}^-h_2) \le 2\,\|(M^-)'\|_{L^{p'}(w)}\,\mathcal{R}^-h_2,$$

so $\mathcal{R}^-h_2 \cdot w \in A_1^+$. Finally, the reverse factorization property noted above implies that $(\mathcal{R}^+h_1)^{1-p_0}\mathcal{R}^-h_2 w \in A_{p_0}^+$.

Given these properties, the remainder of the proof, using duality and Hölder's inequality, follows exactly as in the proof of Theorem 3.9. □

Remark 3.26. As immediate consequences of Theorem 3.25 we can prove a weak type extrapolation theorem and get vector-valued inequalities (as in Corollaries 3.10 and 3.12), and use rescaling to prove the analogs of Corollaries 3.14 and 3.15. The (trivial) details are left to the reader.

Extrapolation for pairs of positive operators

In the proof of Theorem 3.25 we replaced the maximal operator M by a pair of maximal operators, M^+ and M^-. We can do this for arbitrary pairs of operators. To make this precise, we first need to make some definitions. Hereafter, a operator T will be assumed to map its domain $\mathrm{dom}(T)$ into L^1_{loc}. For simplicity we will assume that $\mathrm{dom}(T)$ contains "nice" functions—e.g., simple functions, bounded functions of compact support, Schwartz functions, etc.

An operator T is said to be admissible if it satisfies the following properties:

(a) T is sublinear;

(b) $Tf \ge 0$ for every $f \in \mathrm{dom}(T)$;

(c) for every sequence of non-negative functions $\{f_j\}_j \subset \mathrm{dom}(T)$, if $\sum_j f_j \in \mathrm{dom}(T)$, then $T\big(\sum_j f_j\big) \le \sum_j Tf_j$;

(d) T is positive: if $0 \le f \le g$ and $g \in \mathrm{dom}(T)$, then $f \in \mathrm{dom}(T)$ and $Tf \le Tg$.

Given an admissible operator T and a weight w, we say $w \in A_1(T)$ if $w \in \operatorname{dom}(T)$ and $Tw(x) \le Cw(x)$ for a.e. $x \in \mathbb{R}^n$. Given p, $1 < p < \infty$, we say that $w \in A_p(T)$ if T is bounded on $L^p(w)$. Given $\alpha \in \mathbb{R}$ and p, $1 \le p < \infty$, we say $w \in A_p(T)^\alpha$ if $w = w_1^\alpha$, $w_1 \in A_p(T)$.

Given two admissible operators T_1, T_2, if $w \in A_p(T_1) \cap A_{p'}(T_2)^{1-p}$, then T_1 is bounded on $L^p(w)$ and T_2 is bounded on $L^{p'}(w^{1-p'})$. Further, we say that $w \in A_1(T_1)^{1-p} \cdot A_1(T_2)$ if $w = w_1^{1-p} w_2$ with $w_1 \in A_1(T_1)$ and $w_2 \in A_1(T_2)$. Jawerth [108] used the Rubio de Francia iteration algorithm and ideas from [26] to prove that given admissible operators T_1 and T_2, for every p, $1 < p < \infty$,

$$A_p(T_1) \cap A_{p'}(T_2)^{1-p} \subset A_1(T_1)^{1-p} \cdot A_1(T_2).$$

In other words: every weight w such that T_1 is bounded on $L^p(w)$ and T_2 is bounded on $L^{p'}(w^{1-p'})$ can be factored as $w = w_1^{1-p} w_2$ with $w_1 \in A_1(T_1)$ and $w_2 \in A_1(T_2)$. (See also Hernández [102] and Ruiz and Torrea [198].)

This is a generalization of the difficult half of the Jones factorization theorem for A_p weights. Conversely, if $A_1(T_1)^{1-p} \cdot A_1(T_2) \subset A_p(T_1) \cap A_{p'}(T_2)^{1-p}$, then we say that T_1 and T_2 have the reverse factorization property. Though this property is immediate for A_p weights, in this more general setting it is not necessarily true. For example let $T_1 = M$, the Hardy-Littlewood maximal operator, and let T_2 be the averaging operator

$$T_2 f(x) = \left(\fint_Q |f(y)|\, dy \right) \chi_Q(x),$$

where Q is some fixed cube. Then these operators are admissible. Define $w_1 = 1$ and $w_2 = \chi_Q$. Then $w_1 \in A_1(T_1) = A_1$ and $w_2 \in A_1(T_2)$. Thus $w = w_1^{1-p} w_2 \in A_1(T_1)^{1-p} \cdot A_1(T_2)$, but $w = \chi_Q$ which is not in $A_p(T_1) = A_p$.

In the particular case where the pair of admissible operators consists of a linear operator T and its adjoint T^*, then Jawerth [108] showed that the reverse factorization property holds. (Again, see also [102, 198].)

Remark 3.27. If $T_1 = T_2 = M_{\mathcal{B}}$, then $A_p(M) = A_{p,\mathcal{B}}$, Similarly, if $T_1 = M^+$ and $T_2 = M^-$, then $A_1(M^+) = A_1^-$ and $A_1(M^-) = A_1^+$, and for every $p > 1$, $A_p(M^+) = A_p^+$, $A_p(M^-) = A_p^-$. Note that even though these are sublinear operators, in both cases the reverse factorization property holds.

We can now state our generalized extrapolation theorem.

Theorem 3.28. *Let T_1 and T_2 be admissible operators, and let $\mathcal{F}$ be a family of pairs of functions in* $\operatorname{dom}(T_1) \cap \operatorname{dom}(T_2)$. *Assume that for some p_0, $1 \le p_0 < \infty$, and every $w \in A_1(T_1)^{1-p_0} \cdot A_1(T_2)$,*

$$\int_{\mathbb{R}^n} f(x)^{p_0}\, w(x)\, dx \le C \int_{\mathbb{R}^n} g(x)^{p_0}\, w(x)\, dx \qquad (f,g) \in \mathcal{F}. \tag{3.21}$$

Then for all p, $1 < p < \infty$, and for all $w \in A_p(T_1) \cap A_{p'}(T_2)^{1-p}$,

$$\int_{\mathbb{R}^n} f(x)^{p}\, w(x)\, dx \le C \int_{\mathbb{R}^n} g(x)^{p}\, w(x)\, dx \qquad (f,g) \in \mathcal{F}. \tag{3.22}$$

Furthermore, if T_1 and T_2 satisfy the reverse factorization property, then (3.22) *holds for all* $w \in A_1(T_1)^{1-p} \cdot A_1(T_2)$.

Proof. The proof is essentially the same as the proof of Theorem 3.25, so we sketch the changes. As before, the key changes are in the definition of the iteration algorithms. Fix p, $1 < p < \infty$, and $w \in A_p(T_1) \cap A_{p'}(T_2)^{1-p}$. Then T_1 is bounded on $L^p(w)$ and T_2 is bounded on $L^{p'}(w^{1-p'})$, and so $T_2' f(x) = T_2(f\,w)(x)/w(x)$ is bounded on $L^{p'}(w)$. Given non-negative functions $h_1 \in L^p(w)$ and $h_2 \in L^{p'}(w)$, define the iteration algorithms

$$\mathcal{R}h_1(x) = \sum_{k=0}^{\infty} \frac{T_1^k h_1(x)}{2^k \|T_1\|_{L^p(w)}^k}, \qquad \mathcal{R}'h_2(x) = \sum_{k=0}^{\infty} \frac{(T_2')^k h_2(x)}{2^k \|T_2'\|_{L^{p'}(w)}^k}.$$

They have the following properties whose proofs follow as before:

$$\begin{array}{llll} \text{(A)} & h_1(x) \le \mathcal{R}h_1(x) & \text{(A}'\text{)} & h_2(x) \le \mathcal{R}'h_2(x) \\ \text{(B)} & \|\mathcal{R}h_1\|_{L^p(w)} \le 2\|h_1\|_{L^p(w)} & \text{(B}'\text{)} & \|\mathcal{R}'h_2\|_{L^{p'}(w)} \le 2\|h_2\|_{L^{p'}(w)} \\ \text{(C)} & [\mathcal{R}h_1]_{A_1(T_1)} \le 2\|T_1\|_{L^p(w)} & \text{(C}'\text{)} & [\mathcal{R}h_2\, w]_{A_1(T_2)} \le 2\|T_2'\|_{L^{p'}(w)}. \end{array}$$

The proof now goes through with essentially no change; we only have to note that by (C) and (C′), $(\mathcal{R}h_1)^{1-p_0}\, \mathcal{R}'h_2\, w \in A_1(T_1)^{1-p_0} \cdot A_1(T_2)$, so we can apply hypothesis (3.21). □

If we assume that T_1 and T_2 have the reverse factorization property, then the argument used in Corollary 3.12 extends to this setting to yield vector-valued inequalities. Like the corresponding result for one-sided A_p weights, this is a new result.

Corollary 3.29. *Let T_1 and T_2 be admissible operators, and let $\mathcal{F}$ be a family of pairs of functions in* $\operatorname{dom}(T_1) \cap \operatorname{dom}(T_2)$. *Assume further that T_1 and T_2 have the reverse factorization property. If* (3.21) *holds, then for all p and q, $1 < p, q < \infty$,* $w \in A_1(T_1)^{1-p} \cdot A_1(T_2) = A_p(T_1) \cap A_{p'}(T_2)^{1-p}$, *and sequences* $\{(f_i, g_i)\}_i \subset \mathcal{F}$,

$$\Big\| \Big(\sum_i f_i^q \Big)^{1/q} \Big\|_{L^p(w)} \le C \Big\| \Big(\sum_i g_i^q \Big)^{1/q} \Big\|_{L^p(w)}. \tag{3.23}$$

We can also prove a weak type extrapolation result by arguing as we did in Corollary 3.10. Further, we can use rescaling to obtain A_∞ extrapolation results. In the hypothesis the class of weights is $\cup_{p>1} A_1(T_1)^{1-p} \cdot A_1(T_2)$ and in the conclusion the weights are in the smaller class $\cup_{p>1} A_p(T_1) \cap A_{p'}(T_2)^{1-p}$. More precisely, if w belongs to the latter class, there is $s > 1$ such that $w \in A_s(T_1) \cap A_{s'}(T_2)^{1-s}$and we get estimates on $L^p(w)$ for $0 < p < p_0 s$. If we further assume that for $p < q$, $A_p(T_1) \cap A_{p'}(T_2)^{1-p} \subset A_q(T_1) \cap A_{q'}(T_2)^{1-q}$, then we obtain estimates on $L^p(w)$ for $0 < p < \infty$. The precise statements are left to the reader.

Remark 3.30. The distinction between the factorization and reverse factorization properties also appears in the two-weight case: see Chapter 6 in Part II. However, in this case and unlike in Theorem 3.28, if our initial hypothesis is given in terms of the reverse factorization property (so called "factored weights") then our conclusion is as well. See Section 7.4.

3.7 Limited range extrapolation

In this section we state and prove a version of the Rubio de Francia extrapolation theorem that holds for limited ranges of p. We first introduce the natural generalization of the reverse Hölder inequality: given s, $1 < s < \infty$, we say that $w \in RH_{s,\mathcal{B}}$ if for every $B \in \mathcal{B}$,

$$\left(\fint_B w(x)^s \, dx \right)^{1/s} \leq C \fint_B w(x) \, dx.$$

When $\mathcal{B} = \mathcal{Q}$ we get the class RH_s. We say that $w \in RH_{\infty,\mathcal{B}}$ if for every $B \in \mathcal{B}$ and almost every $x \in B$,

$$w(x) \leq C \fint_B w(y) \, dy.$$

$RH_{\infty,\mathcal{B}}$ is analogous to the class $A_{1,\mathcal{B}}$. The class RH_∞ (i.e., when $\mathcal{B} = \mathcal{Q}$) was introduced in [50].

Theorem 3.31. *Let $\mathcal{B}$ be a Muckenhoupt basis and let $1 \leq p_- < p_+ < \infty$. Suppose that there exists p_0, $p_- \leq p_0 \leq p_+$, such that for every $w_0 \in A_{p_0/p_-,\mathcal{B}} \cap RH_{(p_+/p_0)',\mathcal{B}}$,*

$$\int_{\mathbb{R}^n} f(x)^{p_0} \, w_0(x) \, dx \leq C \int_{\mathbb{R}^n} g(x)^{p_0} \, w_0(x) \, dx, \qquad (f,g) \in \mathcal{F}. \tag{3.24}$$

Then for all p, $p_- < p < p_+$, and for all $w \in A_{p/p_-,\mathcal{B}} \cap RH_{(p_+/p)',\mathcal{B}}$,

$$\int_{\mathbb{R}^n} f(x)^p \, w(x) \, dx \leq C \int_{\mathbb{R}^n} g(x)^p \, w(x) \, dx, \qquad (f,g) \in \mathcal{F}. \tag{3.25}$$

Remark 3.32. Theorem 3.31 remains true if $p_+ = \infty$. In this case the reverse Hölder condition on w is vacuous and we get Corollary 3.14.

Remark 3.33. We can restate the hypotheses of Theorem 3.31 in terms of the $A_{q,\mathcal{B}}$ class of a power of w. To do so, we use the following equivalence: for all s, p, $1 < s, p < \infty$, $w \in A_{p,\mathcal{B}} \cap RH_{s,\mathcal{B}}$ if and only if $w^s \in A_{q,\mathcal{B}}$, where $q = s(p-1)+1$. When $\mathcal{B} = \mathcal{Q}$ this was proved by Johnson and Neugebauer [109], and the same proof works for any Muckenhoupt basis. Given this, we have that $w \in A_{p/p_-,\mathcal{B}} \cap RH_{(p_+/p)',\mathcal{B}}$ if and only if $w^{(p_+/p)'} \in A_{q,\mathcal{B}}$, where

$$q = \left(\frac{p_+}{p}\right)' \left(\frac{p}{p_-} - 1\right) + 1 = \frac{p_+}{p_-} \cdot \frac{p - p_-}{p_+ - p} + 1.$$

Theorem 3.31 was proved in [9] in the classical case $\mathcal{B} = \mathcal{Q}$ using the iteration algorithm, but treating the cases $p > p_0$ and $p < p_0$ separately. This proof can be modified to handle the general case. It was also noted that via rescaling one can prove a limited range, weak type extrapolation theorem similar to Corollary 3.10, and prove vector-valued inequalities as in Corollary 3.18, provided that the range in the ℓ^q norm is also restricted to $p_- < q < p_+$.

Theorem 3.31 has two corollaries. The first is a weak type extrapolation result due to Johnson and Neugebauer [109]; we give its proof below after the proof of Theorem 3.31.

Corollary 3.34. *Given p_0, σ_0, $1 < p_0, \sigma_0 < \infty$, suppose that $T : L^{p_0}(w) \to L^{p_0,\infty}(w)$ for any $w^{\sigma_0} \in A_{p_0}$. Then for every p, $p_0/\sigma_0' + 1/\sigma_0 < p < p_0$, there exists $\sigma = \sigma(p) > \sigma_0$ such that $T : L^p(w) \to L^{p,\infty}(w)$ for every $w^\sigma \in A_p$.*

Remark 3.35. In the limiting case $\sigma_0 = 1$, the hypotheses in Corollary 3.34 are simply that for $w \in A_{p_0}$, T satisfies the weak (p_0, p_0) inequality. These are the same hypotheses as Corollary 3.17, but the conclusion we get here is weaker, since we get a weak type inequality for $1 < p < p_0$ and $w^\sigma \in A_p$ for some $\sigma > 1$.

Remark 3.36. Corollary 3.34 should be compared to another, two-weight, weak type extrapolation theorem due to Neugebauer. See Section 7.3 below.

The second corollary is a result by Duoandikoetxea, *et al.* [70]; we will also give its proof after the proof of Theorem 3.31.

Corollary 3.37. *Given δ, $0 < \delta \le 1$, suppose that for all $w_0 \in A_2$,*

$$\int_{\mathbb{R}^n} f(x)^2\, w_0(x)^\delta\, dx \le C \int_{\mathbb{R}^n} g(x)^2\, w_0(x)^\delta\, dx. \tag{3.26}$$

Then for all p, $\frac{2}{1+\delta} < p < \frac{2}{1-\delta}$, and every $w^{\frac{2}{2-p\,(1-\delta)}} \in A_{\frac{2\,p\,\delta}{2-p\,(1-\delta)}}$,

$$\int_{\mathbb{R}^n} f(x)^p\, w(x)\, dx \le C \int_{\mathbb{R}^n} g(x)^p\, w(x)\, dx. \tag{3.27}$$

Remark 3.38. Though more difficult, the proof of Theorem 3.31 is similar to the proof of off-diagonal extrapolation, Theorem 3.23.

Proof of Theorem 3.31. It suffices to consider the case $p_- = 1$. The general case follows by rescaling: let $\tilde{p}_- = 1$, $\tilde{p}_+ = p_+/p_-$, $\tilde{p}_0 = p_0/p_-$ and consider the new family $\tilde{\mathcal{F}}$ consisting of the pairs $(\tilde{f}, \tilde{g}) = (f^{p_-}, g^{p_-})$ with $(f, g) \in \mathcal{F}$. Then (3.24) gives an estimate for $\tilde{\mathcal{F}}$ with exponent $\tilde{p}_0$ and with weights $w \in A_{\tilde{p}_0/\tilde{p}_-,\mathcal{B}} \cap RH_{(\tilde{p}_+/\tilde{p}_0)',\mathcal{B}}$. We can therefore apply the case for $\tilde{p}_- = 1$ and (3.25) follows from the corresponding estimates for $\tilde{\mathcal{F}}$.

We will need to use the result of Johnson and Neugebauer from Remark 3.33 in the following form: given q, $1 \le q < p_+$, then $w \in A_{q,\mathcal{B}} \cap RH_{(p_+/q)',\mathcal{B}}$ if and only if $w^{(p_+/q)'} \in A_{\tau_q,\mathcal{B}}$ where $\tau_q = (p_+/q)'\,(q-1) + 1$.

We will consider three cases, depending on the initial value p_0.

Case 1: $1 < p_0 < p_+$. Fix p, $p_- = 1 < p < p_+$, and $w \in A_{p,\mathcal{B}} \cap RH_{(p_+/p)',\mathcal{B}}$. Set $s = 1+(p_0-1)\,(p-1)/(p_+-1)$; then $1 < s < \min(p_0, p)$. Also let $r_0 = p_0/s > 1$ and $r = p/s > 1$.

We first define the iteration algorithms. Since $w^{(p_+/p)'} \in A_{\tau_p,\mathcal{B}}$, M is bounded on $L^{\tau_p}(w^{(p_+/p)'})$. Therefore, given non-negative $h_1 \in L^{\tau_p}(w^{(p_+/p)'})$ we define

$$\mathcal{R}h_1(x) = \sum_{k=0}^{\infty} \frac{M_{\mathcal{B}}^k h_1(x)}{2^k \|M_{\mathcal{B}}\|^k_{L^{\tau_p}(w^{(p_+/p)'})}}.$$

$\mathcal{R}h_1$ satisfies properties (A), (B) and (C) in the proof of Theorem 3.9 with the corresponding changes.

Similarly, $w^{(p_+/p)'\,(1-\tau_p')} = w^{1-p'} \in A_{\tau_p',\mathcal{B}}$ and so M is bounded on $L^{\tau_p'}(w^{1-p'})$. Thus for any non-negative $h_2 \in L^{\tau_p'}(w^{1-p'})$ we define

$$\mathcal{R}'h_2(x) = \sum_{k=0}^{\infty} \frac{M_{\mathcal{B}}^k h_2(x)}{2^k \|M_{\mathcal{B}}\|^k_{L^{\tau_p'}(w^{1-p'})}}.$$

(Unlike in the proof of Theorem 3.9, here $\mathcal{R}'$ is composed of iterations of $M_{\mathcal{B}}$ instead of $M_{\mathcal{B}}'$. To distinguish between the two algorithms we abuse our earlier notation and denote this iteration algorithm by $\mathcal{R}'$.) Again, $\mathcal{R}'h_2$ satisfies properties (A), (B) and (C).

We now argue as follows. Fix $(f, g) \in \mathcal{F}$ and assume that both functions are non-trivial and belong to $L^p(w)$. Define

$$h_1(x) = \frac{f(x)}{\|f\|_{L^p(w)}} + \frac{g(x)}{\|g\|_{L^p(w)}};$$

then $\|h_1\|_{L^p(w)} \le 2$. By duality there exists non-negative $h_2 \in L^{r'}(w)$, $\|h_2\|_{L^{r'}(w)} = 1$, such that

$$\|f\|^s_{L^p(w)} = \|f^s\|_{L^r(w)} = \int_{\mathbb{R}^n} f(x)^s\, h_2(x)\, w(x)\, dx.$$

Now define

$$H_1(x) = \mathcal{R}\Big(h_1^{p/\tau_p}\, w^{\frac{1-(p_+/p)'}{\tau_p}}\Big)(x)^{\tau_p/p}\, w(x)^{\frac{(p_+/p)'-1}{p}}$$

and

$$H_2(x) = \mathcal{R}'\Big(h_2^{r'/\tau_p'}\, w^{p'/\tau_p'}\Big)(x)^{\tau_p'/r'}\, w(x)^{-p'/r'}.$$

Then by the properties of the iteration algorithms,

(A_1) $h_1(x) \le H_1(x)$	(A_2) $h_2(x) \le H_2(x)$
(B_1) $\|H_1\|_{L^p(w)} \le C$	(B_2) $\|H_2\|_{L^{r'}(w)} \le C$
(C_1) $H_1^{p/\tau_p}\, w^{\frac{1-(p_+/p)'}{\tau_p}} \in A_{1,\mathcal{B}}$	(C_2) $H_2^{r'/\tau_p'}\, w^{p'/\tau_p'} \in A_{1,\mathcal{B}}$.

By property (A_2) and Hölder's inequality,

$$\begin{aligned}\|f\|_{L^p(w)}^s &\le \int_{\mathbb{R}^n} f(x)^s\, H_2(x)\, w(x)\, dx \\ &= \int_{\mathbb{R}^n} f(x)^s\, H_1(x)^{-s/r_0'}\, H_1(x)^{s/r_0'}\, H_2(x)\, w(x)\, dx \\ &\le \left(\int_{\mathbb{R}^n} f(x)^{p_0}\, H_1(x)^{-(p_0-s)}\, H_2(x)\, w(x)\, dx\right)^{1/r_0} \\ &\qquad \times \left(\int_{\mathbb{R}^n} H_1(x)^s\, H_2(x)\, w(x)\, dx\right)^{1/r_0'} \\ &= I_1^{1/r_0} \times I_2^{1/r_0'}.\end{aligned}$$

We estimate each term separately. By Hölder's inequality, (B_1) and (B_2),

$$I_2 \le \|H_1\|_{L^p(w)}^s\, \|H_2\|_{L^{r'}(w)} \le C.$$

To estimate I_1 using (3.24) we first show that it is finite. Since $f/\|f\|_{L^p(w)} \le h_1 \le H_1$,

$$I_1 \le \|f\|_{L^p(w)}^{p_0} \int_{\mathbb{R}^n} H_1(x)^s\, H_2(x)\, w(x)\, dx = \|f\|_{L^p(w)}^{p_0}\, I_2 \le C\|f\|_{L^p(w)}^{p_0} < \infty.$$

We claim that $W = H_1^{-(p_0-s)}\, H_2\, w \in A_{p_0,\mathcal{B}} \cap RH_{(p_+/p_0)',\mathcal{B}}$. Assuming this for the moment, by (3.24) we have that

$$\begin{aligned}I_1 &\le C \int_{\mathbb{R}^n} g(x)^{p_0}\, H_1(x)^{-(p_0-s)}\, H_2(x)\, w(x)\, dx \\ &\le \|g\|_{L^p(w)}^{p_0} \int_{\mathbb{R}^n} H_1(x)^s\, H_2(x)\, w(x)\, dx = \|g\|_{L^p(w)}^{p_0}\, I_2 \le C\|g\|_{L^p(w)}^{p_0},\end{aligned}$$

where we have used (A_1) to get that $g/\|g\|_{L^p(w)} \le h_1 \le H_1$. Combining the estimates for I_1 and I_2 we get (3.25).

To complete the proof we will show that $W = H_1^{-(p_0-s)}\, H_2\, w \in A_{p_0,\mathcal{B}} \cap RH_{(p_+/p_0)',\mathcal{B}}$. It will suffice to prove that $W^{(p_+/p_0)'} \in A_{\tau_{p_0},\mathcal{B}}$. Fix $B \in \mathcal{B}$. By (C_1) and a lengthy but straightforward computation we get

$$\begin{aligned}&\fint_B W(x)^{(p_+/p_0)'}\, dx \\ &\quad = \fint_B H_1(x)^{-(p_0-s)\,(p_+/p_0)'}\, (H_2(x)\, w(x))^{(p_+/p_0)'}\, dx \\ &\quad \le C \left(\fint_B H_1(x)^{p/\tau_p}\, w(x)^{\frac{1-(p_+/p)'}{\tau_p}}\, dx\right)^{-\frac{\tau_p\,(p_0-s)\,(p_+/p_0)'}{p}}\end{aligned}$$

$$\times \fint_B w(x)^{\frac{(1-(p_+/p)')\,(p_0-s)\,(p_+/p_0)'}{p}}\,(H_2(x)\,w(x))^{(p_+/p_0)'}\,dx$$
$$= \left(\fint_B H_1(x)^{p/\tau_p}\,w(x)^{\frac{1-(p_+/p)'}{\tau_p}}\,dx\right)^{-(1-\tau_{p_0})}$$
$$\times \fint_B H_2(x)^{(p_+/p_0)'}\,w(x)^{p'/\tau_p'}\,dx.$$

Similarly, by (C_2) we have that

$$\begin{aligned}
&\left(\fint_B W(x)^{(p_+/p_0)'\,(1-\tau_{p_0}')}\,dx\right)^{\tau_{p_0}-1}\\
&\quad= \left(\fint_B W(x)^{1-p_0'}\,dx\right)^{\tau_{p_0}-1}\\
&\quad= \left(\fint_B H_1(x)^{\frac{p_0-s}{p_0-1}}\,(H_2(x)\,w(x))^{1-p_0'}\,dx\right)^{\tau_{p_0}-1}\\
&\quad\le C\left(\fint_B H_2(x)^{r'/\tau_p'}\,w(x)^{p'/\tau_p'}\,dx\right)^{-\frac{\tau_p'\,(p_0'-1)\,(\tau_{p_0}-1)}{r'}}\\
&\qquad\times\left(\fint_B H_1(x)^{\frac{p_0-s}{p_0-1}}\,w(x)^{1-p_0'}\,w(x)^{\frac{p'\,(p_0'-1)}{r'}}\,dx\right)^{\tau_{p_0}-1}\\
&\quad= \left(\fint_B H_2(x)^{(p_+/p_0)'}\,w(x)^{p'/\tau_p'}\,dx\right)^{-1}\\
&\qquad\times\left(\fint_B H_1(x)^{p/\tau_p}\,w(x)^{\frac{1-(p_+/p)'}{\tau_p}}\,dx\right)^{\tau_{p_0}-1}.
\end{aligned}$$

Combining the two estimates we conclude that $W^{(p_+/p_0)'} \in A_{\tau_{p_0},\mathcal{B}}$ and our proof is complete.

Case 2: $p_0 = 1$. Set $s = 1$, $r_0 = 1$ and $r = p$. As before we define H_2 associated with $h_2 \in L^{p'}(w)$. (H_1 is not needed.) In this case property (C_2) becomes $(H_2\,w)^{p'/\tau_p'} = (H_2\,w)^{p_+'} \in A_{1,\mathcal{B}}$, and so $W = H_2\,w \in A_{1,\mathcal{B}} \cap RH_{p_+',\mathcal{B}}$. Furthermore, property (B_2) implies that

$$\int_{\mathbb{R}^n} f(x)\,H_2(x)\,w(x)\,dx \le \|f\|_{L^p(w)}\,\|H_2\|_{L^{p'}(w)} \le C\|f\|_{L^p(w)} < \infty.$$

Therefore, we can apply (3.24) to conclude

$$\begin{aligned}
\|f\|_{L^p(w)} &\le \int_{\mathbb{R}^n} f(x)\,H_2(x)\,w(x)\,dx\\
&\le C\int_{\mathbb{R}^n} g(x)\;H_2(x)\,w(x)\,dx \le C\|g\|_{L^p(w)}\,\|H_2\|_{L^{p'}(w)} \le C\|g\|_{L^p(w)}.
\end{aligned}$$

Case 3: $p_0 = p_+$. Let $s = p$, $r_0 = p_+/p$ and $r = 1$ and define h_1 and H_1 as before. (H_2 is not needed.) Then by Hölder's inequality and property (B_1),

$$\begin{aligned}\|f\|_{L^p(w)}^p &= \int_{\mathbb{R}^n} f(x)^p \, H_1(x)^{-p/r_0'} \, H_1(x)^{p/r_0'} \, w(x)\,dx \\ &\le \left(\int_{\mathbb{R}^n} f(x)^{p_+} \, H_1(x)^{-(p_+-p)} \, w(x)\,dx\right)^{1/r_0} \left(\int_{\mathbb{R}^n} H_1(x)^p \, w(x)\,dx\right)^{1/r_0'} \\ &\le C\left(\int_{\mathbb{R}^n} f(x)^{p_+} \, H_1(x)^{-(p_+-p)} \, w(x)\,dx\right)^{1/r_0}.\end{aligned}$$

The last term is finite: by (A_1), $f/\|f\|_{L^p(w)} \le h_1 \le H_1$, so by property ($\mathrm{B}_1$)

$$\int_{\mathbb{R}^n} f(x)^{p_+} \, H_1(x)^{-(p_+-p)} \, w(x)\,dx \le \|f\|_{L^p(w)}^{p_+} \|H_1\|_{L^p(w)}^p \le C\|f\|_{L^p(w)}^{p_+} < \infty.$$

We claim that $W = H_1^{-(p_+-p)} \, w \in A_{p_+,\mathcal{B}} \cap RH_{\infty,\mathcal{B}}$. If we assume this, then we can apply (3.24) and the fact that by (A_1) $g/\|g\|_{L^p(w)} \le h_1 \le H_1$ to get

$$\begin{aligned}\|f\|_{L^p(w)}^p &\le C\left(\int_{\mathbb{R}^n} g(x)^{p_+} \, H_1(x)^{-(p_+-p)} \, w(x)\,dx\right)^{1/r_0} \\ &\le \|g\|_{L^p(w)}^{p_+/r_0} \, \|H_1\|_{L^p(w)}^{p/r_0} \le C\|g\|_{L^p(w)}^p.\end{aligned}$$

Finally, we show that $W \in A_{p_+,\mathcal{B}} \cap RH_{\infty,\mathcal{B}}$. By property ($\mathrm{C}_1$), $W^{1-p_+'} = H_1^{p/\tau_p} \, w^{\frac{1-(p_+/p)'}{\tau_p}} \in A_{1,\mathcal{B}} \subset A_{p_+',\mathcal{B}}$. Therefore, $W \in A_{p_+,\mathcal{B}}$. Furthermore, given $B \in \mathcal{B}$, for a.e. $x \in B$ since $W^{1-p_+'} \in A_{1,\mathcal{B}}$, by Jensen's inequality for the convex function $t \mapsto t^{-(p_+'-1)}$,

$$W(x) \le C\left(\fint_B W(y)^{1-p_+'} \, dy\right)^{-(p_+'-1)} \le C\fint_B W(y)\,dy.$$

Thus, $W \in RH_{\infty,\mathcal{B}}$ and our proof is complete. □

Proof of Corollary 3.34. Let $\tilde{p}_0 = (p_0 - 1)/\sigma_0 + 1$, $p_- = 1$ and $p_+ = \tilde{p}_0 \, \sigma_0'$. Then $p_- < \tilde{p}_0 < p_+$, $\sigma_0 = (p_+/\tilde{p}_0)'$ and $p_0 = (p_+/\tilde{p}_0)'(\tilde{p}_0 - 1) + 1$. Given $\lambda > 0$, let $\mathcal{F}$ be the family of pairs

$$\big((Th)_\lambda^{p_0/\tilde{p}_0}, |h|^{p_0/\tilde{p}_0}\big),$$

where $h \in L^{p_0}(w)$ and $(Th)_\lambda = \lambda \, \chi_{\{x:|Th(x)|>\lambda\}}$. By the observation at the beginning of the proof of Theorem 3.31, if $w \in A_{\tilde{p}_0} \cap RH_{(p_+/\tilde{p}_0)'}$, then $w^{\sigma_0} \in A_{p_0}$, so by assumption $T : L^{p_0}(w) \to L^{p_0,\infty}(w)$. Therefore, for every $(f,g) \in \mathcal{F}$,

$$\|f\|_{L^{\tilde{p}_0}(w)} = \lambda^{p_0/\tilde{p}_0} \, w(\{x : |Th(x)| > \lambda\})^{1/\tilde{p}_0} \le C\|h\|_{L^{p_0}(w)}^{p_0/\tilde{p}_0} = \|g\|_{L^{\tilde{p}_0}(w)}.$$

Hence, we can apply Theorem 3.31 to obtain estimates for the family $\mathcal{F}$ on $L^q(w)$ for every $1 < q < p_+$ and $w \in A_q \cap RH_{(p_+/q)'}$. Equivalently, for every q, $1 < q < p_+$, and $w \in A_q \cap RH_{(p_+/q)'}$,

$$w(\{x : |Th(x)| > \lambda\}) \le C \frac{1}{\lambda^{p_0\, q/\tilde{p}_0}} \int_{\mathbb{R}^n} |h(x)|^{p_0\, q/\tilde{p}_0}\, w(x)\, dx. \tag{3.28}$$

The desired conclusion follows from this inequality. Fix p, $p_0/\sigma_0' + 1/\sigma_0 < p < p_0$. Let $r = (p_0 - 1)/(p-1) > 1$ and $t = (p_0 - p)/(p-1)$. Then $\sigma_0\, t/r < 1$, so there exists δ such that $\sigma_0\, t/r < \delta < 1$. We will fix the precise value of δ below. Define $\sigma(p) = \sigma = \delta\, r\, \sigma_0/(\delta\, r - \sigma_0\, t)$. Fix w such that $w^\sigma \in A_p$. We will show that T is of weak type (p,p) with respect to w. Let $q = p\, \tilde{p}_0/p_0$; then $p = p_0 q/\tilde{p}_0$ and $1 < q < p_+$. Furthermore, by Remark 3.33, $w^\sigma \in A_p$ implies $w \in A_{1+(p-1)/\sigma} \cap RH_\sigma$. We claim that $1 + (p-1)/\sigma < q$ and $\sigma > (p_+/q)'$. Assuming this, then $w \in A_q \cap RH_{(p_+/q)'}$, and (3.28) becomes the desired estimate.

To show that $1 + (p-1)/\sigma < q$, note first that

$$1 + \frac{p-1}{\sigma} = 1 + \frac{p-1}{p_0 - 1}\left(\frac{p_0 - 1}{\sigma_0} - \frac{p_0 - p}{\delta}\right).$$

If we take $\delta = \sigma_0 t/r = \sigma_0 \frac{p_0 - p}{p_0 - 1}$, then the right-hand side equals 1. Therefore, we can fix δ such that $1 + (p-1)/\sigma$ is as close to 1 as desired. Since $q > 1$, we can choose δ such that $1 + (p-1)/\sigma < q$.

Finally, to show that $\sigma > (p_+/q)'$, note that $p < p_0$, and so $q < \tilde{p}_0$. Therefore,

$$\frac{1}{\sigma_1} = \frac{1}{\sigma_0} - \frac{t}{r} < \frac{1}{\sigma_0} = \frac{1}{(p_+/\tilde{p}_0)'} < \frac{1}{(p_+/q)'}. \qquad \square$$

Proof of Corollary 3.37. Let $p_- = \frac{2}{1+\delta}$, $p_+ = \frac{2}{1-\delta}$, and $p_0 = 2$. Then (3.26) becomes

$$\int_{\mathbb{R}^n} f(x)^2\, w_0(x)\, dx \le C \int_{\mathbb{R}^n} g(x)^2\, w_0(x)\, dx, \qquad (f,g) \in \mathcal{F},$$

and this holds for every $w_0^{1/\delta} \in A_2$, which is equivalent to $w_0 \in A_{1+\delta} \cap RH_{1/\delta} = A_{2/p_-} \cap RH_{(p_+/2)'}$. Therefore, by Theorem 3.31 we get estimates for $\mathcal{F}$ in $L^p(w)$ for every $p_- < p < p_+$, and $w \in A_{p/p_-} \cap RH_{(p_+/p)'}$, that is, $w^{\frac{2}{2-p\,(1-\delta)}} \in A_{\frac{2\,p\,\delta}{2-p\,(1-\delta)}}$. $\square$

Remark 3.39. This proof can be modified to get a generalization of Corollary 3.37. Given δ and p_0, $0 < \delta < 1$ and $1 < p_0 < \infty$, if we have weighted estimates for $\mathcal{F}$ in $L^{p_0}(w_0^\delta)$ for every $w_0 \in A_{p_0}$, then we can extrapolate to the range $\frac{p_0}{1+\delta\,(p_0-1)} < p < \frac{p_0}{1-\delta}$ with the class of weights $w^{\frac{p_0}{p_0 - p\,(1-\delta)}} \in A_{\frac{p_0\, p\, \delta}{p_0 - p\,(1-\delta)}}$. Details are left to the reader.

3.8 Applications

The theory of Rubio de Francia extrapolation has found a host of applications in harmonic analysis. We mentioned a few in Section 2.2 and it is not our intention to try to list them all here. The goal of this section is more modest. First, we want to discuss briefly how to apply the theorems in the previous sections; in particular, how to translate from the setting of pairs of functions to weighted norm inequalities for specific operators. Second, we want to describe a few applications that we feel are particularly noteworthy, either because they are new results or are old results whose proofs are considerably simplified by using extrapolation. Additional applications are given in [44, 46]. In Section 4.4 we will discuss applications of extrapolation to Banach function space and modular inequalities.

For brevity, we are going to omit the precise definitions of the classical operators we discuss. Definitions are given in Chapters 9 and 10; also see [68, 88, 91, 92]. Also, we will write A_p instead of $A_{p,\mathcal{Q}}$ for the A_p condition on the basis of cubes.

Norm inequalities for operators

We consider two types of norm inequalities: an operator T is either bounded on $L^p(w)$,

$$\int_{\mathbb{R}^n} |Tf(x)|^p w(x)\,dx \le C \int_{\mathbb{R}^n} |f(x)|^p w(x)\,dx, \tag{3.29}$$

or there exists a second operator, S, such that S and T satisfy a Coifman-Fefferman inequality,

$$\int_{\mathbb{R}^n} |Tf(x)|^p w(x)\,dx \le C \int_{\mathbb{R}^n} |Sf(x)|^p w(x)\,dx. \tag{3.30}$$

To prove inequalities like (3.29), use a family $\mathcal{F}$ consisting of the pairs $(|Tf|, |f|)$; for inequalities like (3.30), take pairs of the form $(|Tf|, |Sf|)$. To insure that the left-hand side of the inequality is finite, restrict f to a suitable dense subset of $L^p(w)$, depending on the operator T. For example, if $w \in A_p$ then one can take f a bounded function of compact support, $f \in C_c^\infty$, $f \in \cap_{p\ge 1} L^p$, etc. For vector-valued inequalities, it suffices to consider pairs consisting of the finite sums

$$\left(\sum_{i=1}^N |Tf_i|^q\right)^{1/q}, \qquad \left(\sum_{i=1}^N |f_i|^q\right)^{1/q},$$

since the general case follows by a limiting argument.

Vector-valued inequalities

An immediate application of our results is that vector-valued inequalities follow directly from weighted norm inequalities in the scalar case. This yields much simpler proofs of known results as well as new results. Though this connection was already noted by Rubio de Francia [195], it has often been overlooked. Further, we want to stress that using our approach of pairs of functions, such inequalities require essentially no extra work.

For the Hardy-Littlewood maximal operator we get a new proof of the vector-valued inequality

$$\int_{\mathbb{R}^n}\Big(\sum_{i=1} Mf_i(x)^q\Big)^{p/q} w(x)\,dx \leq C\int_{\mathbb{R}^n}\Big(\sum_{i=1} |f_i(x)|^q\Big)^{p/q} w(x)\,dx,$$

where $1 < p,\, q < \infty$, $w \in A_p$. This was proved by Fefferman and Stein [75] in the unweighted case using a variant of the Calderón-Zygmund decomposition, and extended to the weighted case by Anderson and John [5].

Vector-valued inequalities for the Hilbert transform date back to the 1930's and the work of Marcinkiewicz and Zygmund [136] and Boas and Bochner [14]. Results for Calderón-Zygmund singular integrals were originally proved by Benedek, Calderón and Panzone [11], and in the weighted case by Rubio de Francia, Ruiz and Torrea [196]. However, our proof has the advantage that we do not have to use the theory of Banach space valued operators. More generally, if T_Ω is a so-called rough singular integral with kernel $\Omega \in L^\infty$, then (using extrapolation) Duoandikoetxea and Rubio de Francia [71] showed that this operator is bounded on $L^p(w)$, $w \in A_p$. By our results we again get vector-valued inequalities; these results are new, and it is not obvious how to prove them directly.

We can also prove vector-valued estimates for commutators. Given an operator T and $b \in BMO$, the commutator $[b,T]$ is defined by

$$[b,T]f(x) = b(x)Tf(x) - T(bf)(x).$$

If T is a Calderón-Zygmund singular integral, then $[b,T]$ is bounded on $L^p(w)$ if $w \in A_p$ (this is implicit in the sharp function estimate due to Strömberg in Janson [107], and proved by Segovia and Torrea [209, 210]). By extrapolation we get vector-valued inequalities. Similarly, if I_α, $0 < \alpha < n$, is the fractional integral operator, then $[b, I_\alpha] : L^p(w^p) \to L^q(w^q)$, $1/p - 1/q = \alpha/n$, $w \in A_{p,q}$. (This is also due to Segovia and Torrea [210]; see also [36].) Again by extrapolation we get vector-valued inequalities.

Remark 3.40. Similar inequalities hold for higher order commutators; details are left to the reader.

Coifman-Fefferman inequalities

The original inequality due to Coifman and Fefferman [25, 24] showed that Calderón-Zygmund singular integrals could be controlled by the Hardy-Littlewood maximal operator: more precisely, given $w \in A_\infty$,

$$\int_{\mathbb{R}^n} |Tf(x)|^p\, w(x)\,dx \leq C\int_{\mathbb{R}^n} Mf(x)^p\, w(x)\,dx. \tag{3.31}$$

Since then, analogous inequalities have been proved for a number of other operators. Originally, such estimates were proved using good-λ inequalities, a method introduced by Burkholder and Gundy [17]. We introduced an alternative approach in [44]. First, by adapting an argument by Lerner [125] and applying A_∞ extrapolation (Corollary 3.15) we prove the Fefferman-Stein inequality [76]: for all p, $0 < p < \infty$, and $w \in A_\infty$,

$$\int_{\mathbb{R}^n} |f(x)|^p w(x)\,dx \leq C \int_{\mathbb{R}^n} M^\# f(x)^p w(x)\,dx, \tag{3.32}$$

where $M^\#$ is the sharp maximal function. We can then deduce a Coifman-Fefferman inequality for any operator T that satisfies a sharp function estimate of the form $M_q^\#(Tf)(x) \leq CMf(x)$, $0 < q < 1$, where $M_q^\# g = M^\#(|g|^q)^{1/q}$. Such inequalities are known for a wide variety of classical operators.

A significant advantage of our approach is that we can prove vector-valued estimates of this kind. For example, we immediately have the vector-valued extension of the original Coifman-Fefferman inequality: for $w \in A_\infty$,

$$\int_{\mathbb{R}^n} \left(\sum_{i=1} |Tf_i(x)|^q\right)^{p/q} w(x)\,dx \leq C \int_{\mathbb{R}^n} \left(\sum_{i=1} Mf_i(x)^q\right)^{p/q} w(x)\,dx.$$

It does not seem to be possible to prove such inequalities using the good-λ inequality.

Sharp function inequalities and the proof of the Fefferman-Stein inequality using Lerner's approach and extrapolation are given in detail (in the two-weight setting) in Section 9.1 below. We also refer the reader to [44, 46] for a complete explanation of this approach and applications to a wide variety of operators, including potential operators, multilinear singular integrals, and multiparameter fractional integrals.

Remark 3.41. There is another approach to Coifman-Fefferman inequalities that yields a larger class of weights. Muckenhoupt [151] introduced the class of weights C_p, $0 < p < \infty$, that satisfy the A_∞-type condition

$$\int_E w(x)\,dx \leq C \left(\frac{|E|}{|Q|}\right)^\epsilon \int_{\mathbb{R}^n} M(\chi_Q)(x)^p w(x)\,dx,$$

where Q is any cube and $E \subset Q$. For each p, A_∞ is a proper subset of C_p: this follows immediately if we restrict the integral on the right-hand side to Q. Muckenhoupt showed that for the Hilbert transform $w \in C_p$ is necessary for inequality (3.31) to hold, and that $w \in C_q$, $q > p$, is sufficient. Sawyer [203] extended this result to Calderón-Zygmund singular integrals. (See also Kahanpää and Mejlbro [112].) Yabuta [237] showed that the same necessary and sufficient conditions hold for the Fefferman-Stein inequality (3.32). If we combine his result with the sharp function estimates we can prove similar results for a wide variety of operators. Details are left to the reader.

Chapter 4

Extrapolation on Function Spaces

In this chapter we extend our theory of extrapolation to get norm inequalities on Banach function spaces starting from inequalities in weighted L^p. Our goal is to find conditions on a Banach function space $\mathbb{X}$ such that if for some p_0, $1 \le p_0 < \infty$, and every $w \in A_{p_0,\mathcal{B}}$,

$$\int_{\mathbb{R}^n} f(x)^{p_0} w(x)\, dx \le C \int_{\mathbb{R}^n} g(x)^{p_0} w(x)\, dx, \qquad (f,g) \in \mathcal{F}, \tag{4.1}$$

then

$$\|f\|_{\mathbb{X}} \le C \|g\|_{\mathbb{X}}, \qquad (f,g) \in \mathcal{F}.$$

Our key assumption is that the maximal operator is bounded on the associate space $(\mathbb{X}^{1/p_0})'$, where $\mathbb{X}^{1/p_0} = \{f : |f|^{1/p_0} \in \mathbb{X}\}$. Hence, in applications the central problem will be determining if a given space $\mathbb{X}$ belongs to a scale of function spaces on whose associate spaces the maximal operator is bounded. If we restrict ourselves to the important class of rearrangement invariant Banach function spaces, then we can establish this immediately. Further, we can extend our extrapolation result to the scale of weighted Banach function spaces $\mathbb{X}(w)$, $w \in A_{p,\mathcal{B}}$. Our proofs follow very closely the proof of our main extrapolation result, Theorem 3.9.

Related to function space inequalities are modular inequalities of the form

$$\int_{\mathbb{R}^n} \Phi(f(x)) w(x)\, dx \le C \int_{\mathbb{R}^n} \Phi(g(x)) w(x)\, dx, \qquad (f,g) \in \mathcal{F},$$

where Φ is a Young function. We show that with appropriate assumptions on the growth of Φ—that both Φ and its conjugate $\bar{\Phi}$ are doubling—such inequalities can be gotten by extrapolating from inequality (4.1). The proof of this result again follows the scheme of the proof of Theorem 3.9. In this case, however, we cannot use duality, and instead use a clever choice of a "dual" function and Young's inequality. (As we noted in Section 2.1, we could avoid duality in the same way in the proof for L^p spaces.)

The idea of using extrapolation to get modular and function space inequalities is a very recent development. It first appeared in [40] where it was used to prove norm inequalities for classical operators on the so-called variable Lebesgue spaces. The theory of A_∞ extrapolation on rearrangement invariant Banach function spaces and modular spaces (extending Corollary 3.15) was developed in [57]. (There the theory was developed in the slightly more general context of rearrangement invariant quasi-Banach spaces.)

The rest of this chapter is organized as follows. In Section 4.1 we give the necessary definitions and preliminary results we need about Banach function spaces and modular spaces. In Section 4.2 we state and prove our extrapolation theorems for general Banach function spaces and for rearrangement invariant Banach function spaces. In Section 4.3 we give our extrapolation theorem for modular spaces. Finally, in Section 4.4 we discuss applications of our extrapolation results. As was the case in Section 3.8 above, our goal is not to give an exhaustive list of applications; rather we want to highlight how many known results are now immediate consequences of extrapolation, and in addition give some new results. In particular, we give two applications in the variable Lebesgue spaces.

For brevity, throughout this chapter we will often abbreviate "rearrangement invariant" by "r.i." and write "function space" instead of "Banach function space."

4.1 Preliminaries

In this section we summarize the definitions and basic properties of Banach function spaces and modular spaces. For complete details we refer the reader to Rao and Ren [189] for Young functions and Orlicz spaces, Kokilashvili and Krbec [115] and Maligranda [135] for modular spaces, and Bennett and Sharpley [12] for function spaces.

Banach function spaces

Given $\Omega \subset \mathbb{R}^n$, let (Ω, μ) be a totally σ-finite, non-atomic measure space. Let $\mathcal{M}_\mu$ be the set of all μ-measurable functions $f : \Omega \to [-\infty, \infty]$ and let $\mathcal{M}_\mu^+ \subset \mathcal{M}_\mu$ be the collection of non-negative μ-measurable functions. If μ is Lebesgue measure, then we will write simply $\mathcal{M}$, etc.

Given a measure space (Ω, μ), and a mapping $\|\cdot\|_{\mathbb{X}} : \mathcal{M}_\mu \to [0, \infty]$, the set

$$\mathbb{X} = \{f \in \mathcal{M}_\mu : \|f\|_{\mathbb{X}} < \infty\},$$

is a Banach function space (that is, a complete normed vector subspace of $\mathcal{M}_\mu$) if $\|\cdot\|_{\mathbb{X}}$ satisfies the following properties for all f, $g \in \mathcal{M}_\mu$:

(a) $\|f\|_{\mathbb{X}} = \||f|\|_{\mathbb{X}}$ and $\|f\|_{\mathbb{X}} = 0$ if and only if $f = 0$ μ-a.e.;

(b) $\|f + g\|_{\mathbb{X}} \leq \|f\|_{\mathbb{X}} + \|g\|_{\mathbb{X}}$;

(c) for all $a \in \mathbb{R}$, $\|af\|_{\mathbb{X}} = |a|\|f\|_{\mathbb{X}}$;

(d) if $|f| \leq |g|$ μ-a.e., then $\|f\|_{\mathbb{X}} \leq \|g\|_{\mathbb{X}}$;

(e) if $\{f_n\} \subset \mathcal{M}_\mu$ is a sequence such that $|f_n|$ increases to $|f|$ μ-a.e., then $\|f_n\|_{\mathbb{X}}$ increases to $\|f\|_{\mathbb{X}}$;

(f) if $E \subset \Omega$ is a measurable set such that $\mu(E) < \infty$, then $\|\chi_E\|_{\mathbb{X}} < \infty$;

(g) $\int_E |f(x)|\, d\mu \leq C_E\|f\|_{\mathbb{X}}$ if $\mu(E) < \infty$, where $C_E < \infty$ depends on E and $\mathbb{X}$, but not on f.

Beginning with $\mathbb{X}$ we can define another Banach function space, $\mathbb{X}'$, the associate space of $\mathbb{X}$, with norm

$$\|f\|_{\mathbb{X}'} = \sup\Big\{ \int_\Omega |f(x)g(x)|\, d\mu : g \in \mathbb{X}, \|g\|_{\mathbb{X}} \leq 1\Big\}.$$

An immediate consequence of this definition is the generalized Hölder's inequality: for all $f \in \mathbb{X}$, $g \in \mathbb{X}'$,

$$\int_\Omega |f(x)g(x)|\, d\mu \leq \|f\|_{\mathbb{X}}\, \|g\|_{\mathbb{X}'}. \tag{4.2}$$

Furthermore, $(\mathbb{X}')' = \mathbb{X}$: i.e., the associate space of $\mathbb{X}'$ is again $\mathbb{X}$. Therefore, we can restate the norm on $\mathbb{X}$ in terms of the norm on $\mathbb{X}'$:

$$\|f\|_{\mathbb{X}} = \sup\Big\{ \int_\Omega |f(x)g(x)|\, d\mu : g \in \mathbb{X}',\ \|g\|_{\mathbb{X}'} \leq 1\Big\}. \tag{4.3}$$

Given a Banach function space $\mathbb{X}$, we define the scale of spaces $\mathbb{X}^r$, $0 < r < \infty$, by

$$\mathbb{X}^r = \big\{f \in \mathcal{M}_\mu : |f|^r \in \mathbb{X}\big\},$$

with the "norm"

$$\|f\|_{\mathbb{X}^r} = \||f|^r\|_{\mathbb{X}}^{1/r}.$$

If $r \geq 1$, then $\|\cdot\|_{\mathbb{X}^r}$ is again an actual norm and $\mathbb{X}^r$ is a Banach function space. However, if $r < 1$, then $\mathbb{X}^r$ need not be a function space. The simplest example is in the scale of Lebesgue spaces: if we let $\mathbb{X} = L^1$, then $(L^1)^r = L^r$, and so $\mathbb{X}^r$ is a Banach space only for $r \geq 1$.

Remark 4.1. If $\mathbb{X}$ satisfies all of the properties given above defining a Banach function space except that (b) is replaced by $\|f + g\|_{\mathbb{X}} < C(\|f\|_{\mathbb{X}} + \|g\|_{\mathbb{X}})$ and (g) is omitted, then $\mathbb{X}$ is referred to as a quasi-Banach function space. For example, if $\mathbb{X}$ is a Banach function space, $\mathbb{X}^r$, $0 < r < 1$, is a quasi-Banach function space. Most of our results extend to this more general case if we assume that $\mathbb{X}$ is a quasi-Banach function space such that $\mathbb{X}^p$ is a Banach function space for some $p > 1$. For further details, see [57].

Remark 4.2. Our notation $\mathbb{X}^r$ is not universal. It is used, for example, in [110], but in [132] the set $\mathbb{X}^r$ consists of the r-th powers of elements in $\mathbb{X}$.

To define a rearrangement invariant function space, let μ_f denote the distribution function of $f \in \mathcal{M}_\mu$:

$$\mu_f(\lambda) = \mu(\{x \in \Omega : |f(x)| > \lambda\}), \qquad \lambda \geq 0.$$

A Banach function space is rearrangement invariant if the norm of a function $f \in \mathbb{X}$ depends only on its distribution function: more precisely, if $\|f\|_{\mathbb{X}} = \|g\|_{\mathbb{X}}$ for every pair of functions $f,\, g \in \mathbb{X}$ such that $\mu_f = \mu_g$. (We say that such functions are equimeasurable.) If $\mathbb{X}$ is an r.i. space, then $\mathbb{X}'$ and $\mathbb{X}^r$, $0 < r < \infty$, are also rearrangement invariant.

Given $f \in \mathcal{M}_\mu$, the decreasing rearrangement of f with respect to μ is the function f_μ^* with domain $[0, \mu(\Omega))$ defined by

$$f_\mu^*(t) = \inf\{\lambda \geq 0 : \mu_f(\lambda) \leq t\}, \qquad 0 \leq t < \mu(\Omega).$$

The function f_μ^* is equimeasurable with f, since

$$\mu(\{x \in \Omega : |f(x)| > \lambda\}) = |\{t \in [0, \mu(\Omega)) : f_\mu^*(t) > \lambda\}|.$$

As a consequence we get the Luxemburg representation theorem: given an r.i. function space $\mathbb{X}$, there exists a unique r.i. function space $\overline{\mathbb{X}}$ on $[0, \mu(\Omega))$ such that $f \in \mathbb{X}$ if and only if $f_\mu^* \in \overline{\mathbb{X}}$, and $\|f\|_{\mathbb{X}} = \|f_\mu^*\|_{\overline{\mathbb{X}}}$. Furthermore, the Luxemburg representation of $\mathbb{X}'$ is the associate space $\overline{\mathbb{X}}'$ of $\overline{\mathbb{X}}$, and so $\|f\|_{\mathbb{X}'} = \|f_\mu^*\|_{\overline{\mathbb{X}}'}$.

If $\mu(\Omega) = \infty$ (e.g., if $(\Omega, \mu) = (\mathbb{R}^n, dx)$), we can use this representation to define the Boyd indices of a function space $\mathbb{X}$. Given $f \in \overline{\mathbb{X}}$, define the dilation operator D_t, $0 < t < \infty$, by $D_t f(s) = f(s/t)$, and define the function

$$h_{\mathbb{X}}(t) = \|D_t\|_{\overline{\mathbb{X}}}.$$

The lower and upper Boyd indices are defined by

$$p_{\mathbb{X}} = \lim_{t\to\infty} \frac{\log t}{\log h_{\mathbb{X}}(t)} = \sup_{1<t<\infty} \frac{\log t}{\log h_{\mathbb{X}}(t)},$$

and

$$q_{\mathbb{X}} = \lim_{t\to 0^+} \frac{\log t}{\log h_{\mathbb{X}}(t)} = \inf_{0<t<1} \frac{\log t}{\log h_{\mathbb{X}}(t)}.$$

We immediately have that $1 \leq p_{\mathbb{X}} \leq q_{\mathbb{X}} \leq \infty$. Given the Boyd indices for $\mathbb{X}$, we can compute the indices of $\mathbb{X}'$: $p_{\mathbb{X}'} = (q_{\mathbb{X}})'$ and $q_{\mathbb{X}'} = (p_{\mathbb{X}})'$. The above definition extends to the spaces $\mathbb{X}^r$, $0 < r < \infty$, and for any $r > 0$, $p_{\mathbb{X}^r} = p_{\mathbb{X}} \cdot r$ and $q_{\mathbb{X}^r} = q_{\mathbb{X}} \cdot r$.

Remark 4.3. Some authors (including [12]) define the Boyd indices as the inverses of the values $p_{\mathbb{X}}$ and $q_{\mathbb{X}}$. We have chosen this definition since it yields that if $\mathbb{X} = L^p$, then $p_{\mathbb{X}} = q_{\mathbb{X}} = p$. The definition of the Boyd indices can also be extended to spaces such that $\mu(\Omega) < \infty$. For simplicity we will not consider this case; details of extending our results to this setting are left to the reader.

We now restrict our attention to rearrangement invariant Banach function spaces defined over the measure space $(\mathbb{R}^n, dx)$. Given such a space $\mathbb{X}$, we want to define a weighted version $\mathbb{X}(w)$ that can be thought of as the corresponding space defined with respect to the measure $d\mu = w\,dx$. These spaces appeared in [57] as an abstract generalization of a number of types of weighted function spaces. However, some care must be taken to make this definition precise since w can be infinite on a set of positive measure and therefore μ will not be totally σ-finite.

To rectify this, we define $\mathbb{X}(w)$ as follows. Given a basis $\mathcal{B}$, let $\Omega_{\mathcal{B}} = \cup_{B\in\mathcal{B}} B$. Fix $w \in A_{\infty,\mathcal{B}}$, and as we did in Chapter 3, let $\Omega^*_{\mathcal{B}} = \{x \in \Omega_{\mathcal{B}} : 0 < w(x) < \infty\}$ and let $w^* = w\chi_{\Omega^*_{\mathcal{B}}}$. Note that w^* is a Lebesgue measurable function and $0 < w^*(x) < \infty$ for almost every $x \in \Omega^*_{\mathcal{B}}$. Define the measure μ by $d\mu = w^*\,dx$. Then the measure space $(\Omega^*_{\mathcal{B}}, \mu)$ is totally σ-finite. To see this, note that $\Omega^*_{\mathcal{B}}$ is open, and so can be written as the union of a countable number of sets $B \in \mathcal{B}$ such that $w(B) < \infty$. (See Proposition 3.7.) Further, to simplify our exposition we will assume that $\mu(\Omega^*_{\mathcal{B}}) = \infty$. In most applications this will not be a problem.

Define the weighted space $\mathbb{X}(w^*)$ by

$$\mathbb{X}(w^*) = \{f \in \mathcal{M}_\mu : \|f^*_\mu\|_{\overline{\mathbb{X}}} < \infty\};$$

this is a Banach function space with norm $\|f\|_{\mathbb{X}(w^*)} = \|f^*_\mu\|_{\overline{\mathbb{X}}}$. We now define the space $\mathbb{X}(w)$ to be equal to $\mathbb{X}(w^*)$. Note that if $\mathbb{X}$ is the Lebesgue space L^p, then $\mathbb{X}(w) = L^p(w^*)$; however, if we define $L^p(w)$ in the usual way, then as we discussed in Section 3.1, if $h \in L^p(w)$, then $h \in L^p(w^*)$ and the norm of h in each space is the same. Therefore, our definition is a natural generalization of this idea. Hereafter we will not mention w^*; it will simply be implicit in the definition.

If we form the weighted space $\mathbb{X}'(w)$ from $\mathbb{X}'$, then since the measure spaces $(\mathbb{R}^n, dx)$ and (Ω^*, μ) are resonant (see [12, Chapter 2]) it follows that $\mathbb{X}'(w) = \mathbb{X}(w)'$. Thus we have a weighted version of (4.3):

$$\|f\|_{\mathbb{X}(w)} = \sup\Big\{\int_\Omega |f(x)g(x)|w(x)\,dx : g \in \mathbb{X}'(w),\ \|g\|_{\mathbb{X}'(w)} \le 1\Big\}. \tag{4.4}$$

Similarly, for $0 < r < \infty$, $\mathbb{X}(w)^r = \mathbb{X}^r(w)$.

Given an operator T such that $\|Tf\|_{\mathbb{X}(w)} \le C\|f\|_{\mathbb{X}(w)}$ for all $f \in \mathbb{X}(w)$, we denote the infimum of all such constants C by $\|T\|_{\mathbb{X}(w)}$.

Examples of function spaces

We give some standard examples of these spaces. In Section 4.4 we will consider the example of the variable Lebesgue spaces; additional examples were given in [57].

Lebesgue spaces

For $1 \leq p \leq \infty$, $\mathbb{X} = L^p$ is an r.i. space. In this case, as we noted above, $p_{\mathbb{X}} = q_{\mathbb{X}} = p$, and $\mathbb{X}(w)$ is just the weighted Lebesgue space $L^p(w)$.

Lorentz spaces

For $0 < p,\, q < \infty$, define

$$\|f\|_{L^{p,q}} = \left(\int_0^\infty f^*(s)^q s^{q/p-1}\, ds\right)^{1/q},$$

and

$$\|f\|_{L^{p,\infty}} = \sup_{0<s<\infty} f^*(s)s^{1/p}.$$

These are only quasi-norms, but when $1 < p < \infty$ and $1 \leq q \leq \infty$ or $p = 1$ and $1 \leq q < \infty$, they are equivalent to norms, $\mathbb{X} = L^{p,q}$ is an r.i. function space and $p_{\mathbb{X}} = q_{\mathbb{X}} = p$. (The space $\mathbb{X} = L^{1,\infty}$ is an r.i. quasi-Banach function space with $p_{\mathbb{X}} = q_{\mathbb{X}} = 1$.) The spaces $\mathbb{X}(w)$ are the weighted Lorentz spaces $L^{p,q}(w)$ gotten by replacing f^* with f_w^* in the definition.

Orlicz spaces

Given a Young function Φ, we define the Orlicz space L^Φ to be the function space with Luxemburg norm

$$\|f\|_{L^\Phi} = \inf\left\{\lambda > 0 : \int_\Omega \Phi\left(\frac{|f(x)|}{\lambda}\right) dx \leq 1\right\}.$$

This is an r.i. function space. The space $\mathbb{X}(w)$ is the weighted Orlicz space defined as above with Lebesgue measure replaced by $w\,dx$.

Clearly the Lebesgue spaces are Orlicz spaces with $\Phi(t) = t^p$. Other examples include the Zygmund spaces $L^p(\log L)^\alpha$, $1 < p < \infty$, $\alpha \in \mathbb{R}$, which are defined using $\Phi(t) \approx t^p \log(e+t)^\alpha$. In this case, $p_{\mathbb{X}} = q_{\mathbb{X}} = p$. The spaces $L^p + L^q$ and $L^p \cap L^q$ can also be treated as Orlicz spaces, with $\Phi(t) \approx \max(t^p, t^q)$ and $\Phi(t) \approx \min(t^p, t^q)$, respectively. In both cases, $p_{\mathbb{X}} = \max(p, q)$ and $q_{\mathbb{X}} = \min(p, q)$. Note that for these and other Orlicz spaces, the Boyd indices can be computed directly from the function Φ—see Remark 4.5 below.

Modular spaces

Let Φ be a Young function, that is, $\Phi : [0,\infty) \to [0,\infty)$ is continuous, convex and strictly increasing, and

$$\lim_{t\to 0^+} \frac{\Phi(t)}{t} = 0 \qquad \text{and} \qquad \lim_{t\to\infty} \frac{\Phi(t)}{t} = \infty.$$

Though $\Phi(t) = t$ is not a Young function, many of our results are still true in this case. We define the complementary function $\bar{\Phi}$ by

$$\bar{\Phi}(t) = \sup_{s>0}\{st - \Phi(s)\}, \qquad t \geq 0.$$

An immediate consequence of this definition is Young's inequality:

$$st \leq \Phi(s) + \bar{\Phi}(t), \qquad s, t \geq 0. \tag{4.5}$$

Furthermore, $\bar{\Phi}$ is also a Young function and we have the relation

$$t \leq \Phi^{-1}(t)\,\bar{\Phi}^{-1}(t) \leq 2\,t, \qquad t \geq 0. \tag{4.6}$$

A Young function Φ satisfies the Δ_2 condition, denoted by $\Phi \in \Delta_2$, if for all $t \geq 0$, $\Phi(2t) \leq C\Phi(t)$. Such functions are also said to be doubling. The conjugate function $\bar{\Phi}$ is in Δ_2 if there exists $r > 1$ such that for all $t \geq 0$, $\Phi(rt) \geq 2r\Phi(t)$.

The growth of a Young function Φ is measured by its upper and lower dilation indices, I_Φ and i_Φ. Intuitively, these allow us to compare Φ with power functions. Let

$$h_\Phi(t) = \sup_{s>0} \frac{\Phi(s\,t)}{\Phi(s)}, \qquad t > 0;$$

then the dilation indices are defined by

$$i_\Phi = \lim_{t\to 0+} \frac{\log h_\Phi(t)}{\log t} = \sup_{0<t<1} \frac{\log h_\Phi(t)}{\log t},$$

and

$$I_\Phi = \lim_{t\to\infty} \frac{\log h_\Phi(t)}{\log t} = \inf_{1<t<\infty} \frac{\log h_\Phi(t)}{\log t}.$$

It follows from this definition that $1 \leq i_\Phi \leq I_\Phi \leq \infty$, $(I_\Phi)' = i_{\bar{\Phi}}$, and $(i_\Phi)' = I_{\bar{\Phi}}$. Furthermore, $\Phi \in \Delta_2$ if and only if $I_\Phi < \infty$. Hence, $1 < i_\Phi \leq I_\Phi < \infty$ if and only if $\Phi, \bar{\Phi} \in \Delta_2$.

Remark 4.4. The dilation indices are often difficult to compute directly from the definition. However, in [79] it was shown that if the limits

$$r_0 = \lim_{t\to 0} \frac{t\Phi'(t)}{\Phi(t)} \qquad \text{and} \qquad r_\infty = \lim_{t\to\infty} \frac{t\Phi'(t)}{\Phi(t)}$$

exist, then $i_\Phi = \min(r_0, r_\infty)$ and $I_\Phi = \max(r_0, r_\infty)$.

Remark 4.5. The dilation indices of a Young function Φ are the same as the Boyd indices of the Orlicz space L^Φ. See [12] and [135].

Given a basis $\mathcal{B}$ and a weight $w \in A_{\infty,\mathcal{B}}$, we define the modular

$$\rho_w^\Phi(f) = \int_{\mathbb{R}^n} \Phi(|f(x)|)\, w(x)\, dx.$$

We define $\rho_w^{\bar{\Phi}}(f)$ similarly. As in the case of the Lebesgue spaces, we will implicitly assume that we have replaced w by w^*—see Section 3.1 for details.

The collection of functions

$$M_w^\Phi = \{f : \rho_w^\Phi(f) < \infty\}$$

is referred to as a modular space. A sublinear operator T satisfies a modular inequality on M_w^Φ—or, more briefly, is (Φ, w)-modular—if there exist constants $C, c > 0$ such that $\rho_w^\Phi(Tf) \le C\rho_w^\Phi(cf)$. If $\Phi \in \Delta_2$, then T is (Φ, w)-modular if

$$\int_{\mathbb{R}^n} \Phi(|Tf(x)|)w(x)\, dx \le C \int_{\mathbb{R}^n} \Phi(|f(x)|)w(x)\, dx.$$

We denote the infimum of all such constants C by $\rho_w^\Phi(T)$.

Examples of modular spaces

We conclude by giving some examples of modular spaces.

- Let $\Phi(t) = t^p$, $1 < p < \infty$. Then the complementary function is $\bar{\Phi}(t) = pt^{p'}/p'$. In this case $i_\Phi = I_\Phi = p$.

- Let $\Phi(t) \approx t^p \log(e+t)^\alpha$, $1 < p < \infty$, and $\alpha \in \mathbb{R}$. Then $\bar{\Phi}(t) \approx t^{p'} \log(e+t)^{\alpha(1-p')}$ and $i_\Phi = I_\Phi = p$.

- Given p, $1 < p < \infty$, let $\Phi(t) \approx t^p$, $0 \le t \le 1$, and $\Phi(t) \approx e^t$, $t \ge 1$. Then $\bar{\Phi}(t) \approx t^{p'}$, $0 \le t \le 1$, and $\bar{\Phi}(t) \approx t\log(e+t)$, $t \ge 1$. In this case, $i_\Phi = p$ and $I_\Phi = \infty$, and Φ is not in Δ_2.

4.2 Extrapolation on Banach function spaces

In this section we state and prove our extrapolation theorems for function spaces. We will first give our results for arbitrary function spaces and then for the special case of rearrangement invariant spaces.

General function spaces

For an arbitrary function space $\mathbb{X}$, we have not defined a weighted space $\mathbb{X}(w)$, and so we are, in effect, proving unweighted estimates. One consequence is that we need a much weaker hypothesis: we only need to assume that weighted norm inequalities hold for weights $w \in A_{1,\mathcal{B}}$ which is the smallest class of weights.

Theorem 4.6. *Let $\mathcal{B}$ be a Muckenhoupt basis and let $\mathbb{X}$ be a Banach function space. Suppose that for some p_0, $0 < p_0 < \infty$, and every $w_0 \in A_{1,\mathcal{B}}$,*

$$\int_{\mathbb{R}^n} f(x)^{p_0} w_0(x)\,dx \le C \int_{\mathbb{R}^n} g(x)^{p_0} w_0(x)\,dx, \qquad (f,g) \in \mathcal{F}. \tag{4.7}$$

If there exists q_0, $p_0 \le q_0 < \infty$, such that $\mathbb{X}^{1/q_0}$ is a Banach function space and $M_{\mathcal{B}}$ is bounded on $(\mathbb{X}^{1/q_0})'$, then

$$\|f\|_{\mathbb{X}} \le C\|g\|_{\mathbb{X}}, \qquad (f,g) \in \mathcal{F}. \tag{4.8}$$

Furthermore, for every p, $p_0/q_0 \le p < \infty$,

$$\|f\|_{\mathbb{X}^p} \le C\|g\|_{\mathbb{X}^p}, \qquad (f,g) \in \mathcal{F}. \tag{4.9}$$

Theorem 4.6 was first proved in [40] in the special case $\mathcal{B} = \mathcal{Q}$ and $\mathbb{X}$ is a variable Lebesgue space (see Section 4.4 for a definition). The proof given below is a direct adaptation of this proof. If $\mathbb{X} = L^p$, then this result was proved in [66] (the original Spanish edition of [68]) as a special case of Rubio de Francia extrapolation; the argument was implicit in [75].

Remark 4.7. We could weaken the hypotheses of Theorem 4.6 by assuming only that $\mathbb{X}$ is a quasi-Banach function space; the key hypothesis is then that some power of $\mathbb{X}$ is a Banach space. Details are left to the interested reader.

If we start with the stronger hypotheses of Theorem 3.9, then we get the following corollary.

Corollary 4.8. *Let $\mathcal{B}$ be a Muckenhoupt basis. Suppose that for some p_0, $1 \le p_0 < \infty$, and every $w_0 \in A_{p_0,\mathcal{B}}$,*

$$\int_{\mathbb{R}^n} f(x)^{p_0} w_0(x)\,dx \le C \int_{\mathbb{R}^n} g(x)^{p_0} w_0(x)\,dx, \qquad (f,g) \in \mathcal{F}. \tag{4.10}$$

If $\mathbb{X}$ is a Banach function space such that $M_{\mathcal{B}}$ is bounded on $\mathbb{X}'$, then for all p, $1 < p < \infty$,

$$\|f\|_{\mathbb{X}^p} \le C\|g\|_{\mathbb{X}^p}, \qquad (f,g) \in \mathcal{F}. \tag{4.11}$$

Furthermore, for every q, $1 < q < \infty$, and sequence $\{(f_i,g_i)\}_i \subset \mathcal{F}$,

$$\Big\|\Big(\sum_i f_i^q\Big)^{1/q}\Big\|_{\mathbb{X}^p} \le C\Big\|\Big(\sum_i g_i^q\Big)^{1/q}\Big\|_{\mathbb{X}^p}. \tag{4.12}$$

Remark 4.9. An equivalent formulation for Corollary 4.8 is the following: if $\mathbb{X}$ is a function space and there exists $1 \le q_0 < \infty$ such that $\mathbb{X}^{1/q_0}$ is a Banach function space and $M_{\mathcal{B}}$ is bounded on $(\mathbb{X}^{1/q_0})'$, then the conclusion holds for $\mathbb{X}^p$ for every p, $1/q_0 < p < \infty$.

Proof of Theorem 4.6. We first show (4.8); the proof is similar to but simpler than the proof of Theorem 3.9 since we do not have to use the dual operator $M'_{\mathcal{B}}$.

Let $\mathbb{Y} = \mathbb{X}^{1/q_0}$. Since $M_{\mathcal{B}}$ is bounded on $\mathbb{Y}'$, we can define the Rubio de Francia iteration algorithm for any non-negative function h:

$$\mathcal{R}h(x) = \sum_{k=0}^{\infty} \frac{M_{\mathcal{B}}^k h(x)}{2^k \|M_{\mathcal{B}}\|_{\mathbb{Y}'}^k}.$$

It is immediate from the definition that

(A) $h(x) \le \mathcal{R}h(x)$,

(B) $\|\mathcal{R}h\|_{\mathbb{Y}'} \le 2\,\|h\|_{\mathbb{Y}'}$,

(C) $[\mathcal{R}h]_{A_{1,\mathcal{B}}} \le 2\,\|M_{\mathcal{B}}\|_{\mathbb{Y}'}$.

We now apply extrapolation to our hypothesis using a rescaling argument. Define the family

$$\mathcal{F}_0 = \{(f^{p_0}, g^{p_0}) : (f,g) \in \mathcal{F}\}.$$

Then we can apply Theorem 3.9 to the family $\mathcal{F}_0$ with initial exponent 1, and conclude that for any $p \ge 1$ and $w \in A_{1,\mathcal{B}} \subset A_{p,\mathcal{B}}$,

$$\int_{\mathbb{R}^n} (f(x)^{p_0})^p w(x)\,dx \le C \int_{\mathbb{R}^n} (g(x)^{p_0})^p w(x)\,dx, \quad (f,g) \in \mathcal{F}.$$

If we let $p = q_0/p_0$ we get

$$\int_{\mathbb{R}^n} f(x)^{q_0} w(x)\,dx \le C \int_{\mathbb{R}^n} g(x)^{q_0} w(x)\,dx, \quad (f,g) \in \mathcal{F}. \tag{4.13}$$

We can now prove the desired inequality. Fix $(f,g) \in \mathcal{F}$. Since $\mathbb{Y}$ is a Banach function space, by (4.3),

$$\|f\|_{\mathbb{X}}^{q_0} = \|f^{q_0}\|_{\mathbb{Y}} = \sup\Big\{ \int_{\mathbb{R}^n} |f(x)^{q_0} h(x)|\,dx : h \in \mathbb{Y}',\ \|h\|_{\mathbb{Y}'} \le 1 \Big\}.$$

Furthermore, since f is non-negative, we may also restrict the supremum to non-negative h. Therefore, it will suffice to fix such a function h and show that

$$\int_{\mathbb{R}^n} f(x)^{q_0} h(x)\,dx \le C \|g\|_{\mathbb{X}}^{q_0}$$

with a constant independent of h.

By (A) we have that

$$\int_{\mathbb{R}^n} f(x)^{q_0}\, h(x)\,dx \le \int_{\mathbb{R}^n} f(x)^{q_0}\, \mathcal{R}h(x)\,dx.$$

By the generalized Hölder's inequality (4.2) and (B),

$$\int_{\mathbb{R}^n} f(x)^{q_0}\, \mathcal{R}h(x)\, dx \le \|f^{q_0}\|_{\mathbb{Y}} \|\mathcal{R}h\|_{\mathbb{Y}'} \le C\|f\|_{\mathbb{X}}^{q_0} \|h\|_{\mathbb{Y}'} < \infty.$$

Therefore, since by (C), $\mathcal{R}h \in A_{1,\mathcal{B}}$, by (4.13) and again by (4.2) and (B),

$$\int_{\mathbb{R}^n} f(x)^{q_0}\, h(x)\, dx \le C \int_{\mathbb{R}^n} g(x)^{q_0}\, \mathcal{R}h(x)\, dx \le C\|g^{q_0}\|_{\mathbb{Y}}\, \|\mathcal{R}h\|_{\mathbb{Y}'} \le 2C\|g\|_{\mathbb{X}}^{q_0}.$$

Since the constant C is independent of h we have proved (4.8).

We now prove (4.9). Fix p, $p_0/q_0 \le p < \infty$, and let $\mathbb{Y} = \mathbb{X}^p$. If we set $\tilde{q}_0 = p\, q_0 \ge p_0$, then $\mathbb{Y}^{1/\tilde{q}_0}$ is a Banach function space and $M_{\mathcal{B}}$ is bounded on $(\mathbb{Y}^{1/\tilde{q}_0})'$. Therefore, we can apply (4.8) to $\mathbb{Y}$ to get the desired inequality. □

Proof of Corollary 4.8. Given p, $1 < p < \infty$, set $\mathbb{Y} = \mathbb{X}^p$; then $\mathbb{Y}$ is a function space such that $\mathbb{Y}^{1/p}$ is also a Banach function space and $M_{\mathcal{B}}$ is bounded on $(\mathbb{Y}^{1/p})'$. If we apply Theorem 3.9 to (4.10), we get that for all $w \in A_{1,\mathcal{B}} \subset A_{p,\mathcal{B}}$,

$$\int_{\mathbb{R}^n} f(x)^p w(x)\, dx \le C \int_{\mathbb{R}^n} g(x)^p w(x)\, dx, \quad (f,g) \in \mathcal{F}.$$

Hence, by Theorem 4.6 with $p_0 = q_0 = p$ we get that for every q, $1 \le q < \infty$,

$$\|f\|_{\mathbb{Y}^q} \le C\|g\|_{\mathbb{Y}^q}, \quad (f,g) \in \mathcal{F}.$$

In particular, if we take $q = 1$ we get the desired estimate for $\mathbb{X}^p$.

To get the vector-valued inequality (4.12), it suffices to use Corollary 3.12 to get the necessary ℓ^q-valued estimates in $L^p(w)$ and then apply Theorem 4.6. □

Rearrangement invariant spaces

In the important special case when $\mathbb{X}$ is an r.i. function space, we can establish the boundedness of the maximal operator via interpolation, so we do not need to make this a hypothesis. This yields an extension of Theorem 3.9 to r.i. function spaces. To state it, recall Definition 3.3: a Muckenhoupt basis $\mathcal{B}$ is $A_{p,\mathcal{B}}$ open if for every p, $1 < p < \infty$, $w \in A_{p,\mathcal{B}}$ implies $w \in A_{p-\varepsilon,\mathcal{B}}$ for some $\epsilon > 0$.

Theorem 4.10. *Let $\mathcal{B}$ be a Muckenhoupt basis that is $A_{p,\mathcal{B}}$ open. Suppose that for some p_0, $1 \le p_0 < \infty$, and every $w_0 \in A_{p_0,\mathcal{B}}$,*

$$\int_{\mathbb{R}^n} f(x)^{p_0} w_0(x)\, dx \le C \int_{\mathbb{R}^n} g(x)^{p_0} w_0(x)\, dx, \qquad (f,g) \in \mathcal{F}. \tag{4.14}$$

If $\mathbb{X}$ is an r.i. function space such that $1 < p_{\mathbb{X}} \le q_{\mathbb{X}} < \infty$, then for all $w \in A_{p_{\mathbb{X}},\mathcal{B}}$,

$$\|f\|_{\mathbb{X}(w)} \le C\|g\|_{\mathbb{X}(w)}, \qquad (f,g) \in \mathcal{F}. \tag{4.15}$$

Furthermore, for every q, $1 < q < \infty$, and sequence $\{(f_i, g_i)\}_i \subset \mathcal{F}$,

$$\Big\|\Big(\sum_i f_i^q\Big)^{1/q}\Big\|_{\mathbb{X}(w)} \le C\Big\|\Big(\sum_i g_i^q\Big)^{1/q}\Big\|_{\mathbb{X}(w)}. \tag{4.16}$$

Remark 4.11. Using the rescaling arguments in Section 3.3 we can prove an A_∞ extrapolation theorem for rearrangement invariant function spaces. If (4.14) holds for some $p_0 > 0$ and every $w \in A_{\infty,\mathcal{B}}$, then for every r.i. invariant function space $\mathbb{X}$ with $q_\mathbb{X} < \infty$, for every $p > 0$, and every $w \in A_{\infty,\mathcal{B}}$, $\|f\|_{\mathbb{X}^p(w)} \le C\|g\|_{\mathbb{X}^p(w)}$. Details are left to the reader. When $\mathcal{B} = \mathcal{Q}$, this was first proved in [57].

The proof of Theorem 4.10 is nearly identical to the proof of Theorem 3.9. The key difference is that to construct the necessary iteration algorithms we must first show that $M_\mathcal{B}$ and $M'_\mathcal{B}$ are bounded. (Recall that for $w \in A_{\infty,\mathcal{B}}$, $M'_\mathcal{B}f(x) = M_\mathcal{B}(f\,w)(x)/w(x)$.)

Lemma 4.12. *Let $\mathcal{B}$ be a Muckenhoupt basis that is $A_{p,\mathcal{B}}$ open. Given an r.i. function space $\mathbb{X}$ with $1 < p_\mathbb{X} \le q_\mathbb{X} < \infty$, and $w \in A_{p_\mathbb{X},\mathcal{B}}$:*

(a) *$M_\mathcal{B}$ is bounded on $\mathbb{X}(w)$;*

(b) *$M'_\mathcal{B}$ is bounded on $\mathbb{X}'(w)$.*

Proof. This lemma follows by adapting the proof of Boyd's interpolation theorem in [12, Chapter 3, Theorem 5.16], and we use the same notation and terminology. Fix $w \in A_{p_\mathbb{X},\mathcal{B}}$. Since $\mathcal{B}$ is $A_{p,\mathcal{B}}$ open, $w \in A_{q,\mathcal{B}}$ for some q, $1 < q < p_\mathbb{X}$. By Hölder's inequality, for any $r > q_\mathbb{X}$, $w \in A_{r,\mathcal{B}}$. Therefore, $M_\mathcal{B}$ is bounded on both $L^q(w)$ and $L^r(w)$, and so is of joint weak type $(q, q; r, r)$; that is, for all $t > 0$,

$$(M_\mathcal{B}f)^*_w(t) \le CS_\sigma(f^*_w)(t),$$

where S_σ is the Calderón operator associated with the interpolation segment $[(1/q, 1/q), (1/r, 1/r)]$. But S_σ is bounded on $\overline{\mathbb{X}}$ because $1 < p_\mathbb{X} \le q_\mathbb{X} < \infty$. Hence,

$$\|M_\mathcal{B}f\|_{\mathbb{X}(w)} = \|(M_\mathcal{B}f)^*_w\|_{\overline{\mathbb{X}}} \le C\|S_\sigma(f^*_w)\|_{\overline{\mathbb{X}}} \le C\|f^*_w\|_{\overline{\mathbb{X}}} = C\|f\|_{\mathbb{X}(w)}.$$

Similarly, since $w \in A_{q,\mathcal{B}}$ and $w \in A_{r,\mathcal{B}}$, $M'_\mathcal{B}$ is bounded on $L^{q'}(w)$ and $L^{r'}(w)$. Furthermore, $p_{\mathbb{X}'} = q'_\mathbb{X} > r'$ and $q_{\mathbb{X}'} = p'_\mathbb{X} < q'$, so by repeating the above argument we have that $M'_\mathcal{B}$ is bounded on $\mathbb{X}'(w)$. □

Remark 4.13. The boundedness of $M_\mathcal{B}$ when $\mathcal{B} = \mathcal{Q}$ was proved in [57] without using Boyd's interpolation theorem.

Proof of Theorem 4.10. As we noted, the proof is very similar to that of Theorem 3.9; here we will sketch the key details.

We first prove (4.15). Since $w \in A_{p_\mathbb{X},\mathcal{B}}$, by Lemma 4.12 we can define the Rubio de Francia iteration algorithms. Given non-negative functions $h_1 \in \mathbb{X}(w)$ and $h_2 \in \mathbb{X}'(w)$, let

$$\mathcal{R}h_1(x) = \sum_{k=0}^\infty \frac{M^k_\mathcal{B}h_1(x)}{2^k\|M_\mathcal{B}\|^k_{\mathbb{X}(w)}}, \qquad \mathcal{R}'h_2(x) = \sum_{k=0}^\infty \frac{(M'_\mathcal{B})^k h_2(x)}{2^k\|M'_\mathcal{B}\|^k_{\mathbb{X}'(w)}}.$$

Then

(A)	$h_1(x) \le \mathcal{R}h_1(x)$	(A′)	$h_2(x) \le \mathcal{R}'h_2(x)$
(B)	$\|\mathcal{R}h_1\|_{\mathbb{X}(w)} \le 2\|h_1\|_{\mathbb{X}(w)}$	(B′)	$\|\mathcal{R}'h_2\|_{\mathbb{X}'(w)} \le 2\|h_2\|_{\mathbb{X}'(w)}$
(C)	$[\mathcal{R}h_1]_{A_{1,\mathcal{B}}} \le 2\|M_{\mathcal{B}}\|_{\mathbb{X}(w)}$	(C′)	$[\mathcal{R}'h_2\, w]_{A_{1,\mathcal{B}}} \le 2\|M'_{\mathcal{B}}\|_{\mathbb{X}'(w)}$.

Fix $(f,g) \in \mathcal{F}$ and define

$$h_1(x) = \frac{f(x)}{\|f\|_{\mathbb{X}(w)}} + \frac{g(x)}{\|g\|_{\mathbb{X}(w)}}.$$

Then $h_1 \in \mathbb{X}(w)$ and $\|h_1\|_{\mathbb{X}(w)} \le 2$. Since $f \in \mathbb{X}(w)$, by (4.4) it will suffice to show that

$$\int_{\mathbb{R}^n} f(x)\, h_2(x)\, w(x)\, dx \le C\|g\|_{\mathbb{X}(w)},$$

where $h_2 \in \mathbb{X}'(w)$, $\|h_2\|_{\mathbb{X}'(w)} = 1$, h_2 is non-negative, and the constant C is independent of h_2.

Fix r, $1 < r < \infty$; then by (A′) and Hölder's inequality with respect to the measure $\mathcal{R}'h_2\, w$,

$$\begin{aligned}
&\int_{\mathbb{R}^n} f(x)h_2(x)w(x)\, dx \\
&\qquad \le \int_{\mathbb{R}^n} f(x)\mathcal{R}'h_2(x)w(x)\, dx \\
&\qquad = \int_{\mathbb{R}^n} f(x)\mathcal{R}h_1(x)^{-1/r'}\mathcal{R}h_1(x)^{1/r'}\mathcal{R}'h_2(x)w(x)\, dx \\
&\qquad \le \left(\int_{\mathbb{R}^n} f(x)^r \mathcal{R}h_1(x)^{1-r}\mathcal{R}'h_2(x)w(x)\, dx\right)^{1/r} \\
&\qquad\qquad \times \left(\int_{\mathbb{R}^n} \mathcal{R}h_1(x)\mathcal{R}'h_2(x)w(x)\, dx\right)^{1/r'}.
\end{aligned}$$

We now estimate each of the integrals in the last term exactly as we did in the proof of Theorem 3.9. We use the above properties of the iteration algorithms and the reverse factorization property (Proposition 3.5). We also apply Theorem 3.9 to show that inequality (4.14) holds with p_0 replaced by r and $w \in A_{r,\mathcal{B}}$.

Finally, to prove (4.16), note that it is an immediate consequence of (4.15): given (4.14), apply Corollary 3.12 and then apply (4.15) to the family $\mathcal{F}_{\ell^q}$ defined in the proof of the corollary. □

Remark 4.14. In the proof of Theorem 4.10 the exponent r is arbitrary and we can take any value. For instance, we could have taken $r = p_0$ (if $p_0 > 1$), thereby eliminating the appeal to extrapolation. The choice of r only affects the final constant; we leave the computation of the best constant to the interested reader.

4.3 Extrapolation on modular spaces

In this section we state and prove our extrapolation theorems for modular spaces. As we did for extrapolation in r.i. spaces, in this section we need to assume the $A_{p,\mathcal{B}}$ openness property.

Theorem 4.15. *Let $\mathcal{B}$ be a Muckenhoupt basis that is $A_{p,\mathcal{B}}$ open. Suppose that for some p_0, $1 \le p_0 < \infty$, and every $w_0 \in A_{p_0,\mathcal{B}}$,*

$$\int_{\mathbb{R}^n} f(x)^{p_0} w_0(x)\,dx \le C \int_{\mathbb{R}^n} g(x)^{p_0} w_0(x)\,dx, \qquad (f,g) \in \mathcal{F}. \tag{4.17}$$

If Φ is any Young function such that $1 < i_\Phi \le I_\Phi < \infty$ (i.e., $\Phi, \bar{\Phi} \in \Delta_2$) and w is any weight in $A_{i_\Phi,\mathcal{B}}$, then

$$\int_{\mathbb{R}^n} \Phi(f(x))w(x)\,dx \le C \int_{\mathbb{R}^n} \Phi(g(x))w(x)\,dx, \quad (f,g) \in \mathcal{F}. \tag{4.18}$$

Further, for every q, $1 < q < \infty$, and sequence $\{(f_i, g_i)\}_i \subset \mathcal{F}$,

$$\int_{\mathbb{R}^n} \Phi\Big(\Big(\sum_i f_i(x)^q\Big)^{1/q}\Big)w(x)\,dx \le C \int_{\mathbb{R}^n} \Phi\Big(\Big(\sum_i g_i(x)^q\Big)^{1/q}\Big)w(x)\,dx. \tag{4.19}$$

Remark 4.16. The assumptions on Φ are not as restrictive as they may appear; in the unweighted cases, for example, the Hilbert transform H satisfies $\rho^\Phi(Hf) \le C\rho^\Phi(f)$ if and only if $1 < i_\Phi \le I_\Phi < \infty$. (See [115].)

Remark 4.17. Using the rescaling arguments in Section 3.3 we can prove an A_∞ extrapolation theorem for modular spaces. If (4.17) holds for some $p_0 > 0$ and every $w \in A_{\infty,\mathcal{B}}$, then for every Young function Φ with $I_\Phi < \infty$, every p, $q > 0$, and every $w \in A_{\infty,\mathcal{B}}$,

$$\int_{\mathbb{R}^n} \Phi(f(x)^p)^q w(x)\,dx \le C \int_{\mathbb{R}^n} \Phi(g(x)^p)^q w(x)\,dx.$$

Details are left to the reader. When $\mathcal{B} = \mathcal{Q}$, this was first proved in [57].

Theorem 4.15 has two corollaries. The first yields weak type modular inequalities; its proof is identical to the proof of Corollary 3.10, but using Theorem 4.15 in place of Theorem 3.9.

Corollary 4.18. *Let $\mathcal{B}$ be a Muckenhoupt basis that is $A_{p,\mathcal{B}}$ open. Suppose that for some p_0, $1 \le p_0 < \infty$, and every $w_0 \in A_{p_0,\mathcal{B}}$,*

$$\|f\|_{L^{p_0,\infty}(w_0)} \le C\|g\|_{L^{p_0}(w_0)}, \qquad (f,g) \in \mathcal{F}.$$

If Φ is any Young function such that $1 < i_\Phi \le I_\Phi < \infty$ and w is any weight in $A_{i_\Phi,\mathcal{B}}$, then

$$\sup_{\lambda>0} \Phi(\lambda) w(\{x \in \mathbb{R}^n : f(x) > \lambda\}) \le C \int_{\mathbb{R}^n} \Phi(g(x))w(x)\,dx, \quad (f,g) \in \mathcal{F}.$$

The second corollary gives estimates that combine modular, vector-valued and function space estimates.

Corollary 4.19. *Let $\mathcal{B}$ be a Muckenhoupt basis that is $A_{p,\mathcal{B}}$ open. Suppose that for some p_0, $1 \le p_0 < \infty$, and every $w_0 \in A_{p_0,\mathcal{B}}$, (4.17) holds. Let Φ be a Young function with $I_\Phi < \infty$. Then:*

(a) *If $\frac{1}{i_\Phi} < p < \infty$ and $w \in A_{p\, i_\Phi,\mathcal{B}}$, then*

$$\int_{\mathbb{R}^n} \Phi(f(x))^p w(x)\, dx \le C \int_{\mathbb{R}^n} \Phi(g(x))^p w(x)\, dx, \quad (f,g) \in \mathcal{F}.$$

(b) *If $\frac{1}{i_\Phi} < p, q < \infty$, $w \in A_{p\, i_\Phi,\mathcal{B}}$ and $\{(f_i, g_i)\}_i \subset \mathcal{F}$,*

$$\Big\| \Big(\sum_i \Phi(f_i)^q \Big)^{1/q} \Big\|_{L^p(w)} \le C \Big\| \Big(\sum_i \Phi(g_i)^q \Big)^{1/q} \Big\|_{L^p(w)}.$$

In particular, if $i_\Phi > 1$ and $w \in A_{i_\Phi,\mathcal{B}}$, then

$$\int_{\mathbb{R}^n} \Big(\sum_i \Phi(f_i(x))^q \Big)^{1/q} w(x)\, dx \le C \int_{\mathbb{R}^n} \Big(\sum_i \Phi(g_i(x))^q \Big)^{1/q} w(x)\, dx.$$

(c) *If $\mathbb{X}$ is a quasi-Banach function space such that $\mathbb{X}^{i_\Phi}$ is an r.i. function space with $\frac{1}{i_\Phi} < p_{\mathbb{X}} \le q_{\mathbb{X}} < \infty$, and $w \in A_{p_{\mathbb{X}}\, i_\Phi,\mathcal{B}}$, then*

$$\|\Phi(f)\|_{\mathbb{X}(w)} \le C \|\Phi(g)\|_{\mathbb{X}(w)}, \quad (f,g) \in \mathcal{F}.$$

In particular, if $i_\Phi > 1$, we have the following weak-weak modular inequality: for all $w \in A_{i_\Phi,\mathcal{B}}$,

$$\sup_{\lambda>0} \Phi(\lambda) w(\{x \in \mathbb{R}^n : f(x) > \lambda\}) \\ \le C \sup_{\lambda>0} \Phi(\lambda) w(\{x \in \mathbb{R}^n : g(x) > \lambda\}), \quad (f,g) \in \mathcal{F}.$$

The proof of Theorem 4.15 follows the same scheme as the proofs of Theorems 3.9 and 4.10. However, we need to make two non-trivial changes. First, we need to define the Rubio de Francia iteration algorithms so that the resulting functions satisfy modular inequalities, which in turn means that we need $M_\mathcal{B}$ and $M'_\mathcal{B}$ to satisfy modular inequalities. We prove this via interpolation. Second, we can no longer use duality and Hölder's inequality; instead we will make a clever choice of the "dual function" and then apply Young's inequality (4.5). (As we noted in our analysis of the proof in Section 2.1, this is a variant of our proof in weighted Lebesgue spaces.) Because we use Young's inequality, we have an "error term" that did not appear in previous proofs but we can reabsorb it into the left-hand side of our inequality.

Lemma 4.20. *Let $\mathcal{B}$ be a Muckenhoupt basis that is $A_{p,\mathcal{B}}$ open. Given a Young function Φ with $1 < i_\Phi \le I_\Phi < \infty$, and $w \in A_{i_\Phi,\mathcal{B}}$:*

(a) *$M_\mathcal{B}$ is (Φ, w)-modular;*

(b) *$M'_\mathcal{B}$ is $(\bar{\Phi}, w)$-modular.*

Proof. The proof is nearly the same as the proof of Lemma 4.12: we replace the Boyd indices $p_\mathbb{X}$, $q_\mathbb{X}$ with the dilation indices i_Φ and I_Φ, and we replace the Boyd interpolation theorem with an interpolation theorem in the scale of modular spaces due to Miyamoto [146] (also see [20]). Since $\mathcal{B}$ is $A_{p,\mathcal{B}}$ open and since $w \in A_{q,\mathcal{B}}$ for any $q > i_\Phi$, there exist p, q such that $1 < p < i_\Phi \le I_\Phi < q < \infty$ and $M_\mathcal{B}$ is bounded on $L^p(w)$ and $L^q(w)$. Therefore, by interpolation,

$$\int_{\mathbb{R}^n} \Phi(M_\mathcal{B} f(x))w(x)\,dx \le C \int_{\mathbb{R}^n} \Phi(|f(x)|)w(x)\,dx$$

provided that for all $t > 0$,

$$\max\left(t^p \int_0^t \frac{\Phi(s)}{s^{p+1}}\,ds,\ t^q \int_t^\infty \frac{\Phi(s)}{s^{q+1}}\,ds\right) \le C\Phi(t).$$

This inequality follows from the definition of the dilation indices: see Maligranda [135]. (Also see [43].)

The same proof works for M', replacing Φ by $\bar{\Phi}$. □

Remark 4.21. That $M_\mathcal{B}$ is (Φ, w)-modular when $\mathcal{B} = \mathcal{Q}$ was proved directly in [57]; also see [114, 115].

Proof of Theorem 4.15. We first prove (4.18). Fix $w \in A_{i_\Phi,\mathcal{B}}$; then by Lemma 4.20, $M_\mathcal{B}$ is (Φ, w)-modular and $M'_\mathcal{B}$ is $(\bar{\Phi}, w)$-modular. Further, we claim that $\rho_w^\Phi(M_\mathcal{B})$, $\rho_w^{\bar{\Phi}}(M'_\mathcal{B}) \ge 1$. The first inequality follows immediately: fix $B \in \mathcal{B}$ such that w is not trivial on B, and let $f = \chi_B$. Then

$$\int_{\mathbb{R}^n} \Phi(f(x))w(x)\,dx = \Phi(1)w(B).$$

On the other hand, for $x \in B$, $M_\mathcal{B} f(x) \ge \fint_B f(y)\,dy = 1$, and so

$$\int_{\mathbb{R}^n} \Phi(M_\mathcal{B} f(x))w(x)\,dx \ge \Phi(1)w(B).$$

To show the second inequality, we first claim that for B as above,

$$\int_B \bar{\Phi}\big(w(x)^{-1}\big)w(x)\,dx < \infty.$$

Let $f = |B|w(B)^{-1}\chi_B$; then for $x \in B$,

$$M'_\mathcal{B} f(x) = w(x)^{-1} M_\mathcal{B}(fw)(x) \ge w(x)^{-1} \fint_B f(y)w(y)\,dy = w(x)^{-1}.$$

Therefore, since $M'_{\mathcal{B}}$ is $(\bar{\Phi}, w)$ modular,

$$\int_B \bar{\Phi}\big(w(x)^{-1}\big)w(x)\,dx \le \int_B \bar{\Phi}(M'_{\mathcal{B}}f(x))w(x)\,dx$$
$$\le C\int_{\mathbb{R}^n} \bar{\Phi}(f(x))w(x)\,dx = C\bar{\Phi}(|B|w(B)^{-1})w(B) < \infty.$$

Now let $f = w^{-1}\chi_B$. Then

$$\int_{\mathbb{R}^n} \bar{\Phi}(f(x))w(x)\,dx = \int_B \bar{\Phi}\big(w(x)^{-1}\big)w(x)\,dx < \infty;$$

but since for $x \in B$, $M'_{\mathcal{B}}f(x) = w(x)^{-1}M_{\mathcal{B}}(\chi_B)(x) \ge w(x)^{-1}$, we have that

$$\int_{\mathbb{R}^n} \bar{\Phi}(M'_{\mathcal{B}}f(x))w(x)\,dx \ge \int_B \bar{\Phi}\big(w(x)^{-1}\big)w(x)\,dx.$$

We now define the Rubio de Francia iteration algorithms.
Let $\Lambda = (2\rho_w^{\Phi}(M_{\mathcal{B}}))^{-1}$; then $\Lambda \le 1/2$ and $\Lambda\,\rho_w^{\Phi}(M_{\mathcal{B}}) = 1/2$. Further,

$$\sum_{k=0}^{\infty}(1-\Lambda)\,\Lambda^k = 1.$$

Similarly, if we let $\Lambda' = (2\rho_w^{\bar{\Phi}}(M'_{\mathcal{B}}))^{-1}$, then analogous estimates hold. Given non-negative $h_1 \in M_w^{\Phi}$ and $h_2 \in M_w^{\bar{\Phi}}$, let

$$\mathcal{R}h_1(x) = \sum_{k=0}^{\infty} \Lambda^k M_{\mathcal{B}}^k h_1(x), \qquad \mathcal{R}'h_2(x) = \sum_{k=0}^{\infty}(\Lambda')^k(M'_{\mathcal{B}})^k h_2(x).$$

(The slightly different definitions of $\mathcal{R}$ and $\mathcal{R}'$ are for technical reasons that will be made clear below.) Then the following are true:

(A) $h_1(x) \le \mathcal{R}h_1(x)$ (A′) $h_2(x) \le \mathcal{R}'h_2(x)$

(B) $\rho_w^{\Phi}(\mathcal{R}h_1) \le 2\rho_w^{\Phi}(2h_1)$ (B′) $\rho_w^{\bar{\Phi}}(\mathcal{R}'h_2) \le 2\,\rho_w^{\bar{\Phi}}(2h_2)$

(C) $[\mathcal{R}h_1]_{A_{1,\mathcal{B}}} \le 2\,\rho_w^{\Phi}(M_{\mathcal{B}})$ (C′) $[\mathcal{R}'h_2\cdot w]_{A_{1,\mathcal{B}}} \le 2\rho_w^{\bar{\Phi}}(M'_{\mathcal{B}})$.

Properties (A), (A′), (C) and (C′) are proved exactly as in the proof of the corresponding properties in Theorem 3.9. We will prove (B); the proof of (B′) is identical.

Define $\widetilde{h}_1(x) = h_1(x)/(1-\Lambda)$; then, since Φ is convex and $M_{\mathcal{B}}$ is (Φ, w)-modular,

$$\rho_w^{\Phi}(\mathcal{R}h_1) = \int_{\mathbb{R}^n} \Phi\Big(\sum_{k=0}^{\infty}(1-\Lambda)\Lambda^k M_{\mathcal{B}}^k\widetilde{h}_1(x)\Big)w(x)\,dx$$

$$\begin{aligned}
&\le \sum_{k=0}^{\infty}(1-\Lambda)\Lambda^k \int_{\mathbb{R}^n} \Phi\big(M_{\mathcal{B}}^k \widetilde{h}_1(x)\big)w(x)\,dx \\
&\le (1-\Lambda)\sum_{k=0}^{\infty}\Lambda^k\,\rho_w^{\Phi}(M_{\mathcal{B}})^k \int_{\mathbb{R}^n} \Phi\big(\widetilde{h}_1(x)\big)w(x)\,dx \\
&= 2(1-\Lambda)\int_{\mathbb{R}^n} \Phi\big(h_1(x)/(1-\Lambda)\big)w(x)\,dx \\
&\le 2\rho_w^{\Phi}(2\,h_1).
\end{aligned}$$

Now let $(f,g)\in\mathcal{F}$ be such that $\rho_w^{\Phi}(f)<\infty$; we may also assume that $\rho_w^{\Phi}(g)<\infty$ since otherwise there is nothing to prove. Fix δ, $0<\delta<1/2$, and define $h_1(x)=\delta\,f(x)+(1-\delta)\,g(x)$. Since Φ is convex,

$$\rho_w^{\Phi}(h_1)\le\delta\rho_w^{\Phi}(f)+(1-\delta)\rho_w^{\Phi}(g)<\infty.$$

Define $h_2(x)=\Phi\big(f(x)\big)/f(x)$ when $f(x)\neq 0$ and 0 otherwise. (The function h_2 takes the place of the function gotten by duality in the proof of Theorem 3.9.) If we replace t by $\Phi(t)$ in (4.6) we get that $\bar{\Phi}\big(\Phi(t)/t\big)\le\Phi(t)$, and so $\bar{\Phi}\big(h_2(x)\big)\le\Phi\big(f(x)\big)$. Therefore,

$$\rho_w^{\bar{\Phi}}(h_2)\le\rho_w^{\Phi}(f)<\infty. \tag{4.20}$$

Fix r, $1<r<\infty$; then by (A′) and Hölder's inequality,

$$\begin{aligned}
\rho_w^{\Phi}(f) &= \int_{\mathbb{R}^n} f(x)h_2(x)w(x)\,dx \\
&\le \int_{\mathbb{R}^n} f(x)\mathcal{R}'h_2(x)w(x)\,dx \\
&= \int_{\mathbb{R}^n} f(x)\mathcal{R}h_1(x)^{-1/r'}\mathcal{R}h_1(x)^{1/r'}\mathcal{R}'h_2(x)w(x)\,dx \\
&\le \left(\int_{\mathbb{R}^n} f(x)^r\mathcal{R}h_1(x)^{1-r}\mathcal{R}'h_2(x)w(x)\,dx\right)^{1/r} \\
&\qquad\times\left(\int_{\mathbb{R}^n} \mathcal{R}h_1(x)\mathcal{R}'h_2(x)w(x)\,dx\right)^{1/r'} \\
&= I_1^{1/r}\times I_2^{1/r'}.
\end{aligned}$$

We estimate I_1 and I_2 separately, beginning with I_2. Given ε, $0<\varepsilon<1$ (the exact value to be fixed below), by Young's inequality (4.5),

$$I_2\le\rho_w^{\Phi}(\varepsilon^{-1}\mathcal{R}h_1)+\rho_w^{\bar{\Phi}}(\varepsilon\mathcal{R}'h_2)=J_1+J_2.$$

Since $\Phi\in\Delta_2$ there exist constants C_Φ, $D>0$ such that $\Phi(\lambda t)\le C_\Phi\lambda^D\Phi(t)$ for every $t>0$ and $\lambda>1$. (This follows from the definition of I_Φ: it suffices to take any $D>I_\Phi$.) Therefore, by property (B) and by the convexity of Φ,

$$J_1 \le C_\Phi \varepsilon^{-D} \rho_w^\Phi(\mathcal{R}h_1) \le 2C_\Phi \varepsilon^{-D} \rho_w^\Phi(2\,h_1)$$
$$\le 2^{D+1} C_\Phi^2 \varepsilon^{-D} \rho_w^\Phi(h_1) \le C\varepsilon^{-D}\big(\delta\,\rho_w^\Phi(f) + (1-\delta)\rho_w^\Phi(g)\big).$$

Similarly, if we use that $\bar{\Phi}$ is convex and in Δ_2, property (B′), and (4.20), we get that

$$J_2 \le \varepsilon \rho_w^{\bar{\Phi}}(\mathcal{R}'h_2) \le \varepsilon 2 \rho_w^{\bar{\Phi}}(2\,h_2) \le C\varepsilon\, \rho_w^{\bar{\Phi}}(h_2) \le C\varepsilon\, \rho_w^\Phi(f).$$

Combining these two estimates we get that for a fixed constant C_0 (independent of ϵ),

$$I_2 \le C_0(\delta\varepsilon^{-D} + \varepsilon)\rho_w^\Phi(f) + C_0 \varepsilon^{-D} \rho_w^\Phi(g).$$

In particular, we have that $I_2 < \infty$.

To estimate I_1 we first show that it is finite. By property (A), $\delta\, f(x) \le h_1(x) \le \mathcal{R}h_1$, and so

$$I \le \delta^{-r} \int_{\mathbb{R}^n} \mathcal{R}h_1(x)\mathcal{R}'h_2(x)w(x)\,dx = \delta^r I_2 < \infty.$$

By properties (C) and (C′) and by the reverse factorization property (Proposition 3.5), $\mathcal{R}h_1(x)^{1-r}\,\mathcal{R}'h_2(x)\,w(x) \in A_{r,\mathcal{B}}$. Therefore, we can apply Theorem 3.9 with initial hypothesis (4.17) to conclude that

$$\begin{aligned} I_1 &\le C \int_{\mathbb{R}^n} g(x)^r \mathcal{R}h_1(x)^{1-r}\mathcal{R}'h_2(x)w(x)\,dx \\ &\le C(1-\delta)^{-r} \int_{\mathbb{R}^n} \mathcal{R}h_1(x)\mathcal{R}'h_2(x)w(x)\,dx \\ &\le CI_2; \end{aligned}$$

the second inequality holds since by property (A), $(1-\delta)g \le h_1 \le \mathcal{R}h_1$; the third since $0 < \delta < 1/2$.

We now combine the estimates for I_1 and I_2, and choose ε and δ such that $0 < \varepsilon < 1/(C_0\,4)$, $0 < \delta < \min(1/2, \varepsilon^D/(C_0\,4))$. This yields

$$\begin{aligned} \rho_w^\Phi(f) &\le I_1^{1/r} \times I_2^{1/r'} \\ &\le CI_2 \\ &\le C_0(\delta\varepsilon^{-D} + \varepsilon)\rho_w^\Phi(f) + C_0\varepsilon^{-D}\rho_w^\Phi(g) \\ &\le \frac{1}{2}\rho_w^\Phi(f) + C_0\varepsilon^{-D}\rho_w^\Phi(g). \end{aligned}$$

Since $\rho_w^\Phi(f) < \infty$ we can rearrange terms to get inequality (4.18).

To complete the proof we note that (4.19) follows immediately from Corollary 3.12 and (4.18). □

Remark 4.22. As in the proof of Theorem 4.10, the exact value of r chosen only affects the size of the constant.

Proof of Corollary 4.19. To prove part (a) we could simply define the Young function $\Psi(t) = \Phi(t)^p$; since $i_\psi = p\, i_\Phi$ and $I_\psi = p\, I_\Phi$, by Theorem 4.15 we get the desired inequality.

We can also prove (a) using a rescaling argument; this has the advantage that it adapts to prove (b) and (c). By our hypotheses, (4.18) holds. Define a new family $\mathcal{F}_\Phi$ consisting of the pairs

$$(F,G) = \big(\Phi(f)^{1/i_\Phi}, \Phi(g)^{1/i_\Phi}\big), \qquad (f,g) \in \mathcal{F}.$$

Then (4.18) is equivalent to the following: for every $w \in A_{i_\Phi,\mathcal{B}}$,

$$\int_{\mathbb{R}^n} F(x)^{i_\Phi} w(x)\,dx \le C \int_{\mathbb{R}^n} G(x)^{i_\Phi} w(x)\,dx \qquad (F,G) \in \mathcal{F}_\Phi. \tag{4.21}$$

Therefore, we can apply Theorem 3.9 with $p_0 = i_\Phi$ to get that if $w \in A_{p\,i_\Phi,\mathcal{B}}$,

$$\int_{\mathbb{R}^n} F(x)^{p i_\Phi} w(x)\,dx \le C \int_{\mathbb{R}^n} G(x)^{p i_\Phi} w(x)\,dx \qquad (F,G) \in \mathcal{F}_\Phi,$$

which is equivalent to (a).

To prove the first inequality in (b), we apply Corollary 3.12 beginning with (4.21) to get that

$$\Big\| \Big(\sum_i F^{i_\Phi q}\Big)^{1/(i_\Phi q)} \Big\|_{L^{p i_\Phi}(w)} \le C \Big\| \Big(\sum_i G^{i_\Phi q}\Big)^{1/(i_\Phi q)} \Big\|_{L^{p i_\Phi}(w)},$$

which is equivalent to the first inequality. We get the second by taking $p = 1$.

Finally, the first inequality in (c) is gotten the same way except that we use Theorem 4.10. The second inequality is gotten by taking the special case $\mathbb{X} = L^{1,\infty}$ which is a r.i. quasi-Banach function space with $p_\mathbb{X} = q_\mathbb{X} = 1$. □

4.4 Applications

In this section we consider some applications of extrapolation to Banach function spaces and modular spaces. We first briefly consider applications to modular spaces and rearrangement invariant Banach function spaces, and then give two applications to a particular (non-rearrangement invariant) function space, the variable Lebesgue space $L^{p(\cdot)}$.

Remark 4.23. To apply Theorem 4.6 and Corollary 4.8 to a given (non rearrangement invariant) Banach function space $\mathbb{X}$, we need information on the boundedness of the maximal operator on $(\mathbb{X}^r)'$, $r > 0$. Even for the Hardy-Littlewood maximal operator (i.e., when $\mathcal{B} = \mathcal{Q}$) this is often a non-trivial question. Two recent papers [128, 131] consider the boundedness of the maximal operator on abstract function spaces.

Modular spaces and r.i. function spaces

Clearly, given an operator T that is bounded on $L^{p_0}(w)$ for all $w \in A_{p_0}$, we can immediately use extrapolation to deduce function space and modular space inequalities for it. On the other hand, such inequalities also follow for rearrangement invariant function spaces from the classical Rubio de Francia extrapolation (Theorem 1.4) and Boyd's interpolation theorem (see [12]). Similarly, modular inequalities can be gotten by using modular interpolation (see [95, 118]).

However, even for these spaces our approach yields new results or elementary proofs of known results. In particular, we can deduce vector-valued inequalities and Coifman-Fefferman type inequalities. We sketch a few illustrative examples. Let $\mathbb{X}$ be an r.i. function space such that $1 < p_{\mathbb{X}} \le q_{\mathbb{X}} < \infty$. Then for any $w \in A_{p_{\mathbb{X}}}$, and any q, $1 < q < \infty$,

$$\Big\|\Big(\sum_i Mf_i(x)^q\Big)^{1/q}\Big\|_{\mathbb{X}(w)} \le C\Big\|\Big(\sum_i |f_i(x)|^q\Big)^{1/q}\Big\|_{\mathbb{X}(w)}.$$

Similarly, if Φ is a Young function such that $1 < i_\Phi \le I_\Phi < \infty$, and $w \in A_{i_\Phi}$, then

$$\int_{\mathbb{R}^n} \Phi\Big(\Big(\sum_i Mf_i(x)^q\Big)^{1/q}\Big)w(x)\,dx \le \int_{\mathbb{R}^n} \Phi\Big(\Big(\sum_i |f_i(x)|^q\Big)^{1/q}\Big)w(x)\,dx.$$

In the unweighted case, the modular inequality is proved in Kokilashvili and Krbec [115]; the weighted modular inequality and the function space inequality appear in [57].

Similarly, if T is a Calderón-Zygmund singular integral, by inequality (3.31) and Remarks 4.11 and 4.17 we have that for all $w \in A_\infty$,

$$\|Tf\|_{\mathbb{X}(w)} \le C\|Mf\|_{\mathbb{X}(w)},$$
$$\int_{\mathbb{R}^n} \Phi(|Tf(x)|)w(x)\,dx \le C\int_{\mathbb{R}^n} \Phi(Mf(x))w(x)\,dx.$$

These inequalities can be found in [57]. Both inequalities can be proved using the good-λ inequality of Coifman and Fefferman [25]; as we noted in Section 3.8 above, extrapolation gives us an alternative approach.

Variable Lebesgue spaces

The variable Lebesgue spaces and the associated variable Sobolev spaces are of interest in their own right, and also have applications to partial differential equations and the calculus of variations. In the past decade there has been a great deal of work in this area; we refer the reader to the survey papers by Diening, Hästö and Nekvinda [63] and Samko [199] for details and further references. It was in this context of these spaces that extrapolation to Banach function spaces was first considered in [40].

We define these spaces and sketch their basic properties; for complete information we refer the reader to Fan and Zhao [73], Kováčik and Rákosník [116], and Diening [61]. Given a measurable function $p(\cdot) : \mathbb{R}^n \to [1, \infty]$, let $\Omega_\infty = \{x \in \mathbb{R}^n : p(x) = \infty\}$, and define

$$\rho_{p(\cdot)}(f) = \int_{\mathbb{R}^n \setminus \Omega_\infty} |f(x)|^{p(x)}\, dx + \|f\|_{L^\infty(\Omega_\infty)}.$$

Then the variable Lebesgue space $L^{p(\cdot)}$ is the collection of all measurable functions f such that $\rho_{p(\cdot)}(f/\lambda) < \infty$ for some $\lambda > 0$. It is a Banach function space with respect to the norm

$$\|f\|_{p(\cdot)} = \inf\{\lambda > 0 : \rho_{p(\cdot)}(f/\lambda) \leq 1\}.$$

These spaces generalize the classical Lebesgue spaces: if $p(x) \equiv p_0$, then $L^{p(\cdot)}$ equals L^{p_0}. (Here and below we write $p(\cdot)$ instead of p to emphasize that the exponent is a function and not a constant.) They have many properties in common with the standard L^p spaces, except that they are not rearrangement invariant. For use below we highlight the fact that the associate space of $L^{p(\cdot)}$ is $L^{p'(\cdot)}$, where the conjugate exponent function $p'(\cdot)$ is defined by

$$\frac{1}{p(x)} + \frac{1}{p'(x)} = 1$$

with $1/\infty = 0$.

To apply Theorem 4.6 and Corollary 4.8 to $\mathbb{X} = L^{p(\cdot)}$ we need to understand the boundedness of the Hardy-Littlewood maximal operator on variable Lebesgue spaces. To do so we make some definitions. Let

$$p_- = \operatorname*{ess\,inf}_{x \in \mathbb{R}^n} p(x), \qquad p_+ = \operatorname*{ess\,sup}_{x \in \mathbb{R}^n} p(x).$$

A function $p(\cdot)$ is locally log-Hölder continuous if there exists a constant $C_0 > 0$ such that

$$|p(x) - p(y)| \leq \frac{C_0}{-\log(|x-y|)}, \qquad |x-y| \leq 1/2.$$

Similarly, $p(\cdot)$ is log-Hölder continuous at infinity if there exist constants p_∞, $0 \leq p_\infty < \infty$, and $C_\infty > 0$ such that

$$|p(x) - p_\infty| \leq \frac{C_\infty}{\log(e + |x|)}.$$

If $p(\cdot)$ satisfies both these conditions we say that it is log-Hölder continuous and write $p(\cdot) \in LH$.

If the exponent function is log-Hölder continuous, then the Hardy-Littlewood maximal function is bounded on $L^{p(\cdot)}$. Since $p(\cdot)$ can be unbounded or even equal infinity on a set of positive measure, continuity is given in terms of $1/p(\cdot)$ (where again we let $1/\infty = 0$).

Theorem 4.24. *Given an exponent $p(\cdot)$, suppose that $p_- > 1$ and $1/p(\cdot) \in LH$. Then the maximal operator is bounded on $L^{p(\cdot)}$:*

$$\|Mf\|_{p(\cdot)} \leq C\|f\|_{p(\cdot)}.$$

Theorem 4.24 was proved in stages and a number of authors contributed to the final result. We refer the reader to the following papers: [18, 34, 41, 42, 59, 61, 62, 162]. The condition $p_- > 1$ is necessary for M to be bounded; see [41]. Examples given in [41, 185] show that log-Hölder continuity is sharp: if the modulus of continuity is weakened, the maximal operator may not be bounded. On the other hand, it is not necessary, as is shown by the examples of discontinuous exponent functions given in [126]. Diening [60] has given a necessary and sufficient condition for the maximal operator to be bounded on $L^{p(\cdot)}$ in terms of families of averaging operators.

As a consequence of Theorem 4.6 and Corollary 4.8 we get the following extrapolation result on variable Lebesgue spaces that was first proved in [40].

Theorem 4.25. *Suppose that for some p_0, $0 < p_0 < \infty$, and every $w \in A_1$,*

$$\int_{\mathbb{R}^n} f(x)^{p_0} w(x)\,dx \leq C \int_{\mathbb{R}^n} g(x)^{p_0} w(x)\,dx, \qquad (f,g) \in \mathcal{F}. \tag{4.22}$$

If the exponent $p(\cdot)$ is such that $p_0 \leq p_- \leq p_+ < \infty$ and $p(\cdot) \in LH$, then

$$\|f\|_{p(\cdot)} \leq C\|g\|_{p(\cdot)}, \qquad (f,g) \in \mathcal{F}. \tag{4.23}$$

If in addition $p_- > 1$, $p_0 \geq 1$ and (4.22) holds for every $w \in A_{p_0}$, then (4.23) again holds and for every q, $1 < q < \infty$, and sequence $\{(f_i, g_i)\}_i \subset \mathcal{F}$,

$$\Big\| \Big(\sum_i f_i^q\Big)^{1/q} \Big\|_{p(\cdot)} \leq \Big\| \Big(\sum_i g_i^q\Big)^{1/q} \Big\|_{p(\cdot)}. \tag{4.24}$$

Proof. By the definition of the norm on variable Lebesgue spaces, if $\mathbb{X} = L^{p(\cdot)}$, then $\mathbb{X}^{1/p_-} = L^{p(\cdot)/p_-}$ is again a Banach function space. Since $p(\cdot) \in LH$ and $p_+ < \infty$, $1/p(\cdot) \in LH$. Therefore, $p_-/p(\cdot) \in LH$ and so $1/(p(\cdot)/p_-)' \in LH$. Further, since $p_+ < \infty$, $[(p(\cdot)/p_-)']_- > 1$. Hence, by Theorem 4.24, the Hardy-Littlewood maximal operator is bounded on $L^{(p(\cdot)/p_-)'}$, and so by Theorem 4.6 we get (4.23).

If we assume that $p_- > 1$, then we can fix p, $1 < p < p_-$ and let $\mathbb{X} = L^{p(\cdot)/p}$. Then we can repeat the above argument to get that the maximal operator is bounded on $\mathbb{X}' = L^{(p(\cdot)/p)'}$, and so the desired conclusion follows from Corollary 4.8 since $\mathbb{X}^p = L^{p(\cdot)}$. □

Remark 4.26. In [40] we proved a more general result. Instead of assuming a regularity condition on $p(\cdot)$, we assumed that $p_- > 1$ and that M was bounded on $L^{p(\cdot)}$. Then, via a deep characterization of such exponents due to Diening [60] we get that the maximal operator is bounded on the necessary associate spaces. We refer the reader to the original proof for details.

In [40] Theorem 4.25 was used to prove that a number of classical operators from harmonic analysis—singular integral operators, fractional integral operators, square functions—were bounded on variable Lebesgue spaces. The advantage of using extrapolation is that these results are immediate consequences of known weighted norm inequalities. (By way of comparison, consider the original proof [64] that singular integrals are bounded on variable Lebesgue spaces.) Also, since $L^{p(\cdot)}$ is not rearrangement invariant, we cannot prove these results via interpolation.

Here we give two new applications of extrapolation to variable Lebesgue spaces. First, we show that $L^{p(\cdot)}$ can be characterized in terms of wavelets. Second, we prove the Sobolev embedding theorem for variable Sobolev spaces.

Wavelets

We begin with the definition of wavelets; we refer the reader to Meyer [145] and Hernández and Weiss [103] for the construction and properties of wavelets. Enumerate the dyadic cubes as

$$\mathcal{D} = \{Q_{j,k} = 2^{-j}([0,1)^n + k) : j \in \mathbb{Z},\, k \in \mathbb{Z}^n\}.$$

A finite collection of functions $\Psi = \{\psi^1, \dots, \psi^L\} \subset L^2(\mathbb{R}^n)$ is an orthonormal wavelet family if the system

$$\{\psi_Q^l\} = \{\psi_{Q_{j,k}}^l(x) = 2^{j\,n/2}\psi^l(2^j\,x - k) : j \in \mathbb{Z},\, k \in \mathbb{Z}^n,\, 1 \le l \le L\}$$

is an orthonormal basis of $L^2(\mathbb{R}^n)$. Define

$$\mathcal{W}_\Psi f = \Big(\sum_{l=1}^{L}\sum_{Q\in\mathcal{D}} |\langle f, \psi_Q^l\rangle|^2\, |Q|^{-1}\, \chi_Q\Big)^{1/2};$$

a wavelet family Ψ is admissible if for some p_0, $1 < p_0 < \infty$, and every $w_0 \in A_{p_0}$, there exists a constant $C > 1$ such that

$$C^{-1}\|f\|_{L^{p_0}(w_0)} \le \|\mathcal{W}_\Psi f\|_{L^{p_0}(w_0)} \le C\|f\|_{L^{p_0}(w_0)}. \tag{4.25}$$

Admissible wavelets in $\mathbb{R}$ include the Haar system [113], spline wavelets [85], Daubechies wavelets [124], compactly supported wavelets [236], and smooth wavelets in the class $\mathcal{R}^1$ [87]. In [2] it is shown that MRA wavelets in $\mathbb{R}^n$ are admissible.

One consequence of being an admissible wavelet is that it implies that the wavelet family is an unconditional basis for $L^p(w)$, $1 < p < \infty$, $w \in A_p$. This was shown on the real line in [87]; the arguments there can readily be adapted to higher dimensions.

By extrapolation we can immediately extend these results to variable Lebesgue spaces.

Theorem 4.27. *Given an exponent $p(\cdot)$, suppose $1 < p_- \le p_+ < \infty$ and $p(\cdot) \in LH$. If Ψ is an admissible orthonormal wavelet family, then for every $f \in L^{p(\cdot)}$,*

$$\|f\|_{p(\cdot)} \approx \left\| \left(\sum_{l=1}^{L} \sum_{Q \in \mathcal{D}} |\langle f, \psi_Q^l \rangle|^2 \, |Q|^{-1} \, \chi_Q \right)^{1/2} \right\|_{p(\cdot)}. \tag{4.26}$$

Furthermore, the wavelet Ψ is an unconditional basis for $L^{p(\cdot)}(\mathbb{R}^n)$.

A version of Theorem 4.27 was proved independently by Izuki [106].

Proof. We first prove (4.26) for bounded functions of compact support. Define the two families

$$\mathcal{F}_1 = \{(|f|, \mathcal{W}_\Psi f) : f \in L_c^\infty\},$$
$$\mathcal{F}_2 = \{(\mathcal{W}_\Psi f, |f|) : f \in L_c^\infty\}.$$

Since $L_c^\infty \subset L^{p_0}(w_0)$, $p_0 > 1$, and Ψ is an admissible family, (4.25) holds for some p_0. We can therefore take this as our hypothesis (since $A_1 \subset A_{p_0}$) and apply Theorem 4.25 to the families $\mathcal{F}_1$ and $\mathcal{F}_2$.

Now fix $f \in L^{p(\cdot)}$. Since $p_+ < \infty$, L_c^∞ is dense in $L^{p(\cdot)}$ (see [116]). Therefore, there exists a sequence $\{f_k\} \subset L_c^\infty$ such that $f_k \to f$ in norm. Since convergence in $L^{p(\cdot)}$ norm implies convergence in measure (see [39, 73]), by passing to a subsequence we may assume the sequence converges pointwise almost everywhere.

The definition of admissibility implies that $\psi_Q^l \subset \cap_{p>1} L^p$. To see this, first fix a dyadic cube Q and index l and define

$$T_Q^l f = |\langle f, \psi_Q^l \rangle| |Q|^{-1/2} \chi_Q.$$

Then $0 \le T_Q^l f(x) \le \mathcal{W}_\Psi f(x)$, and so by admissibility, for all $w \in A_{p_0}$, $\|T_Q^l f\|_{L^{p_0}(w)} \le C \|f\|_{L^{p_0}(w)}$. Therefore, by extrapolation (Theorem 3.9) we have that for all p, $1 < p < \infty$,

$$\|T_Q^l f\|_p \le C \|f\|_p$$

whenever $T_Q f \in L^p$. Fix p and fix $f \in L_c^\infty$ such that $\|f\|_{p'} \le 1$. Since $f, \psi_Q^l \in L^2$, we have that $|\langle f, \psi_Q^l \rangle| < \infty$, and so

$$\|T_Q f\|_{p'} = |\langle f, \psi_Q^l \rangle| |Q|^{1/p' - 1/2} < \infty.$$

Therefore,

$$|\langle f, \psi_Q^l \rangle| |Q|^{1/p' - 1/2} = \|T_Q f\|_{p'} \le C \|f\|_{p'} \le C,$$

and so by duality, $\psi_Q^l \in L^p$ and $\|\psi_Q^l\|_p \le C |Q|^{1/2 - 1/p'}$.

Since $\psi_Q^l \in \cap_{p>1} L^p$, $\psi_Q^l \in L^{p'(\cdot)}$ (again see [116]). Therefore, by Hölder's inequality for variable Lebesgue spaces,

$$|\langle f - f_k, \psi_Q^l \rangle| \le C \|f - f_k\|_{p(\cdot)} \|\psi_Q^l\|_{p'(\cdot)},$$

and so $\langle f_k, \psi_Q^l\rangle \to \langle f, \psi_Q^l\rangle$. Hence, for almost every x,

$$\mathcal{W}_\Psi f(x) \le \liminf_{k\to\infty} \mathcal{W}_\Psi f_k(x).$$

Therefore, by Fatou's lemma for variable Lebesgue spaces (which is true in any Banach function space—see [12]),

$$\|\mathcal{W}_\Psi f\|_{p(\cdot)} \le \liminf_{k\to\infty} \|\mathcal{W}_\Psi f_k\|_{p(\cdot)} \le C \liminf_{k\to\infty} \|f_k\|_{p(\cdot)} = C\|f\|_{p(\cdot)}.$$

To prove the reverse inequality, note first that the above estimate proves that $\mathcal{W}_\Psi f_k \to \mathcal{W}_\Psi f$ in $L^{p(\cdot)}$, since

$$\|\mathcal{W}_\Psi f - \mathcal{W}_\Psi f_k\|_{p(\cdot)} \le \|\mathcal{W}_\Psi (f - f_k)\|_{p(\cdot)} \le C\|f - f_k\|_{p(\cdot)}.$$

Therefore, again by Fatou's lemma,

$$\|f\|_{p(\cdot)} \le \liminf_{k\to\infty} \|f_k\|_{p(\cdot)} \le C \liminf_{k\to\infty} \|\mathcal{W}_\Psi f_k\|_{p(\cdot)} = C\|\mathcal{W}_\Psi f\|_{p(\cdot)}.$$

Finally, the fact that the wavelet system is an unconditional basis for $L^{p(\cdot)}$ follows from the argument in [87]. $\square$

Remark 4.28. By Theorem 4.10, for every r.i. function space $\mathbb{X}$ with $1 < p_{\mathbb{X}} \le q_{\mathbb{X}} < \infty$ and for every $w \in A_{p_{\mathbb{X}}}$,

$$\|f\|_{\mathbb{X}(w)} \approx \|\mathcal{W}_\Psi f\|_{\mathbb{X}(w)}.$$

In the unweighted case this was proved in [211] using Boyd's interpolation theorem. A similar argument works in the weighted case, since we can use Theorem 3.9 to extend (4.25) to $L^p(w)$, $1 < p < \infty$ and $w \in A_p$. We stress, however, that Theorem 4.27 cannot be derived in this way.

Sobolev embedding theorem

Given an exponent function $p(\cdot)$, define the variable Sobolev space $W^{1,p(\cdot)}$ to be the set of all functions f such that f, $\nabla f \in L^{p(\cdot)}$, where ∇f is understood in terms of weak derivatives. The norm on this space is $\|f\|_{W^{1,p(\cdot)}} = \|f\|_{p(\cdot)} + \|\nabla f\|_{p(\cdot)}$. If $p_+ < n$, define the Sobolev exponent $p^*(\cdot)$ by

$$\frac{1}{p(\cdot)} - \frac{1}{p^*(\cdot)} = \frac{1}{n}.$$

Theorem 4.29. *Given $p(\cdot)$ such that $1 \le p_- \le p_+ < n$ and $p(\cdot) \in LH$, then $W^{1,p(\cdot)} \subset L^{p^*(\cdot)}$ and*

$$\|f\|_{p^*(\cdot)} \le C\|\nabla f\|_{p(\cdot)}.$$

With the additional assumption that $p_- > 1$ (and replacing $\mathbb{R}^n$ by a bounded domain), Theorem 4.29 was proved by several authors: see [18] and the references it contains. When $p_- = 1$ this was proved by Harjulehto and Hästö [99] for bounded domains and extended to all of $\mathbb{R}^n$ by Hästö [100].

To prove Theorem 4.29 we need two lemmas that are interesting results in their own right. First, we need a generalization of Theorem 4.25 to off-diagonal extrapolation. This was originally proved in [40]; the proof is much the same as the proof of Theorem 4.6, modified according to the ideas in the proof of Theorem 3.23. Details are left to the interested reader.

Lemma 4.30. *Suppose that for some p_0, q_0, $1 \le p_0 < q_0 < \infty$, and every $w_0 \in A_1$,*

$$\left(\int_{\mathbb{R}^n} f(x)^{q_0} w_0(x)\, dx\right)^{1/q_0} \le C\left(\int_{\mathbb{R}^n} g(x)^{p_0} w_0(x)^{p_0/q_0}\, dx\right)^{1/p_0}, \qquad (f,g) \in \mathcal{F}. \tag{4.27}$$

Given an exponent $p(\cdot)$ such that $p(\cdot) \in LH$ and $p_0 \le p_- \le p_+ < \frac{p_0 q_0}{q_0 - p_0}$, define $q(\cdot)$ by

$$\frac{1}{p(x)} - \frac{1}{q(x)} = \frac{1}{p_0} - \frac{1}{q_0}.$$

Then

$$\|f\|_{q(\cdot)} \le C\|g\|_{p(\cdot)}, \qquad (f,g) \in \mathcal{F}.$$

The second is a weighted version of the Sobolev embedding theorem.

Lemma 4.31. *For all p, $1 \le p < n$, $w \in A_1$, and $f \in C_c^\infty$,*

$$\left(\int_{\mathbb{R}^n} |f(x)|^{p^*} w(x)\, dx\right)^{1/p^*} \le C\left(\int_{\mathbb{R}^n} |\nabla f(x)|^p w(x)^{p/p_*}\, dx\right)^{1/p}.$$

For $p > 1$, Lemma 4.31 follows at once from the well-known formula $|f(x)| \le c_n I_1(|\nabla f|)(x)$ and the weighted norm inequalities for the fractional integral operator [154]. When $p = 1$, this is due to Mazya [144] (also see Turesson [225]). The proof we give is implicit in [80] (see also [177]) and depends in a fundamental way on an argument of Long and Nie [133] that is based in turn on ideas of Mazya [144, p. 110].

Proof. Fix $f \in C_c^\infty$. For each $j \in \mathbb{Z}$, define

$$\Omega_j = \{x \in \mathbb{R}^n : 2^j < |f(x)| \le 2^{j+1}\},$$

and the function f_j by

$$f_j(x) = \begin{cases} |f(x)| - 2^j & x \in \Omega_j, \\ 2^j & x \in \Omega_i,\ i > j, \\ 0 & \text{otherwise}. \end{cases}$$

It follows immediately from this definition that $|\nabla f_j(x)| = |\nabla f(x)|\chi_{\Omega_j}$. Further, by a well-known inequality (see, for example, [222]) we have that if $x \in \Omega_j$, then

$$c_n I_1(|\nabla f_{j-1}|)(x) \geq |f_{j-1}(x)| = 2^{j-1}, \tag{4.28}$$

where I_1 is the fractional integral operator

$$I_1 h(x) = \int_{\mathbb{R}^n} \frac{h(y)}{|x-y|^{n-1}}\, dy.$$

Since $w \in A_1$, $w \in A_{1+p^*/p'}$, which is equivalent to $w^{1/p^*} \in A_{p,p^*}$. (See Section 3.5 above.) Thus, the fractional integral operator satisfies a weak type inequality (see [154]):

$$w(\{x \in \mathbb{R}^n : |I_1 h(x)| > \lambda\}) \leq C\left(\frac{1}{\lambda^p}\int_{\mathbb{R}^n} |h(x)|^p w(x)^{p/p^*}\, dx\right)^{p^*/p}.$$

Therefore, we can estimate as follows:

$$\begin{aligned}
\int_{\mathbb{R}^n} |f(x)|^{p^*} w(x)\, dx &= \sum_j \int_{\Omega_j} |f(x)|^{p^*} w(x)\, dx \\
&\leq \sum_j \int_{\Omega_j} 2^{(j+1)p^*} w(x)\, dx \\
&= 4^{p^*} c_n^{p^*} \sum_j \int_{\Omega_j} \left[c_n^{-1} 2^{j-1}\right]^{p^*} w(x)\, dx \\
&\leq C \sum_j \int_{\{x\in\mathbb{R}^n : I_1(|\nabla f_{j-1}|)(x) > c_n^{-1}2^{j-1}\}} \left[c_n^{-1} 2^{j-1}\right]^{p^*} w(x)\, dx \\
&\leq C \sum_j \left(\int_{\mathbb{R}^n} |\nabla f_{j-1}(x)|^p w(x)^{p/p_*}\, dx\right)^{p^*/p} \\
&\leq C \left(\sum_j \int_{\Omega_{j-1}} |\nabla f(x)|^p w(x)^{p/p_*}\, dx\right)^{p^*/p} \\
&\leq C \left(\int_{\mathbb{R}^n} |\nabla f(x)|^p w(x)^{p/p_*}\, dx\right)^{p^*/p}. \qquad \square
\end{aligned}$$

Proof of Theorem 4.29. We first prove this for $f \in C_c^\infty$. Let $p_0 = 1$ and $q_0 = 1^* = n/(n-1)$. Then $p_+ < n = p_0 q_0/(q_0 - p_0)$, and $1/p(x) - 1/p^*(x) = 1/p_0 - 1/q_0$. Define the family $\mathcal{F} = \{(|f|, |\nabla f|) : f \in C_c^\infty\}$. Then we can extrapolate using Lemma 4.30 (since $p(\cdot) \in LH$) starting from Lemma 4.31 to conclude that for all $f \in C_c^\infty$, $\|f\|_{p^*(\cdot)} \leq C\|\nabla f\|_{p(\cdot)}$.

To get the full result, fix $f \in W^{1,p(\cdot)}$. Since $p(\cdot) \in LH$, C_c^∞ is dense in $W^{1,p(\cdot)}$ (see [37]), so there exists $\{f_k\} \subset C_c^\infty$ such that $f_k \to f$ in $W^{1,p(\cdot)}$ norm. Since convergence in $L^{p(\cdot)}$ norm implies convergence in measure (see [39, 73]), by passing to a subsequence (twice), we may assume it also converges pointwise. Hence, by Fatou's lemma for variable Lebesgue spaces (see [12]),

$$\|f\|_{p^*(\cdot)} \leq \liminf_{k\to\infty} \|f_k\|_{p^*(\cdot)} \leq C \liminf_{k\to\infty} \|\nabla f_k\|_{p(\cdot)} \leq C\|\nabla f\|_{p(\cdot)}. \qquad \square$$

Part II

Two-Weight Factorization and Extrapolation

Chapter 5

Preliminary Results

In this chapter we gather together many of the basic facts we will use throughout Part II: Orlicz spaces, A_p-type conditions, maximal operators, and fractional maximal operators. Fractional maximal operators are treated separately even though a unified presentation is possible; there is some duplication but we believe that the exposition is made clearer this way. Some of the results we present are not new and we give complete references. In some cases we give the proofs, either for completeness or because we can improve upon the original.

5.1 Weights

Throughout Part II we will work with pairs of weights (u, v). Hereafter we will always assume the following: u and v are non-negative; $u > 0$ on a set of positive measure and $u < \infty$ almost everywhere; $v > 0$ almost everywhere and $v < \infty$ on a set of positive measure. Further, in A_p type conditions for pairs of weights we will use the standard convention that $0 \cdot \infty = 0$.

5.2 Orlicz spaces

Central to our approach to two-weight norm inequalities is the theory of Orlicz spaces. Here we summarize some basic facts; we refer the reader to Bennett and Sharpley [12], Krasnosel'skiĭ and Rutickiĭ [117], Maligranda [135], or Rao and Ren [189] for further information and detailed references.

A function $B : [0, \infty) \to [0, \infty)$ is a Young function if it is continuous, convex and strictly increasing, if $B(0) = 0$, and if $B(t)/t \to \infty$ as $t \to \infty$. $B(t) = t$ is not a Young function, but many of the results below apply to it. We will make it explicit if what we say does not apply to this limiting case.

Remark 5.1. The definition of Young functions given here is slightly different than the one in Section 4.1: we do not assume that $B(t)/t \to 0$ as $t \to 0^+$. We can weaken the definition in this way since we are only concerned with Orlicz spaces defined on domains with finite measure.

We will generally use capital Roman letters A, $B,\ldots$ to denote Young functions. In some cases (primarily for the Orlicz maximal operators defined in the next section) we will use capital Greek letters Φ, $\Psi,\ldots$.

If A, B are Young functions, we write $A(t) \approx B(t)$ if there are constants $c_1, c_2 > 0$ such that $c_1A(t) \leq B(t) \leq c_2A(t)$ for $t \geq t_0 > 0$. Also, we say that B dominates A, and denote this by $A(t) \preceq B(t)$, if there exists $c > 0$ such that for all $t \geq t_0 > 0$, $A(t) \leq B(ct)$. (In practice we can always assume that $t_0 = 1$.) For all Young functions B, $t \preceq B(t)$. Further, if for some $c > 1$, $A(t) \leq cB(t)$, then by convexity, $A(t) \leq B(ct)$.

The relative size of a Young function is gotten by comparing it to power functions. Given p, $1 \leq p < \infty$, we say that a Young function B is a p-Young function, if $\Phi(t) = B(t^{1/p})$ is a Young function. (For convenience, we will also say this if $B(t) = t^p$–i.e., $\Phi(t) = t$.) If B is a p-Young function, then $t^p \preceq B(t)$ and $B(t)/t^p$ is non-decreasing. A function F is said to be quasi-increasing if there exists $c > 1$ such that for all $t < s$, $F(t) \leq cF(s)$. Similarly, F is quasi-decreasing if there exists $c > 1$ such that $F(s) \leq cF(t)$ for all $t < s$.

A Young function B is said to be doubling if there exists a positive constant C such that $B(2t) \leq CB(t)$ for all $t > 0$; in the context of Orlicz and modular spaces this is usually referred to as the Δ_2 condition—see Section 4.1. B is called submultiplicative if $B(st) \leq CB(s)B(t)$ for all $s, t > 0$. Clearly $B(t) = t^r$, $r \geq 1$, is submultiplicative. A straightforward computation shows that $B(t) = t^a[\log(e+t)]^b$, $a \geq 1$, is doubling, and if $b \geq 0$, then it is also submultiplicative. (For simplicity, hereafter we will omit the brackets and write $t^a \log(e + t)^b$.)

Given a Young function B and a cube Q, we define the normalized Luxemburg norm of f on Q by

$$\|f\|_{B,Q} = \inf\left\{\lambda > 0 : \fint_Q B\left(\frac{|f(x)|}{\lambda}\right) dx \leq 1\right\}. \tag{5.1}$$

When $B(t) = t^p$, $1 \leq p < \infty$,

$$\|f\|_{B,Q} = \left(\fint_Q |f(x)|^p \, dx\right)^{1/p} = \|f\|_{p,Q},$$

so the Luxemburg norm coincides with the normalized L^p norm.

The Luxemburg norm has two rescaling properties which we will use repeatedly. Given any Young function A, for all $r > 0$,

$$\|f^r\|_{A,Q} = \|f\|^r_{B,Q},$$

where $B(t) = A(t^r)$. In particular, if A is a p-Young function, then

$$\|f^{1/p}\|_{A,Q} = \|f^{1/r}\|^{r/p}_{B,Q},$$

where $B(t) = A(t^{r/p})$ is an r-Young function. This follows by a change of variables in the definition. By convexity, we have that for all $\tau > 1$ and all cubes Q,

$$\|f\|_{A,Q} \le \tau^n \|f\|_{A,\tau Q}. \tag{5.2}$$

If $A(t)/t^p$ is quasi-increasing for some $p > 1$, we can sharpen this inequality, replacing τ^n by $\tau^{n/p}$.

If $A \preceq B$, then there exists a constant C, depending on A and B, such that for all cubes Q and functions f, $\|f\|_{A,Q} \le C\|f\|_{B,Q}$. This follows from the definition and convexity; we stress that because this is the normalized Luxemburg norm, the constant C is independent of Q and only depends on the relative behavior of A and B for large values of t.

Given a Young function B, the complementary Young function $\bar{B}$ is defined by

$$\bar{B}(t) = \sup_{s>0}\{st - B(s)\}, \qquad t > 0. \tag{5.3}$$

B and $\bar{B}$ satisfy the following inequality:

$$t \le B^{-1}(t)\bar{B}^{-1}(t) \le 2t. \tag{5.4}$$

Hölder's inequality can be generalized to the scale of Orlicz spaces.

Lemma 5.2. *Given a Young function B, then for all functions f and g and all cubes Q,*

$$\fint_Q |f(x)g(x)|\,dx \le 2\|f\|_{B,Q}\|g\|_{\bar{B},Q}. \tag{5.5}$$

More generally, if A, B and C are Young functions such that for all $t \ge t_0 > 0$,

$$B^{-1}(t)C^{-1}(t) \le cA^{-1}(t),$$

then

$$\|fg\|_{A,Q} \le K\|f\|_{B,Q}\|g\|_{C,Q}. \tag{5.6}$$

5.3 Orlicz maximal operators

Given a Young function Φ, we define the Orlicz maximal operator

$$M_\Phi f(x) = \sup_{Q\ni x} \|f\|_{\Phi,Q}, \tag{5.7}$$

where the supremum is taken over all cubes Q containing x. When $\Phi(t) = t$, M_Φ is the classical Hardy-Littlewood maximal operator. When $\Phi(t) = t^r$, $r > 1$, $M_\Phi f = M_r f = M(|f|^r)^{1/r}$. If $\Phi(t) \preceq \Psi(t)$, then for all $x \in \mathbb{R}^n$, $M_\Phi f(x) \le CM_\Psi f(x)$.

Remark 5.3. Hereafter, whenever we deal with an Orlicz maximal operator M_Φ, we will always implicitly assume that the functions f are such that $M_\Phi f(x) < \infty$ for almost every $x \in \mathbb{R}^n$. This condition implies both that the norms $\|f\|_{\Phi,Q}$ are finite for every cube Q, and that they are bounded as $|Q| \to \infty$. In some settings we need stronger hypotheses; these will either be clear from the context or will be stated explicitly.

Remark 5.4. We can define the dyadic Orlicz maximal operator M_Φ^d by restricting the supremum in (5.7) to dyadic cubes containing x. All the results for Orlicz maximal operators remain true for dyadic maximal operators. In some cases sharper results hold, and we will point these out as the need arises.

Orlicz maximal operators satisfy a modular weak type inequality that generalizes the weak $(1,1)$ inequality for the Hardy-Littlewood maximal operator. This was originally proved in [10, 174]; a proof is given below in Appendix A, Remark A.3.

Theorem 5.5. *Given a Young function Φ, for all f satisfying $\|f\|_{\Phi,Q} \to 0$ as $|Q| \to \infty$, and all $\lambda > 0$,*

$$|\{x \in \mathbb{R}^n : M_\Phi f(x) > \lambda\}| \le 3^n \int_{\{x\in\mathbb{R}^n:|f(x)|>\lambda/2\}} \Phi\left(\frac{2\cdot 4^n|f(x)|}{\lambda}\right) dx.$$

If Φ is doubling, then we can replace $2 \cdot 4^n|f(x)|$ by $|f(x)|$ on the right-hand side at the cost of a larger constant depending on Φ on the outside. Further, if $\operatorname{supp}(f) \subset Q_0$ for some cube Q_0, then Φ doubling implies that the growth condition on f holds, since $\|f\|_{\Phi,\tau Q_0} \le \tau^{-n/D}\|f\|_{\Phi,Q_0}$ for $\tau > 1$, for some $D > 0$.

For $p > 1$, the Orlicz maximal function M_Φ is weak type (p,p) provided that Φ satisfies a natural growth condition.

Proposition 5.6. *Given a Young function Φ and p, $1 < p < \infty$, M_Φ satisfies the weak (p,p) inequality*

$$|\{x \in \mathbb{R}^n : M_\Phi f(x) > \lambda\}| \le \frac{C}{\lambda^p}\int_{\mathbb{R}^n} |f(x)|^p\, dx \tag{5.8}$$

if and only if $\Phi(t) \preceq t^p$.

Proof. The sufficiency of this condition follows by replacing M_Φ by M_p and using the weak $(1,1)$ inequality for the Hardy-Littlewood maximal operator. To see necessity, fix $t > 1$ and two cubes $Q_0 \subset Q_1$ such that $|Q_0| = 1$, $|Q_1| = t$. Let $f = \chi_{Q_0}$; then for $x \in Q_1$, $M_\Phi f(x) > \|f\|_{\Phi,Q_1} = \Phi^{-1}(t)^{-1} = \lambda$. On the other hand, $\|f\|_p^p = 1$. Therefore, (5.8) yields $\Phi^{-1}(t) \ge ct^{1/p}$, so $\Phi(t) \preceq t^p$. □

The Orlicz maximal M_Φ is bounded on L^p if Φ is not too large; for example, if $\Phi(t) = t^q$, $1 \le q < p$, then it is immediate that $M_\Phi = M_q$ is bounded on L^p; on the other hand, if $\Phi(t) = t^p$, then M_Φ is not bounded on L^p. It is very surprising that there exist Orlicz functions Φ satisfying $t^q \preceq \Phi(t)$ for all $q < p$ such that M_Φ is bounded on L^p. To state the precise result we need the following definition which we will use repeatedly in this and the subsequent chapters.

Definition 5.7. Given p, $1 < p < \infty$, the Young function A is said to be in B_p (denoted by $A \in B_p$) if for some $c > 0$,

$$\int_c^\infty \frac{A(t)}{t^p}\frac{dt}{t} < \infty. \tag{5.9}$$

Simple examples of functions in B_p are

$$A(t) = \frac{t^p}{\log(e+t)^{1+\nu}}, \quad A(t) = \frac{t^p}{\log(e+t)\log\log(e^e+t)^{1+\nu}}, \quad \nu > 0.$$

Remark 5.8. If (5.9) is finite for some value of c, then it is finite for every $c > 0$.

Remark 5.9. It follows from the definition that if $A \in B_p$, then $A(t^s) \in B_{ps}$ for every $s > 1$.

Proposition 5.10. *If A and $\bar{A}$ are doubling Young functions, then $A \in B_p$ if and only if*

$$\int_c^\infty \left(\frac{t^{p'}}{\bar{A}(t)}\right)^{p-1}\frac{dt}{t} < \infty. \tag{5.10}$$

Proof. Since A is convex, A' is increasing; therefore,

$$A(t) = \int_0^t A'(s)\,ds \le tA'(t).$$

Since A is doubling there is a constant $C > 1$ such that for all $t > 0$,

$$CA(t) \ge A(2t) \ge \int_t^{2t} A'(s)\,dx \ge tA'(t).$$

Thus, $A(t) \approx tA'(t)$. The same argument also shows that $\bar{A}(t) \approx t\bar{A}'(t)$.

We now make the change of variables $s = A(t)$ and use (5.4) to get

$$\begin{aligned}\int_c^\infty \frac{A(t)}{t^p}\frac{dt}{t} &= \int_{A(c)}^\infty \frac{s}{A^{-1}(s)^p}\frac{ds}{A^{-1}(s)A'(A^{-1}(s))}\\ &\approx \int_{A(c)}^\infty \frac{s}{A^{-1}(s)^p}\frac{ds}{A(A^{-1}(s))} = \int_{A(c)}^\infty \frac{ds}{A^{-1}(s)^p} \approx \int_{A(c)}^\infty \frac{\bar{A}^{-1}(s)^p}{s^p}\,ds.\end{aligned}$$

We can now reverse this argument, making the change of variables $s = \bar{A}(t)$ to get

$$\begin{aligned}\int_c^\infty \frac{A(t)}{t^p}\frac{dt}{t} &\approx \int_{\bar{A}^{-1}(A(c))}^\infty \frac{t^p}{\bar{A}(t)^p}\bar{A}'(t)\,dt\\ &\approx \int_{\bar{A}^{-1}(A(c))}^\infty \frac{t^p}{\bar{A}(t)^{p-1}}\frac{dt}{t} = \int_{\bar{A}^{-1}(A(c))}^\infty \left(\frac{t^{p'}}{\bar{A}(t)}\right)^{p-1}\frac{dt}{t}.\end{aligned}$$

The desired conclusion follows immediately. □

Remark 5.11. Proposition 5.10 is very useful since we will often make use of conditions of the form $\bar{A} \in B_p$, which can then be stated strictly in terms of A. In particular, we will be able to do this if $A(t) = t^p \log(e+t)^a$.

Remark 5.12. If $\bar{A} \in B_p$, then A is a p'-Young function. By (5.9), $\bar{A}(t) \leq ct^p$ for $t \geq 1$. Hence,

$$c^{-1/p} t^{1/p} \leq \bar{A}^{-1}(t) \leq \frac{2t}{A^{-1}(t)},$$

so $A^{-1}(t) \leq 2c^{1/p} t^{1/p'}$. Thus $t^{p'} \preceq A(t)$.

Theorem 5.13. *Given p, $1 < p < \infty$, then $M_\Phi : L^p \to L^p$ if and only if $\Phi \in B_p$.*

For a proof, see [174]. In the original statement it is assumed that Φ is doubling, but an examination of the proof shows that this hypothesis is not necessary: doubling was only used to invoke Proposition 5.10.

Of critical importance in what follows is the two-weight generalization of Theorem 5.13.

Theorem 5.14. *Given p, $1 < p < \infty$, let Φ, B, and C be Young functions such that $B^{-1}(t)C^{-1}(t) \leq c\Phi^{-1}(t)$, $t \geq t_0 > 0$, and $C \in B_p$. If (u, v) is a pair of weights such that for every cube Q,*

$$\|u^{1/p}\|_{p,Q} \|v^{-1/p}\|_{B,Q} \leq K < \infty, \tag{5.11}$$

then for every function $f \in L^p(v)$,

$$\int_{\mathbb{R}^n} M_\Phi f(x)^p u(x)\, dx \leq C \int_{\mathbb{R}^n} |f(x)|^p v(x)\, dx. \tag{5.12}$$

For a proof, see [53, 174]. Again in the original statement of this result it is assumed that C is doubling, but this is not necessary: this hypothesis was only to use Theorem 5.13 which, as we noted, does not require this condition.

Remark 5.15. In the special case of the Hardy-Littlewood maximal operator (i.e., $\Phi(t) = t$) we can state the hypotheses more simply as (5.11) holds for some B such that $\bar{B} \in B_p$. This condition is sharp: given B, if (5.12) holds for every pair (u, v) that satisfies (5.11), then $\bar{B} \in B_p$.

It is implicit in the hypotheses of Theorem 5.14 that B is a p'-Young function: $\Phi^{-1}(t) \leq t$, so the inequality $B^{-1}(t)C^{-1}(t) \leq C\Phi^{-1}(t)$ implies that $\bar{C}(t) \preceq B(t)$, and by Remark 5.12, $\bar{C}$ is a p'-Young function.

We can also characterize the weights governing the weak (p,p) inequality for M_Φ with a condition similar to (5.11). This result is new.

Proposition 5.16. *Given p, $1 < p < \infty$, let Φ and B be Young functions such that $B^{-1}(t)t^{1/p} \leq c\Phi^{-1}(t)$, $t \geq t_0 > 0$. If (u, v) is a pair of weights such that for every cube Q,*

$$\|u^{1/p}\|_{p,Q} \|v^{-1/p}\|_{B,Q} \leq K < \infty,$$

then for every function $f \in L^p(v)$,

$$u(\{x \in \mathbb{R}^n : M_\Phi f(x) > \lambda\}) \leq \frac{C}{\lambda^p} \int_{\mathbb{R}^n} |f(x)|^p v(x)\, dx.$$

Proof. The proof is standard, so we only sketch the details. Fix $\lambda > 0$. Then by Proposition A.1 there exists a countable, disjoint set of dyadic cubes $\{Q_j\}$ such that $\|f\|_{\Phi,Q_j} > 4^{-n}\lambda$ and $\{3Q_j\}$ is a cover of the set $\{x \in \mathbb{R}^n : M_\Phi f(x) > \lambda\}$. By the generalized Hölder's inequality, Lemma 5.2,

$$\begin{aligned} &u(\{x \in \mathbb{R}^n : M_\Phi f(x) > \lambda\}) \\ &\quad \leq \frac{C}{\lambda^p} \sum_j \|u^{1/p}\|_{p,3Q_j}^p |Q_j| \|f\|_{\Phi,Q_j}^p \\ &\quad \leq \frac{C}{\lambda^p} \sum_j \|u^{1/p}\|_{p,3Q_j}^p \|v^{-1/p}\|_{B,3Q_j}^p |Q_j| \|fv^{1/p}\|_{p,Q_j}^p \\ &\quad \leq \frac{C}{\lambda^p} \int_{\mathbb{R}^n} |f(x)|^p v(x)\, dx. \end{aligned}$$

□

Remark 5.17. If $\Phi(t) = t\log(e+t)$, then we can take $B(t) = t^{p'} \log(e+t)^{p'}$.

Remark 5.18. Unless $\Phi(t) = t$ there is no corresponding weak $(1,1)$ inequality. For the Hardy-Littlewood maximal operator we have that for all weights u,

$$u(\{x \in \mathbb{R}^n : Mf(x) > \lambda\}) \leq \frac{C}{\lambda} \int_{\mathbb{R}^n} |f(x)| Mu(x)\, dx. \tag{5.13}$$

For a proof, see [68, 88].

5.4 Generalizations of the A_p condition

Given a pair of weights (u,v), we say that (u,v) satisfy the two-weight Muckenhoupt condition A_p, $1 < p < \infty$, denoted by $(u,v) \in A_p$, if for every cube Q,

$$⨍_Q u(x)\, dx \left(⨍_Q v(x)^{1-p'}\, dx \right)^{p-1} \leq K < \infty.$$

When $p = 1$, $(u,v) \in A_1$ if $Mu(x) \leq Cv(x)$ for almost every x; in particular, all pairs (u, Mu) are in A_1. We can restate the A_p condition in terms of normalized L^p and $L^{p'}$ norms: $(u,v) \in A_p$ if

$$\|u^{1/p}\|_{p,Q} \|v^{-1/p}\|_{p',Q} \leq K < \infty. \tag{5.14}$$

In the one-weight case the A_p condition provides a sufficient (and often necessary and sufficient) condition for a large number of operators to be bounded on a weighted L^p space. In the two-weight case, the A_p condition is necessary and sufficient for the weak type inequalities $M : L^p(v) \to L^{p,\infty}(u)$ and $M : L^{p'}(u^{1-p'}) \to L^{p',\infty}(v^{1-p'})$, but it is not sufficient for the maximal operator to be strong type (p,p). (See [88] and Remark 6.6 below.) A necessary and sufficient condition was given by Sawyer [201] (see also [88]), but it involves the maximal operator itself and has none of the structural properties associated with the A_p condition.

To adapt our proof of the Rubio de Francia extrapolation theorem to the two-weight case, we need conditions on the weights (u,v) such that the maximal operator satisfies $M : L^p(v) \to L^p(u)$ and/or $M : L^{p'}(u^{1-p'}) \to L^{p'}(v^{1-p'})$. We get such conditions from Theorem 5.14 above: we will consider pairs of weights that satisfy conditions of the form

$$\|u^{1/p}\|_{A,Q}\|v^{-1/p}\|_{B,Q} \leq K < \infty, \tag{5.15}$$

where A and B are Young functions such that $\bar{A} \in B_{p'}$ and/or $\bar{B} \in B_p$, and Q is any cube in $\mathbb{R}^n$. In order that condition (5.15) imply the two-weight A_p condition, we will also assume that A is a p-Young function and B is a p'-Young function. (If $\bar{A} \in B_{p'}$ or $\bar{B} \in B_p$, this follows from Remark 5.12.) In some instances we will explicitly restrict ourselves to the dyadic version of this condition and assume that Q is any dyadic cube.

Informally, we refer to (5.15) as an "A_p bump" condition. There are many possible choices for A and B; generically we will refer to them as "Orlicz bumps." If $\bar{A} \in B_{p'}$ and/or $\bar{B} \in B_p$, then we will refer to them as B_p bumps.

For many operators, we believe that B_p bumps yield optimal sufficient conditions for weak and strong type inequalities. These conjectures (discussed in detail in Chapters 9 and 10 below) are variations of the conjecture due to Muckenhoupt and Wheeden [152] discussed in Section 1.1. Since the B_p condition is so general, we feel that our conjecture is morally, though not necessarily logically, equivalent to theirs. (It is tempting to speculate that if $M : L^p(v) \to L^p(u)$, then (u,v) satisfy the hypotheses of Theorem 5.14 for some Young function B such that $\bar{B} \in B_p$, but there is no evidence to warrant making this conjecture.)

In many cases, particularly for singular integral operators, assuming that (u,v) satisfy a B_p bump condition has not been enough, and we were forced to assume that the Young functions belong to particular families that satisfy the B_p condition. Here we will consider four such families: log bumps, log-log bumps, exponential log bumps, and power bumps. Of these the most important are the log bumps, for which we have proved a number of important results. We include the others both to compare and contrast to the log bumps, and also because we are able to prove easily using extrapolation a number of known results that are stated in terms of these bump conditions.

Remark 5.19. Bump conditions can be defined in terms of the norm on an arbitrary Banach function X. The analog of Theorem 5.14 is true provided that the maximal operator M_X defined using the norm on X, is bounded on L^p. This is the case, for example, if X is a Lorentz space. See [171, 174] for these and related results.

Log bumps

If $A(t) = t^p \log(e+t)^a$ or if $B(t) = t^{p'} \log(e+t)^b$, then we say that (5.15) holds with log bumps. The associated Orlicz space norms are connected to the Zygmund spaces of classical analysis (cf. Bennett and Sharpley [12]).

An important special case are log bumps of the form

$$A(t) = t^p \log(e+t)^{p-1+\delta}, \quad B(t) = t^{p'} \log(e+t)^{p'-1+\delta}, \quad \delta > 0,$$

since $\bar{A} \in B_{p'}$ and $\bar{B} \in B_p$. To show this we estimate the complementary function. If $\Phi(t) = t\log(e+t)^a$, $a \in \mathbb{R}$, then a straightforward computation shows that $\Phi^{-1}(t) \approx t\log(e+t)^{-a}$. It then follows (for instance) that

$$A^{-1}(t) \approx \frac{t^{1/p}}{\log(e+t)^{1/p'+\delta/p}},$$

and so

$$\bar{A}^{-1}(t) \approx t^{1/p'} \log(e+t)^{1/p'+\delta/p}$$

and

$$\bar{A}(t) \approx \frac{t^{p'}}{\log(e+t)^{1+(p'-1)\delta}}.$$

One drawback to working with log bumps is that the B_p condition is not preserved by rescaling. If we let $A(t) = t^{p_0} \log(e+t)^{p_0-1+\delta}$, then $A(t^{p/p_0}) \approx t^p \log(e+t)^{p_0-1+\delta}$, and for $p > p_0$ sufficiently large (depending on δ), $A(t^{p/p_0})$ is not in B_p.

As a consequence we lose an important property of A_p weights. By Hölder's inequality, if $(u, v) \in A_{p_0}$, then $(u, v) \in A_p$ for all $p > p_0$. This holds because the first norm on the left-hand side of (5.14) is invariant under rescaling. The same is no longer true for weights (u, v) that satisfy (5.15) with A a log bump.

However, a weaker version of this property holds for log bumps, and this has great importance in applications since we can use it to prove many results for log bumps that we conjecture are true in general.

Proposition 5.20. *Given p_0, $1 < p_0 < \infty$, and $A_0(t) = t^{p_0} \log(e+t)^{p_0-1+\delta_0}$, $\delta_0 > 0$, suppose that*

$$\|u^{1/p_0}\|_{A_0,Q} \|v^{-1/p_0}\|_{B_0,Q} \le K < \infty,$$

where B_0 is a p_0'-Young function. Then there exists $p > p_0$ such that

$$\|u^{1/p}\|_{A,Q} \|v^{-1/p}\|_{B,Q} \le K < \infty$$

where $A(t) = t^p \log(e+t)^{p-1+\delta}$ (so $\bar{A} \in B_{p'}$) and $B(t) = B_0(t^{p'/p_0'})$ is a p'-Young function.

Proof. This follows immediately by rescaling. There exist $p > p_0$ and $\delta > 0$ such that $p - 1 + \delta = p_0 - 1 + \delta_0$. Let $A(t) = t^p \log(e+t)^{p-1+\delta}$ and $B(t) = B_0(t^{p'/p_0'})$. Then $\bar{A} \in B_{p'}$ and B is a p'-Young function. Since $p'/p_0' < p/p_0$, $B(t) \preceq B_0(t^{p/p_0})$. Therefore, by rescaling,

$$\left(\|u^{1/p}\|_{A,Q}\|v^{-1/p}\|_{B,Q}\right)^{p/p_0} \leq C\|u^{1/p_0}\|_{A_0,Q}\|v^{-1/p_0}\|_{B_0,Q}.$$

□

The heart of the proof of Proposition 5.20 is the fact that for any $\delta > 0$, if $A(t) = t^p \log(e+t)^{p-1+\delta}$, then $\bar{A} \in B_{p'}$. We can exploit this fact in another way that we will use in Chapter 6.

Proposition 5.21. *Given p, $1 < p < \infty$, and $\delta > 0$, let $B(t) = t^{p'} \log(e+t)^{p'-1+\delta}$. Then for all ϵ, $0 < \epsilon < \delta/p'$, if $\Phi(t) = t\log(e+t)^\epsilon$, then $M_\Phi : L^p(v) \to L^p(u)$ for any pair (u, v) that satisfies*

$$\|u^{1/p}\|_{p,Q}\|v^{-1/p}\|_{B,Q} \leq K < \infty.$$

Proof. Fix ϵ such that $0 < \epsilon < \delta/p'$ and fix $\nu > 0$ such that

$$\epsilon = \frac{\delta}{p'} - \frac{\nu}{p}.$$

Let $C(t) = t^p \log(e+t)^{-(1+\nu)}$; then C is in B_p and

$$B^{-1}(t)C^{-1}(t) \approx \frac{t^{1/p'}}{\log(e+t)^{1/p+\delta/p'}}\, t^{1/p}\log(e+t)^{1/p+\nu/p}$$
$$= \frac{t}{\log(e+t)^\epsilon} \approx \Phi^{-1}(t).$$

Therefore, by Theorem 5.14, $M_\Phi : L^p(v) \to L^p(u)$. □

Log-log bumps

The log-log bumps are of the form

$$A(t) = t^p \log(e+t)^{p-1} \log\log(e^e+t)^{p-1+\delta}, \qquad \delta > 0.$$

They are B_p bumps since $\bar{A} \in B_{p'}$; we omit the details of this calculation. (See [165].) They are of interest since they are smaller than the log bumps, but they do not have any rescaling properties. If we replace the exponent $p-1$ on the log term by any smaller value, then $\bar{A}$ is no longer in $B_{p'}$. It follows that Propositions 5.20 and 5.21 do not hold for these bump conditions.

More generally, we can define increasingly fine scales of bump conditions by iterating the logarithm. For instance, if we let

$$A(t) = t^p \log(e+t)^{p-1} \log\log(e^e+t)^{p-1} \log\log\log(e^{e^e}+t)^{p-1+\delta}, \quad \delta > 0,$$

then $\bar{A} \in B_{p'}$. Details are left to the reader.

We have many proofs of two-weight norm inequalities assuming log bump conditions, but none for log-log bumps except for those results we have proved for arbitrary B_p bumps. We suspect that the "gap" between log bumps and log-log bumps is so large that any proof that works for log-log bumps will yield a proof for B_p bumps.

Exponential log bumps

The exponential log bumps are Young functions of the form

$$A(t) = t^p\big[\exp\big(\log(1+t^p)^r\big) - 1\big], \qquad 0 < r < 1. \tag{5.16}$$

For t small it is immediate that $A(t) \approx t^{p(1+r)}$; for t large, by L'Hôpital's rule we have that $A(t) \approx t^p e^{\log(t^p)^r}$. For brevity, hereafter we will use the latter approximate expression as shorthand for the precise definition in (5.16).

Exponential log bumps were introduced in [35], where it was shown that they have a crucial measure theoretic property in common with power bumps (defined in the next section). This allowed many two-weight norm inequalities proved for weights satisfying power bump conditions to be extended to weights satisfying exponential log bump conditions.

When $0 < r \leq 1/2$ we can compute A^{-1} and $\bar{A}$. This restriction on r does not have a practical effect, since if a bump condition holds with an exponential log bump A with $r > 1/2$, then it holds for all $r \leq 1/2$.

Proposition 5.22. *If $0 < r \leq 1/2$, then for t large,*

$$A^{-1}(t) \approx t^{1/p} e^{-(1/p)\log(t)^r}, \qquad \bar{A}(t) \approx t^{p'} e^{-(p'/p)\log(t^{p'})^r},$$

and $\bar{A} \in B_{p'}$.

Proof. Throughout the proof we will assume $t > e$. Fix $r \leq 1/2$ and $a \in \mathbb{R}$, $a \neq 0$, and define $\Phi(t) = te^{a\log(t)^r}$. We will first show that if $\Psi(t) = te^{-a\log(t)^r}$, then $\Phi^{-1}(t) \approx \Psi(t)$. It will suffice to show that $\Psi(\Phi(t)) \approx t$, since then by symmetry we have that $\Phi(\Psi(t)) \approx t$ as well. But

$$\Psi(\Phi(t)) = t\exp\big[a\log(t)^r - a\big(\log(t) + a\log(t)^r\big)^r\big];$$

the exponential term is clearly bounded above, so it will suffice to show that it is bounded below. Let $x^{-1} = \log(t)$; it will then suffice to show that

$$\lim_{x\to 0+} x^{-r} - (x^{-1} + ax^{-r})^r = \lim_{x\to 0+} \frac{1 - (1 + ax^{1-r})^r}{x^r}$$

is finite. But by L'Hôpital's rule the right-hand limit is equal to

$$\lim_{x\to 0+} \frac{-a\,(1-r)(1+ax^{1-r})^{r-1}}{x^{2r-1}} = c,$$

where $c = -a/2$ if $r = 1/2$ and $c = 0$ if $0 < r < \frac{1}{2}$.

If we now let $a = 1$, then $A(t) \approx \Phi(t^p)$, so by the convexity of A and Φ we get that $A^{-1}(t) \approx \Phi^{-1}(t)^{1/p} \approx t^{1/p} e^{-(1/p)\log(t)^r}$ and $\bar{A}^{-1}(t) \approx t^{1/p'} e^{(1/p)\log(t)^r}$. The desired formula for $\bar{A}(t)$ then follows from the formula for Φ^{-1} with $a = p'/p$ and the fact that $\Psi(t)/t$ is decreasing

Finally, to see that $\bar{A} \in B_{p'}$, note that there exists $C > 0$ such that $e^{-cx^r} \leq Cx^{-2}$, $x > 0$, so we have that

$$\int_e^\infty \frac{\bar{A}(t)}{t^{p'}} \frac{dt}{t} \leq C \int_e^\infty \frac{dt}{t \log(t)^2} < \infty.$$

□

In passing we note three additional properties of exponential log bumps. First, A and $\bar{A}$ are both doubling. Clearly, $A(2t)/A(t)$ is bounded for t small, so it will suffice to show that it is bounded for t large. (By Proposition 5.22 the same argument holds for $\bar{A}$.) A straight-forward computation shows that

$$\frac{A(2t)}{A(t)} \approx 2^p \exp\left[\log((2t)^p)^r - \log(t^p)^r\right].$$

If we let $x^{-1} = \log(t^p)$ and $a = \log(2^p)$, then the quantity in square brackets equals

$$\frac{(ax+1)^r - 1}{x^r};$$

by L'Hôpital's rule this tends to 0 as $x \to 0^+$, which shows that $A(2t)/A(t)$ is bounded for t large.

Second, we have that the inclusion $\bar{A} \in B_{p'}$ is invariant under rescaling. More precisely, if for any $s > 1/p$ we define $B(t) = A(t^s)$, then by Proposition 5.22 it follows that $\bar{B} \in B_{(sp)'}$.

Third, exponential log bumps satisfy the rescaling property given in Proposition 5.23 below. Details are left to the reader.

Power bumps

Bumps that yield Orlicz space norms in the scale of Lebesgue spaces, $A(t) = t^{rp}$, $B(t) = t^{rp'}$, $r > 1$, are referred to as power bumps. These were the first bump conditions to be considered, appearing in a paper by Neugebauer [163].

Power bumps, have very strong rescaling properties analogous to those in the one-weight case. The next result corresponds to the property that if $w \in A_p$, then $w \in A_q$ for all $q > p$. The proof is essentially the same as the proof of Proposition 5.20 and so is omitted.

Proposition 5.23. *Given p_0, $1 < p_0 < \infty$, suppose the pair of weights (u, v) satisfy*

$$\|u^{1/p_0}\|_{A_0,Q} \|v^{-1/p_0}\|_{B_0,Q} \leq K < \infty, \tag{5.17}$$

where $A_0(t) = t^{rp_0}$, $r > 1$, and B is a p_0'-Young function. Then for all $p > p_0$,

$$\|u^{1/p}\|_{A,Q}\|v^{-1/p}\|_{B,Q} \le K < \infty,$$

where $A(t) = t^{rp}$ and $B(t) = B_0(t^{p'/p_0'})$ is a p'-Young function.

The next result corresponds to the property that if $w \in A_p$, $p > 1$, there exists $\epsilon > 0$ such that $w \in A_{p-\epsilon}$.

Proposition 5.24. *Given p_0, $1 < p_0 < \infty$, if (u, v) satisfy (5.17) with $\bar{A}_0 \in B_{p_0'}$, $\bar{A}_0$ doubling, and $B_0(t) = t^{rp_0'}$, $r > 1$, then for all $1 < p < p_0$ such that*

$$p \ge \frac{p_0 - 1}{r} + 1,$$

(u, v) satisfy

$$\|u^{1/p}\|_{A,Q}\|v^{-1/p}\|_{B,Q} \le K < \infty, \tag{5.18}$$

where $B(t) = t^{r_p p'}$,

$$r_p = r\frac{p-1}{p_0 - 1} \ge 1.$$

and $\bar{A} \in B_{p'}$.

That the pair (u, v) satisfies (5.18) follows by rescaling and our choice of r_p. That $\bar{A} \in B_{p'}$ is a consequence of the following lemma.

Lemma 5.25. *Let A_0 be a Young function such that $\bar{A}_0$ is doubling and $\bar{A}_0 \in B_{p_0'}$ for some $p_0 > 1$. Then for all p, $1 < p < p_0$, if $A(t) = A_0(t^{p/p_0})$, then $\bar{A} \in B_{p'}$.*

Proof. The proof is very similar to the proof of Proposition 5.10, so we only give the most important details. Since $\bar{A}$ is convex, for all $t > 0$, $\bar{A}(t) \le t\bar{A}'(t)$. By inequality (5.4),

$$\bar{A}^{-1}(s) \ge \frac{s}{A^{-1}(s)} = \frac{s}{A_0^{-1}(s)^{p_0/p}} \ge \frac{s^{1-p_0/p}}{2^{p_0/p}}\bar{A}_0^{-1}(s)^{p_0/p}.$$

Therefore, if we make the change of variables $\bar{A}(t) = s$ we get that

$$\begin{aligned}\int_c^\infty \frac{\bar{A}(t)}{t^{p'}}\frac{dt}{t} &= \int_{\bar{A}(c)}^\infty \frac{s}{\bar{A}^{-1}(s)^{p'}}\frac{ds}{\bar{A}^{-1}(s)\bar{A}'(\bar{A}^{-1}(s))} \\ &\le \int_{\bar{A}(c)}^\infty \frac{s}{\bar{A}^{-1}(s)^{p'}}\frac{ds}{s} \le C\int_{\bar{A}(c)}^\infty \left(\frac{s}{\bar{A}_0^{-1}(s)^{p_0'}}\right)^{\frac{p_0-1}{p-1}}\frac{ds}{s}.\end{aligned}$$

Since $\bar{A}_0$ is doubling, $\bar{A}_0(t) \ge t\bar{A}_0'(t)$. Since $\bar{A}_0 \in B_{p_0'}$, $\bar{A}_0(t)/t^{p_0'}$ is bounded for t bounded away from 0. Therefore, if we make the change of variables $t = \bar{A}_0^{-1}(s)$, we get that

$$\int_{\bar{A}(c)}^\infty \left(\frac{s}{\bar{A}_0^{-1}(s)^{p_0'}}\right)^{\frac{p_0-1}{p-1}}\frac{ds}{s} = \int_{\bar{A}_0^{-1}(\bar{A}(c))}^\infty \left(\frac{\bar{A}_0(t)}{t^{p_0'}}\right)^{\frac{p_0-1}{p-1}}\frac{\bar{A}_0'(t)dt}{\bar{A}_0(t)}$$

$$\leq C \int_{\bar{A}_0^{-1}(\bar{A}(c))}^{\infty} \left(\frac{\bar{A}_0(t)}{t^{p_0'}}\right)^{\frac{p_0-1}{p-1}} \frac{dt}{t} \leq C \int_{\bar{A}_0^{-1}(\bar{A}(c))}^{\infty} \frac{\bar{A}_0(t)}{t^{p_0'}} \frac{dt}{t} < \infty. \qquad \square$$

5.5 The composition of maximal operators

The composition $M^k = M \circ M \circ M \circ \cdots \circ M$ of k copies of the Hardy-Littlewood maximal operator can be treated as a single operator, the Orlicz maximal function M_Φ, where $\Phi(t) = t\log(e+t)^{k-1}$. More precisely, for all functions f such that both sides are finite, $M^k f \approx M_\Phi f$. The proof of this uses the weak $(1,1)$ inequality for the maximal operator and a "reverse" weak $(1,1)$ inequality due to Stein [215]. This equivalence was first observed in [173, 174] (but see also [123]) and a proof can be found in [57].

This equivalence can be generalized to the composition of two arbitrary Orlicz maximal operators M_Φ and M_Ψ. This is due to Carozza and Passarelli di Napoli [19]. In Chapter 9 below we will need one half of this equivalence, and will also need an analogous result for fractional maximal operators (discussed in Section 5.6 below). Therefore, here we will state and prove the precise result we need; our proof is a revised version of their proof.

Through the remainder of this section we assume without loss of generality that all functions f are non-negative.

Theorem 5.26. *Given two Young functions Φ and Ψ, for $t > 1$ define*

$$G(t) = \int_1^t \Phi(t/\lambda)\Psi'(\lambda)\,d\lambda.$$

If Γ is any Young function such that $G(t) \preceq \Gamma(t)$, then there exists a constant c such that for $x \in \mathbb{R}^n$,

$$M_\Psi(M_\Phi f)(x) \leq c M_\Gamma f(x). \tag{5.19}$$

Remark 5.27. In Theorem 5.26 we can take $\Phi(t) = t$ and/or $\Psi(t) = t$. In [19], G is defined with lower limit of integration 0, and in this case the integral does not converge if $\Phi(t) = \Psi(t) = t$. With our definition of G we can avoid this problem.

The order of composition in Theorem 5.26 does not matter. By integration by parts and a change of variables,

$$G(t) \leq c\Psi(t) + \int_1^t \Psi(t/\lambda)\Phi'(\lambda)\,d\lambda = c\Psi(t) + \tilde{G}(t).$$

In the proof of Theorem 5.26 we show that $\Phi(t) \preceq G(t)$; the same argument shows that $\Psi(t) \preceq \tilde{G}(t)$. Hence $G(t) \preceq \tilde{G}(t)$, and so proceeding as before we get the reverse inequality.

The proof of Theorem 5.26 requires a localization lemma for Orlicz maximal operators.

Lemma 5.28. *Given a Young function Φ, then for every Q and $x \in Q$ we have*

$$M_\Phi(f\,\chi_Q)(x) = \sup_{x\in P\subset Q} \|f\|_{\Phi,P}, \tag{5.20}$$

and

$$M_\Phi(f\,\chi_{\mathbb{R}^n\setminus 3Q})(x) \approx \sup_{P\supset Q} \|f\,\chi_{\mathbb{R}^n\setminus 3Q}\|_{\Phi,P}. \tag{5.21}$$

Remark 5.29. Lemma 5.28 remains true for the dyadic maximal operator. Further, we can strengthen (5.21) by replacing $3Q$ by Q, Q dyadic, and the equivalence by equality. Details are left to the reader.

The identity (5.20) shows that if the support of f is contained in a cube Q, then for $x \in Q$ it suffices to take cubes contained in P in the definition of the maximal operator. On the other hand, (5.21) shows that $M_\Phi(f\,\chi_{\mathbb{R}^n\setminus 3Q})$ is essentially constant on Q.

Proof. Fix a function f. We first prove (5.20). Clearly the supremum on the right-hand side is less than or equal to the left-hand side, so it will suffice to prove the opposite inequality.

Fix $x \in Q$ and a cube $P \not\subset Q$. There are two cases. If $|P| < |Q|$ then by translating P we can find another cube $\tilde{P}$ containing x such that $\tilde{P} \subset Q$, $Q \cap P \subset Q \cap \tilde{P}$, and $|\tilde{P}| = |P|$. But then,

$$\|f\chi_Q\|_{\Phi,P} \le \|f\|_{\Phi,\tilde{P}} \le \sup_{x\in P\subset Q} \|f\|_{\Phi,P}.$$

Now suppose $|P| \ge |Q|$. Then for all $\lambda > 0$,

$$\fint_P \Phi\left(\frac{f\,\chi_Q}{\lambda}\right) dx \le \fint_Q \Phi\left(\frac{f\,\chi_Q}{\lambda}\right) dx.$$

Therefore,

$$\|f\chi_Q\|_{\Phi,P} \le \|f\|_{\Phi,Q} \le \sup_{x\in P\subset Q} \|f\|_{\Phi,P}.$$

If we take the supremum over all P we get (5.20).

We now prove (5.21). As before, the supremum on the right-hand side is less than or equal to the left-hand side, so it will suffice to prove that the opposite inequality holds up to a constant. Fix $x \in Q$ and P_0 containing x such that $(\mathbb{R}^n \setminus 3Q) \cap P_0 \neq \emptyset$. Then $\ell(Q) \le \ell(P_0)$ and so $Q \subset 3P_0$. Therefore, by (5.2),

$$\|f\,\chi_{\mathbb{R}^n\setminus 3Q}\|_{\Phi,P_0} \le 3^n\,\|f\,\chi_{\mathbb{R}^n\setminus 3Q}\|_{\Phi,3P_0} \le 3^n \sup_{P\supset Q} \|f\,\chi_{\mathbb{R}^n\setminus 3Q}\|_{\Phi,P}.$$

Taking the supremum over all such cubes P_0 yields the desired estimate. □

Proof of Theorem 5.26. Fix a function f. We will first show that there exists constants c_0, $c_1 > 1$ such that for any cube Q,

$$\int_Q \Psi(M_\Phi(f\chi_{3Q})(y))\,dy \le c_0|Q| + c_0\int_{3Q}\Gamma(c_1 f(y))\,dy. \tag{5.22}$$

By Theorem 5.5 and Fubini's theorem,

$$\begin{aligned}
&\int_Q \Psi(M_\Phi(f\chi_{3Q})(y))\,dy \\
&\qquad = \int_0^\infty \Psi'(\lambda)|\{x\in Q : M_\Phi(f\chi_{3Q})(x) > \lambda\}|\,d\lambda \\
&\qquad \le c_0|Q| + c_0\int_1^\infty \Psi'(\lambda)\int_{\{x\in 3Q: f(x)>\lambda/2\}} \Phi\left(\frac{cf(x)}{\lambda}\right)dx\,d\lambda \\
&\qquad \le c_0|Q| + c_0\int_{\{x\in 3Q: f(x)>1/2\}}\int_1^{2f(x)} \Phi\left(\frac{cf(x)}{\lambda}\right)\Psi'(\lambda)\,d\lambda\,dx \\
&\qquad = c_0|Q| + c_0\int_{\{x\in 3Q: f(x)>1/2\}} G(cf(x))\,dx \\
&\qquad \le c_0|Q| + c_0\int_{3Q}\Gamma(c_1 f(x))\,dx.
\end{aligned}$$

We will now prove (5.19). Fix x and any cube Q containing x. Then it will suffice to show that $\|M_\Phi f\|_{\Psi,Q} \le CM_\Gamma f(x)$. But

$$\|M_\Phi f\|_{\Psi,Q} \le \|M_\Phi(f\chi_{3Q})\|_{\Psi,Q} + \|M_\Phi(f\chi_{\mathbb{R}^n\setminus 3Q})\|_{\Psi,Q} = I_1 + I_2.$$

We estimate each term separately. To estimate I_2, first note that for $\lambda \ge 1$, $\Psi'(\lambda) \ge \Psi'(1) > 0$. (If Ψ is a Young function this follows from convexity; if $\Psi(t) = t$ then equality holds.) Therefore, for $t > 1$,

$$G(2t) \ge c\int_1^2 \Phi(t/\lambda)\,d\lambda \ge c\Phi(t/2).$$

Hence, $\Phi(t) \preceq G(t) \preceq \Gamma(t)$, so by (5.21),

$$I_2 \le c\sup_{P\supset Q}\|f\,\chi_{\mathbb{R}^n\setminus 3Q}\|_{\Phi,P} \le cM_\Phi f(x) \le cM_\Gamma f(x).$$

To estimate I_1 we use the definition of the norm and (5.22), which holds if we replace f by f/λ for any $\lambda > 0$. By convexity,

$$\begin{aligned}
&(3^n c_0 + c_0)^{-1}\|M_\Phi(f\chi_{3Q})\|_{\Psi,Q} \\
&\qquad \le \inf\left\{\lambda > 0 : \fint_Q \Psi\left(\frac{M_\Phi(f\chi_{3Q})(y)}{\lambda}\right)dy \le 3^n c_0 + c_0\right\}
\end{aligned}$$

$$\begin{aligned} &\leq \inf\left\{\lambda > 0 : \fint_{3Q} \Gamma\left(c_1 \frac{f(y)}{\lambda}\right) dy \leq 1\right\} \\ &= c_1 \|f\|_{\Gamma,3Q} \\ &\leq c_1 M_\Gamma f(x). \end{aligned}$$

This completes the proof. □

For the classes of Orlicz bumps we are interested in—log bumps, power bumps and exponential log bumps—we can estimate G and find specific functions Γ such that Theorem 5.26 holds.

Example 5.30. If $\Phi(t) = \Psi(t) = t$, then $G(t) = t\log(t)$, so we can take $\Gamma(t) = t\log(e+t)$. More generally, if $\Phi(t) = t\log(e+t)^a$, $a > 0$, then

$$G(t) = \int_1^t \Phi(t/\lambda)\, d\lambda = t\int_1^t \Phi(u)u^{-2}\, du \leq t\log(e+t)^a \int_1^t \frac{du}{u} \leq Ct\log(e+t)^{a+1}.$$

Thus by induction we get the inequality $M^k f(x) \leq CM_\Gamma f(x)$, where $\Gamma(t) = t\log(e+t)^{k-1}$.

Example 5.31. Let $\Phi(t)$ be an r-Young function, $r > 1$; then by definition $\Phi(t)/t^r$ is non-decreasing. (The simplest example of such a function Φ is $\Phi(t) = t^r$.) If $\Psi(t) = t^s$, $1 \leq s < r$, then

$$G(t) = s\int_1^t \Phi(t/\lambda)\lambda^{s-1}\, d\lambda \leq C\Phi(t)\int_1^t \lambda^{s-r-1}\, d\lambda \leq C\Phi(t).$$

Therefore, we can take $\Gamma(t) = \Phi(t)$.

As an immediate corollary to this example we get a generalization of a result due to Coifman and Rochberg [27] that lets us readily find examples of A_1 weights. (Also see [68, 88].)

Proposition 5.32. *Let $\Phi(t)$ be an r-Young function, $r > 1$. Then for $0 \leq s < r$, $(M_\Phi f)^s \in A_1$, and the A_1 constant depends only on r, s, Φ and n. More generally, given any Young function Φ, for all δ, $0 < \delta < 1$, $(M_\Phi f)^\delta \in A_1$.*

Note that in the second half of the proposition we can take $\Phi(t) = t$.

Proof. If $w \in A_1$, then it follows from the definition and Hölder's inequality that $w^s \in A_1$ for $0 < s < 1$. Therefore, we may assume that $1 \leq s < r$. But then by Example 5.31,

$$M((M_\Phi f)^s) = M_s(M_\Phi f)^s \leq C(M_\Phi f)^s.$$

To prove the second part of this result, let $r = 1/\delta$ and define $\Psi(t) = \Phi(t^r)$. Since Φ is convex, Ψ is an r-Young function and $\Psi(t)/t^r$ is quasi-increasing. Then by rescaling and by the first part with $s = 1$ we have that $(M_\Phi f)^\delta = M_\Psi(|f|^\delta) \in A_1$. □

Remark 5.33. Recall that a weight w is in the class RH_∞ if there exists a constant C such that for every cube Q and almost every $x \in Q$,

$$w(x) \leq C \fint_Q w(y)\,dy.$$

The class RH_∞ is the counterpart of the A_1 class and plays a central role in the structure of A_p weights. (See [50] for more information.) As a consequence of Proposition 5.32 we can also construct examples of RH_∞ weights: given any Young function Φ and any $r > 0$, $(M_\Phi f)^{-r} \in RH_\infty$. To see this, fix $s > r$; then $(M_\Phi f)^{r/s} \in A_1$, so by the definition and by Hölder's inequality, for any cube Q and almost every $x \in Q$,

$$(M_\Phi f)(x)^{-r/s} \leq c\left(\fint_Q M_\Phi f(y)^{r/s}\,dy\right)^{-1} \leq c\left(\fint_Q M_\Phi f(y)^{-r}\,dy\right)^{1/s}.$$

Example 5.34. Suppose for $t > e$, $\Phi(t) \approx te^{\log(t)^r}$, $0 < r < 1$, and $\Psi(t) = t$. Fix s, $r < s < 1$; then by a change of variables

$$\begin{aligned} G(t) = \int_1^t \Phi(t/\lambda)\,d\lambda = t\int_1^t \frac{\Phi(u)}{u^2}\,du \\ \leq te^{\log(t)^r}\int_1^t \frac{du}{u} = t\log(t)e^{\log(t)^r} \leq Cte^{\log(t)^s}. \end{aligned}$$

Therefore, we can take $\Gamma(t) \approx te^{\log(t)^s}$.

5.6 Orlicz fractional maximal operators

Given a Young function Φ and α, $0 < \alpha < n$, define the Orlicz fractional maximal operator by

$$M_{\Phi,\alpha}f(x) = \sup_{Q \ni x} |Q|^{\alpha/n}\|f\|_{\Phi,Q}, \tag{5.23}$$

where the supremum is taken over all cubes containing x. When $\Phi(t) = t$ this reduces to the classical fractional maximal operator M_α introduced by Muckenhoupt and Wheeden [154]. If we take $\alpha = 0$, then these become the Orlicz maximal operators.

Remark 5.35. As we did for Orlicz maximal operators, hereafter we will always assume that the functions f are such that $M_{\Phi,\alpha}f(x) < \infty$ for almost every $x \in \mathbb{R}^n$. In particular, this means that $|Q|^{\alpha/n}\|f\|_{\Phi,Q}$ is bounded for every cube Q and is uniformly bounded as $|Q| \to \infty$.

Remark 5.36. We can also define the dyadic Orlicz fractional maximal operator $M^d_{\Phi,\alpha}$ by restricting the supremum in (5.23) to dyadic cubes containing x. All the results for Orlicz fractional maximal operators remain true for their dyadic counterparts. In some cases sharper results hold and we will point these out when necessary.

Fractional maximal operators have many properties similar to those of the Orlicz maximal operators. In addition, they satisfy "off-diagonal" estimates, mapping L^p into L^q when $p \le q$. The key two-weight inequality is the following generalization of Theorem 5.14. It was first proved in [36], where the Orlicz fractional maximal operator was introduced.

Theorem 5.37. *Given p, q, $1 < p \le q < \infty$, and α, $0 < \alpha < n$, let Φ, B, and C be Young functions such that $B^{-1}(t)C^{-1}(t) \le c\Phi^{-1}(t)$, $t > t_0 > 0$, and $C \in B_p$. If (u,v) is a pair of weights such that for every cube Q,*

$$|Q|^{\alpha/n+1/q-1/p}\|u^{1/q}\|_{q,Q}\,\|v^{-1/p}\|_{B,Q} \le K < \infty, \tag{5.24}$$

then for every function $f \in L^p(v)$,

$$\left(\int_{\mathbb{R}^n} M_{\Phi,\alpha}f(x)^q u(x)\,dx\right)^{1/q} \le C\left(\int_{\mathbb{R}^n} |f(x)|^p v(x)\,dx\right)^{1/p}.$$

Remark 5.38. If $\Phi(t) = t$, then the hypotheses simplify to $\bar{B} \in B_p$ (see [171]). When $p = q$ and $\alpha = 0$, this theorem reduces to Theorem 5.14.

Condition (5.24) is a generalization of the $A^\alpha_{p,q}$ condition. A pair $(u,v) \in A^\alpha_{p,q}$ if

$$|Q|^{\alpha/n+1/q-1/p}\|u^{1/q}\|_{q,Q}\|v^{-1/p}\|_{p',Q} \le K < \infty. \tag{5.25}$$

When $p = 1$ the $A^\alpha_{1,q}$ condition for $1 \le q < \infty$ is $|Q|^{\alpha/n+1/q-1}\|u^{1/q}\|_{q,Q} \le K\,v(x)$ for a.e. $x \in Q$.

The two-weight, weak type inequalities for the fractional maximal operator M_α are governed by the class $A^\alpha_{p,q}$. In particular, when $p = q = 1$ we get the analog of (5.13):

$$u(\{x \in \mathbb{R}^n : M_\alpha f(x) > \lambda\}) \le \frac{C}{\lambda}\int_{\mathbb{R}^n} |f(x)|M_\alpha u(x)\,dx. \tag{5.26}$$

The condition $A^\alpha_{p,q}$ and its relationship to the two-weight, weak (p,q) inequalities for M_α do not seem to have been stated explicitly in the literature, but the proof is identical to that for the analogous result for the Hardy-Littlewood maximal operator. (See [88]; also see [86] for related results in non-homogeneous spaces that can be adapted trivially to the Euclidean case.)

Remark 5.39. The $A^\alpha_{p,q}$ condition is a generalization of the one-weight $A_{p,q}$ condition for fractional operators due to Muckenhoupt and Wheeden [154] and discussed in Section 3.5 in Part I. To compare the two conditions, replace (u,v) by the pair

(u^q, v^p), and let $\alpha/n = 1/p - 1/q$. Written this way we see that if $u = v$, then the $A^{\alpha}_{p,q}$ condition is vacuous unless $1/p - 1/q = \alpha/n$. Otherwise, since $p \le q$, by Hölder's inequality,

$$|Q|^{\alpha/n+1/q-1/p} \le |Q|^{\alpha/n+1/q-1/p} \|u\|_{q,Q} \|u^{-1}\|_{p',Q} \le K,$$

and this is impossible as $|Q|$ gets very large or very small.

The condition (5.24) with $p = q$ leads naturally to our definition of fractional A_p bump conditions. Given α, $0 < \alpha < n$, and p, $1 < p < \infty$, we will consider pairs of weights that satisfy conditions of the form

$$|Q|^{\alpha/n} \|u^{1/p}\|_{A,Q} \|v^{-1/p}\|_{B,Q} \le K < \infty,$$

where A is a p-Young function and B is a p'-Young function. Besides the result above, conditions of this type were first considered when A and B were power bumps by Sawyer and Wheeden [208]; general B_p bumps were introduced in [171]. Also see [35, 55, 137].

Remark 5.40. We could also define fractional bump conditions associated with off-diagonal inequalities; however, hereafter we will only consider (p,p) inequalities, so we leave such generalizations to the interested reader.

5.7 Composition of fractional maximal operators

In this section we consider the composition of fractional maximal operators and maximal operators. In this case the order of composition makes a significant difference as our two main results show. For instance, we have that $M(M_\alpha f) \approx M_\alpha f$ (as $M_\alpha f \in A_1$ by Remark 5.44), but $M_\alpha(Mf) \le CM_{\Phi,\alpha} f$, where $\Phi(t) = t\log(e+t)$ (see Example 5.42). Throughout this section we will assume without loss of generality that all functions f are non-negative.

Theorem 5.41. *Given two Young functions Φ and Ψ, for $t > 1$ define*

$$G(t) = \int_1^t \Phi(t/\lambda)\Psi'(\lambda)\, d\lambda.$$

If Γ is any Young function such that $G(t) \preceq \Gamma(t)$, then there exists a constant c such that for all α, $0 < \alpha < n$, and $x \in \mathbb{R}^n$,

$$M_{\Psi,\alpha}(M_\Phi f)(x) \le cM_{\Gamma,\alpha} f(x).$$

The proof is essentially the same as the proof of Theorem 5.26, requiring only an additional factor of $|Q|^{\alpha/n}$ in the final estimates, and using Lemma 5.47 below instead of Lemma 5.28.

All of the examples given after the proof of Theorem 5.26 apply to Theorem 5.41 as well. In Chapter 9 we will need the analog of Example 5.30, so we state it here.

Example 5.42. If $\Phi(t) = t\log(e+t)^\epsilon$, $\epsilon \geq 0$, and $\Psi(t) = t\log(e+t)^{1+\epsilon}$, then $M_\alpha(M_\Phi f) \leq CM_{\Psi,\alpha} f$.

A stronger version of Proposition 5.32 holds for fractional maximal operators.

Theorem 5.43. *Given α, $0 < \alpha < n$, let Φ be a Young function such that $\Phi(t)/t^{n/\alpha}$ is quasi-decreasing and $\Phi(t)/t^{n/\alpha} \to 0$ as $t \to \infty$. If there exists r, $1 \leq r < n/\alpha$ such that $\Phi(t)/t^r$ is quasi-increasing, then $(M_{\Phi,\alpha} f)^s \in A_1$ for every $0 < s < r\,n/(n-\alpha)$. The A_1 constant depends only on Φ, s and n.*

Remark 5.44. Since Φ is convex, $\Phi(t)/t$ is increasing, so we can always take $r = 1$ in Theorem 5.43. If $\Phi(t) = t$, then $(M_\alpha f)^s \in A_1$, $0 < s < n/(n-\alpha)$. This special case is due to Sawyer [202].

Remark 5.45. Arguing as we did in Remark 5.33, we have that given a Young function Φ satisfying the hypotheses of Theorem 5.43, $0 < \alpha < n$, and $r > 0$, $(M_{\Phi,\alpha} f)^{-r} \in RH_\infty$.

The proof of Theorem 5.43 requires several lemmas. The first two are analogs of Lemma 5.28.

Lemma 5.46. *Given α, $0 < \alpha < n$, let Φ be a Young function so that $\Phi(t)/t^{n/\alpha}$ is quasi-decreasing. Then for every Q and $x \in Q$,*

$$M_{\Phi,\alpha}(f\,\chi_Q)(x) \approx \sup_{x\in P\subset Q} |P|^{\alpha/n}\,\|f\|_{\Phi,P}. \tag{5.27}$$

Proof. The proof is nearly identical to the proof of (5.20) in Lemma 5.28, up to the additional factor $|Q|^{\alpha/n}$, so we give here only the details of the step that is different.

Having fixed a cube P as in the proof of (5.20), suppose $|P| \geq |Q|$. Let $s = |P|/|Q| \geq 1$. Since $\Phi(t)/t^{n/\alpha}$ is quasi-decreasing, for all $\lambda > 0$,

$$⨍_P \Phi\left(\frac{f(x)\chi_Q}{\lambda}\right) dx \leq ⨍_Q s^{-1}\Phi\left(\frac{f(x)}{\lambda}\right) dx \leq ⨍_Q \Phi\left(\frac{Cf(x)}{s^{\alpha/n}\lambda}\right) dx.$$

Therefore,

$$\begin{aligned} |P|^{\alpha/n}\|f\chi_Q\|_{\Phi,P} &\leq Cs^{-\alpha/n}|P|^{\alpha/n}\|f\|_{\Phi,Q} \\ &= C|Q|^{\alpha/n}\|f\|_{\Phi,Q} \leq C \sup_{x\in P\subset Q} |P|^{\alpha/n}\,\|f\chi_Q\|_{\Phi,P}. \end{aligned}$$

The proof now finishes as before. □

Lemma 5.47. *Given α, $0 < \alpha < n$, and a Young function Φ, then for every Q and for every $x \in Q$,*

$$M_{\Phi,\alpha}(f\,\chi_{\mathbb{R}^n\setminus 3Q})(x) \approx \sup_{P\supset Q} |P|^{\alpha/n}\,\|f\,\chi_{\mathbb{R}^n\setminus 3Q}\|_{\Phi,P}.$$

The proof of Lemma 5.47 is identical to the proof of (5.21) in Lemma 5.28 (with the additional factor of $|Q|^{\alpha/n}$) and so is omitted.

Lemma 5.48. *Given α, $0<\alpha<n$, suppose the Young function Φ is such that $\Phi(t)/t^{n/\alpha}$ is quasi-decreasing and $\Phi(t)/t^{n/\alpha}\to 0$ as $t\to\infty$. If* $\operatorname{supp}(f)\subset Q_0$ *for some cube Q_0, then $|Q|^{\alpha/n}\|f\|_{\Phi,Q}\to 0$ as $|Q|\to\infty$.*

Proof. Since $\|f\|_{\Phi,Q}\le\|f+\chi_{Q_0}\|_{\Phi,Q}$, we may assume without loss of generality that $f(x)\ge 1$. Let $\|f\|_{\Phi,Q_0}=M$; then

$$\int_{Q_0}\Phi\left(\frac{f(x)}{M}\right)dx\le|Q_0|<\infty.$$

Since $\operatorname{supp}(f)\subset Q_0$, it follows that the integrand is in L^1.

Fix $\epsilon>0$; we need to show that there exists $N>0$ such that if $|Q|>N$, then $|Q|^{\alpha/n}\|f\|_{\Phi,Q}\le\epsilon$; to obtain this it suffices to see that

$$\fint_Q\Phi\left(\frac{f(x)|Q|^{\alpha/n}}{\epsilon}\right)dx\le 1. \tag{5.28}$$

Since $f(x)\ge 1$ and since $\Phi(t)/t^{n/\alpha}$ is quasi-decreasing, for almost every $x\in Q_0$ and for $|Q|$ sufficiently large,

$$|Q|^{-1}\Phi\left(\frac{f(x)|Q|^{\alpha/n}}{\epsilon}\right)\le C\epsilon^{-n/\alpha}\Phi\left(\frac{f(x)}{M}\right)\in L^1.$$

Since $\Phi(t)/t^{n/\alpha}\to 0$,

$$|Q|^{-1}\Phi\left(\frac{f(x)|Q|^{\alpha/n}}{\epsilon}\right)=\left(\epsilon^{-1}f(x)\right)^{n/\alpha}\frac{\Phi\left(\epsilon^{-1}f(x)|Q|^{\alpha/n}\right)}{\left(\epsilon^{-1}f(x)|Q|^{\alpha/n}\right)^{n/\alpha}}\to 0$$

as $|Q|\to\infty$. Therefore, by the dominated convergence theorem,

$$\fint_Q\Phi\left(\frac{f(x)|Q|^{\alpha/n}}{\epsilon}\right)dx\to 0,$$

so there exists $N>0$ such that if $|Q|>N$, (5.28) holds. □

The next lemma is a modular inequality analogous to Theorem 5.5. Similar results were proved in [36, 38], and our proof is adapted from theirs.

Lemma 5.49. *Given α, $0<\alpha<n$, let Φ be a Young function such that $\Phi(t)/t^{n/\alpha}$ is quasi-decreasing and $\Phi(t)/t^{n/\alpha}\to 0$ as $t\to\infty$. Then for all cubes Q with $|Q|=1$, and every $\lambda>0$,*

$$\begin{aligned}&|\{x\in Q: M_{\Phi,\alpha}(f\chi_Q)(x)>\lambda\}|^{(n-\alpha)/n}\\ &\qquad\le C\int_{\{x\in Q:f(x)\ge\lambda/c\}}\Phi\left(\frac{f(x)}{\lambda}\right)dx.\end{aligned} \tag{5.29}$$

Proof. We will first show that $M_{\Phi,\alpha}(\chi_Q)(x) \leq C$ for every $x \in Q$. By Lemma 5.46, it suffices to show that for all cubes $P \subset Q$, $|P|^{\alpha/n}\|\chi_Q\|_{\Phi,P} \leq C$. Fix such a P; then $|P| \leq |Q| = 1$ and by the definition of the norm,

$$|P|^{\alpha/n}\,\|\chi_Q\|_{\Phi,P} \leq \|\chi_Q\|_{\Phi,P} \leq C.$$

Now write $f\chi_Q$ as $f_1 + f_2$, where $f_1 = f\,\chi_{\{x\in Q: f\leq 1\}}$. By the above observation, if $x \in Q$, $M_{\Phi,\alpha}f_1(x) \leq M_{\Phi,\alpha}(\chi_Q)(x) \leq C$. By Lemma 5.46 it follows that $M_{\Phi,\alpha}f_2(x) \leq C\,M^Q_{\Phi,\alpha}f_2(x)$, where

$$M^Q_{\Phi,\alpha}f_2(x) = \sup_{x\in P\subset Q} |P|^{\alpha/n}\,\|f_2\|_{\Phi,P}.$$

Therefore, there exists a constant $C_0 > 0$ such that

$$\begin{aligned}\{x \in Q : M_{\Phi,\alpha}(f\chi_Q)(x) > 2\,C_0\} &\subset \{x \in Q : M_{\Phi,\alpha}f_2(x) > C_0\}\\ &\subset \{x \in Q : M^Q_{\Phi,\alpha}f_2(x) > 1\} = E.\end{aligned}$$

For each $x \in E$, there exists a cube $P_x \subset Q$ containing x such that $|P_x|^{\alpha/n}\,\|f_2\|_{\Phi,P_x} > 1$. By Lemma 5.48, $|Q|^{\alpha/n}\|f_2\|_{\Phi,Q} \to 0$ as $|Q| \to \infty$. Therefore, we can adapt the proof of the fractional Calderón-Zygmund decomposition in Proposition A.7 in Appendix A to show that there exists a collection of disjoint dyadic cubes $\{P_j\}_j$ such that $\ell(P_j) \leq 2$, $E \subset \cup_j 3P_j$, and $|P_j|^{\alpha/n}\,\|f_2\|_{\Phi,P_j} > \beta > 0$ for some $\beta < 1$.

Since Φ is convex, for each cube P_j,

$$1 \leq \fint_{P_j} \Phi\left(\frac{|P_j|^{\alpha/n}\,f_2(x)}{\beta}\right) dx \leq \frac{1}{2^\alpha|P_j|^{1-\alpha/n}} \int_{P_j} \Phi\left(\frac{2^\alpha f_2(x)}{\beta}\right) dx.$$

Therefore, since $\Phi(t)/t^{n/\alpha}$ is quasi-decreasing and the P_j's are disjoint,

$$\begin{aligned}|\{x \in Q : M_{\Phi,\alpha}(f\chi_Q)(x) > 2\,C_0\}|^{(n-\alpha)/n} &\leq |E|^{(n-\alpha)/n}\\ \leq \sum_j |3P_j|^{(n-\alpha)/n} &\leq C\sum_j \int_{P_j} \Phi\left(\frac{2^\alpha f_2(x)}{\beta}\right) dx\\ &\leq C\int_{\{x\in Q: f\geq 1\}} \Phi\left(\frac{f(x)}{2\,C_0}\right) dx.\end{aligned}$$

Inequality (5.29) now follows by homogeneity: replace f by $2\,C_0\,f/\lambda$ to get the desired result. □

Proof of Theorem 5.43. We will first show that if Q is a cube such that $|Q| = 1$, then for any $x \in Q$,

$$\fint_Q M_{\Phi,\alpha}(f\chi_Q)(x)^s\,dx \leq C\,\|f\|^s_{\Phi,Q}. \tag{5.30}$$

By homogeneity we may assume $\|f\|_{\Phi,Q} = 1$, and so, in particular, that

$$\int_Q \Phi(f(x))\,dx = \fint_Q \Phi(f(x))\,dx \le 1.$$

Therefore, by Lemma 5.49 and because $\Phi(t)/t^r$ is quasi-increasing and $0 < s < r\,n/(n-\alpha)$,

$$\begin{aligned}
&\int_Q M_{\Phi,\alpha}(f\chi_Q)(x)^s\,dx \\
&\quad = \int_0^\infty s\,\lambda^s\,|\{x\in Q : M_{\Phi,\alpha}(f\chi_Q)(x) > \lambda\}|\,\frac{d\lambda}{\lambda} \\
&\quad \le c^s + C\int_c^\infty \lambda^s\left(\int_{Q\cap\{f\ge\lambda/c\}} \Phi\left(\frac{f(x)}{\lambda}\right)dx\right)^{n/(n-\alpha)}\frac{d\lambda}{\lambda} \\
&\quad \le C + C\int_c^\infty \lambda^s\left(\int_Q \frac{1}{\lambda^r}\,\Phi(f(x))\,dx\right)^{n/(n-\alpha)}\frac{d\lambda}{\lambda} \\
&\quad \le C.
\end{aligned}$$

This yields (5.30).

We will now prove via a homogeneity argument that (5.30) extends to arbitrary cubes: for all Q,

$$\fint_Q M_{\Phi,\alpha}(f\chi_Q)(x)^s\,dx \le C\,|Q|^{s\,\alpha/n}\,\|f\|_{\Phi,Q}^s. \tag{5.31}$$

Fix a cube Q; by translation invariance we may assume without loss of generality that Q is centered at the origin. Let $l = \ell(Q)$, and let $f_l(x) = f(lx)$. If P is any cube contained in Q with center x_P, let P_l be the cube centered at x_P/l with side-length $\ell(P)/l$. Note that $|Q_l| = 1$ and every cube contained in Q_l is of the form P_l for some $P \subset Q$. Therefore, if we make the change of variables $x = ly$, we get

$$\begin{aligned}
|P|^{\alpha/n}\|f\|_{\Phi,P} &= |P|^{\alpha/n}\inf\left\{\lambda>0 : \frac{1}{|P|}\int_P \Phi\left(\frac{f(x)}{\lambda}\right)dx \le 1\right\} \\
&= l^\alpha |P_l|^{\alpha/n}\inf\left\{\lambda>0 : \frac{1}{|P_l|}\int_{P_l} \Phi\left(\frac{f(ly)}{\lambda}\right)dy \le 1\right\} \\
&= l^\alpha |P_l|^{\alpha/n}\|f_l\|_{\Phi,P_l}.
\end{aligned}$$

Since $x\in P$ if and only if $x/l \in P_l$, this identity combined with Lemma 5.46 shows that $M_{\Phi,\alpha}(f\chi_Q)(x) \le C l^\alpha M_{\Phi,\alpha}(f_l\chi_{Q_l})(x/l)$. Hence, if we make the change of variables $y = x/l$, it follows from (5.30) that

$$\fint_Q M_{\Phi,\alpha}(f\chi_Q)(x)^s\,dx \le C l^{s\,\alpha}\fint_Q M_{\Phi,\alpha}(f_l\chi_{Q_l})(x/l)^s\,dx$$

$$
\begin{aligned}
&= C l^{s\,\alpha} \frac{1}{|Q_l|} \int_{Q_l} M_{\Phi,\alpha}(f_l \chi_{Q_l})(y)^s \, dy \\
&\leq C l^{s\,\alpha} \|f_l\|_{\Phi,Q_l}^s = C|Q|^{s\,\alpha/n} \, \|f\|_{\Phi,Q}^s .
\end{aligned}
$$

We can now finish the proof. Fix any cube Q. By (5.31) and Lemma 5.47, for every $y \in Q$,

$$
\begin{aligned}
&\fint_Q M_{\Phi,\alpha} f(x)^s \, dx \\
&\qquad \leq C \fint_{3Q} M_{\Phi,\alpha}(f\chi_{3Q})(x)^s \, dx + \fint_Q M_{\Phi,\alpha}(f\chi_{\mathbb{R}^n\setminus 3Q})(x)^s \, dx. \\
&\qquad \leq C\,|3Q|^{s\,\alpha/n} \, \|f\|_{\Phi,3Q}^s + C \Big(\sup_{P \supset Q} |P|^{\alpha/n} \, \|f \, \chi_{\mathbb{R}^n\setminus 3Q}\|_{\Phi,P} \Big)^s \\
&\qquad \leq C M_{\Phi,\alpha} f(y)^s .
\end{aligned}
$$

This completes the proof. □

Chapter 6

Two-Weight Factorization

Our primary goal in this chapter is to define the appropriate analogs of A_1 weights and prove a "reverse factorization" property for pairs of weights that satisfy the A_p bump conditions defined in Chapter 5. As we discussed in Chapter 2, reverse factorization is an essential tool in the proof of the Rubio de Francia extrapolation theorem; in the two-weight case these ideas play a similar but less direct role, as we will discuss in Chapter 7.

To put our results into context, we begin by recalling the Jones factorization theorem for A_p weights (see Theorem 1.3): for all p, $1 < p < \infty$, $w \in A_p$ if and only if there exist $w_1, w_2 \in A_1$ such that $w = w_1 w_2^{1-p}$. The factorization also includes information on the reverse Hölder inequality: if $w \in RH_s$, then $w_1 \in A_1 \cap RH_s$ and $w_2^{1-p} \in A_p \cap RH_\infty$. (For this extension, see [50]; for the definition of RH_∞, see Remark 5.33.) Furthermore, since $w_1, w_2 \in A_1$, $w_1 \approx Mw_1$ and $w_2 \approx Mw_2$, so $w \in A_p$ if and only if $w \approx Mw_1(Mw_2)^{1-p}$.

As in the one-weight case, a reverse factorization theorem follows almost immediately from the definition of the A_p bump conditions. We define a new class of weights, the factored weights, that are of the form

$$(\tilde{u}, \tilde{v}) = \big(w_1(M_\Psi w_2)^{1-p}, (M_\Phi w_1)w_2^{1-p}\big);$$

by rescaling Φ and Ψ we determine which bump condition these weights satisfy. We also consider the converse (and more difficult) question: when can a pair of weights be "factored" and replaced by a pair of factored weights? This problem is of interest in its own right and the solution requires a variation of the Rubio de Francia iteration algorithm discussed in Chapter 2. Furthermore, as we will show in Chapters 9 and 10, factored weights are well suited to proving certain two-weight norm inequalities, so for applications it is of interest to know when a pair of weights can be replaced by a pair of factored weights.

Two-weight factorization theorems have been considered by a number of authors: implicitly in Bloom [13], Neugebauer [163] and Segovia and Torrea [209], and explicitly by Hernández [102] and Ruiz and Torrea [198]. These authors all started with the assumption that a pair of weights (u,v) were such that the maximal operator satisfied $M : L^p(v) \to L^p(u)$, and $M : L^{p'}(u^{1-p'}) \to L^{p'}(v^{1-p'})$, and then give a related pair of factored weights defined in terms of the Hardy-Littlewood maximal operator. (Bloom and Hernández generalize this by replacing M with any bounded sublinear operator.) The problem with this approach, as we will see below, is that the factored weights obtained are not, *a priori*, such that the maximal operator has the same boundedness properties. Hernández [102] also considered the problem of reverse factorization, but found results only for linear operators.

Our approach to factorization is different and is motivated by our focus on A_p bump conditions. We instead begin with weights that satisfy an A_p bump condition (which implies that M has the requisite boundedness properties) and then construct a pair of factored weights that satisfy a related A_p bump condition. By controlling the bump condition we can deduce sharper results; in particular, we can specify bump conditions on the original pair of weights so that the maximal operator is again bounded with respect to the new factored weights.

We discuss reverse factorization in Section 6.1 and factorization in Section 6.2. In Section 6.3 we consider a related problem first posed by Neugebauer [163]: given a pair of weights (u,v), when can you "insert" an A_∞ weight between them—i.e., when can you find $w \in A_\infty$ such that $c_1 u \leq w \leq c_2 v$? Finally, in Section 6.4 we extend our factorization results to the pairs of weights associated with fractional operators. As in Chapter 5, we could have treated all of these results in one general theorem, but for clarity we have split them up.

Remark 6.1. All of the results in this chapter hold if we everywhere replace cubes and maximal operators by dyadic cubes and the corresponding dyadic maximal operators.

6.1 Reverse factorization and factored weights

Given a Young function Φ and a weight u, the pair of weights $(u, M_\Phi u)$ is called an A_1 pair associated with Φ or simply a pair of A_1 weights. When $\Phi(t) = t$ we get the pairs (u, Mu) which are the classic A_1 weights in the two-weight case. More generally, if we let $\Phi(t) = t\log(e+t)^{k-1}$, $k \geq 1$, we get the pairs $(u, M_\Phi u) \approx (u, M^k u)$. (See Section 5.5.) A_1 pairs appear frequently in applications: see, for instance, [28, 54, 75] and Chapters 9 and 10 below.

Closely related to the A_1 pairs associated with Φ are the pairs of weights $((M_\Phi u)^{1-p}, u^{1-p})$, $1 < p < \infty$. These are are referred to as the dual A_1 pairs associated with Φ or dual A_1 weights. The reason for this name is that if a linear operator T satisfies

$$\int_{\mathbb{R}^n} |Tf(x)|^p u(x)\,dx \leq C \int_{\mathbb{R}^n} |f(x)|^p M_\Phi u(x)\,dx,$$

then by duality,

$$\int_{\mathbb{R}^n} |T^* f(x)|^{p'} M_\Phi u(x)^{1-p'}\,dx \le C \int_{\mathbb{R}^n} |f(x)|^{p'} u(x)^{1-p'}\,dx.$$

The converse is also true. Thus, norm inequalities for A_1 weights are equivalent to norm inequalities for dual A_1 weights.

A_1 and dual A_1 pairs are closely connected to the one-weight theory: they are "almost" A_1 and A_p weights, respectively.

Proposition 6.2. *Given a Young function Φ and a weight u, then:*

(a) *for any $0 < \delta < 1$, $(M_\Phi u)^\delta \in A_1$. If $\Phi(t) = t^r$, $r > 1$, then we can take $\delta = 1$;*

(b) *given $1 < p < \infty$, for all $q > p$, $(M_\Phi u)^{1-p} \in A_q \cap RH_\infty$.*

The first part of Proposition 6.2 is just a restatement of Proposition 5.32, and the second follows from the definition of A_p and Remark 5.33.

A_1 pairs and dual A_1 pairs play the role of the A_1 weights in the Jones factorization theorem. To state and prove the reverse factorization property we first make a definition.

Definition 6.3. Given Young functions Φ and Ψ and non-negative functions w_1 and w_2 such that w_1, w_2 are positive on sets of positive measure, and $M_\Phi w_1$, $M_\Psi w_2 < \infty$ almost everywhere, the pair

$$(\tilde{u}, \tilde{v}) = \big(w_1 (M_\Psi w_2)^{1-p}, (M_\Phi w_1) w_2^{1-p}\big) \tag{6.1}$$

is called a pair of factored weights.

Theorem 6.4. *Given p, $1 < p < \infty$, let A be a p-Young function and B a p'-Young function, and define $\Phi(t) = A(t^{1/p})$ and $\Psi(t) = B(t^{1/p'})$. Then the pair of factored weights* (6.1) *satisfies the A_p bump condition*

$$\|\tilde{u}^{1/p}\|_{A,Q} \|\tilde{v}^{-1/p}\|_{B,Q} \le 1. \tag{6.2}$$

Proof. By the definition of the Orlicz maximal operators and by rescaling

$$\|\tilde{u}^{1/p}\|_{A,Q} = \|w_1 (M_\Psi w_2)^{1-p}\|_{\Phi,Q}^{1/p} \le \|w_1\|_{\Phi,Q}^{1/p} \|w_2\|_{\Psi,Q}^{-1/p'},$$
$$\|\tilde{v}^{-1/p}\|_{B,Q} = \|(M_\Phi w_1)^{1-p'} w_2\|_{\Psi,Q}^{1/p'} \le \|w_1\|_{\Phi,Q}^{-1/p} \|w_2\|_{\Psi,Q}^{1/p'}.$$

Condition (6.2) follows immediately. □

If we take either $w_2 = 1$ or $w_1 = 1$, then we get that the A_1 pairs $(w_1, M_\Phi w_1)$ and dual A_1 pairs $\big((M_\Psi w_2)^{1-p}, w_2^{1-p}\big)$ satisfy condition (6.2). Note that the A_1 weights depend only on the bump on the left, and the dual A_1 weights depend only on the bump on the right. Intuitively, therefore, in factored weights we have separated the effects of each Orlicz bump.

An immediate corollary of Theorems 5.14 and 6.4, one which we will use repeatedly, is the following.

Corollary 6.5. *Fix p, $1 < p < \infty$, and let $\bar{B} \in B_p$. Given any weights w_1, w_2, define the pair*

$$(\tilde{u}, \tilde{v}) = \big(w_1(M_\Psi w_2)^{1-p}, (Mw_1)w_2^{1-p}\big),$$

where $\Psi(t) = B(t^{1/p'})$. Then $M : L^p(\tilde{v}) \to L^p(\tilde{u})$.

Remark 6.6. Corollary 6.5 sheds light on the classic example to show that the two-weight A_p condition is not sufficient for the Hardy-Littlewood maximal operator to satisfy a strong (p,p) inequality. For $p = 2$, Muckenhoupt and Wheeden [155] noted that the pair $u(x) = -x\log(x)\chi_{(0,1/2)}(x)$ and $v(x) = x\log(x)^2/\chi_{(0,1/2)}(x)$ is in A_2 but M is not bounded from $L^2(v)$ to $L^2(u)$. But on $(0,1/2)$ we have that $(u,v) \approx \big((Mw)^{-1}, w^{-1}\big)$, where $w = [x\log(x)^2]^{-1}$. Hence, this pair does not satisfy the sharp bump condition in Theorem 5.14 (see Remark 5.15). Their other examples can be gotten in a similar manner; details are left to the reader.

6.2 Factorization of weights

In this section we consider the converse of Theorem 6.4. Our goal is to factor pairs of weights in terms of A_1 weights and dual A_1 weights. Given a pair of weights (u,v) that satisfies the A_p bump condition

$$\|u^{1/p}\|_{A,Q}\|v^{-1/p}\|_{B,Q} \le K < \infty, \tag{6.3}$$

where A is a p-Young function and B a p'-Young function, the strongest analog of the Jones factorization theorem would be that there exist w_1, w_2 such that

$$(u,v) \approx \big(w_1(M_\Psi w_2)^{1-p}, (M_\Phi w_1)w_2^{1-p}\big),$$

with $\Phi(t) = A(t^{1/p})$ and $\Psi(t) = B(t^{1/p'})$. However, since we can make u arbitrarily smaller and v arbitrarily larger and (6.3) still holds, this seems implausible.

For application to weighted norm inequalities, a weaker but still useful version of the factorization theorem is given by the following definition.

Definition 6.7. Given a pair of weights (u,v), $u \le Cv$, if there exists a pair of factored weights $(\tilde{u},\tilde{v})$ such that $c_1 u \le \tilde{u} \le \tilde{v} \le c_2 v$, then we say that we can insert the factored pair $(\tilde{u},\tilde{v})$ between (u,v).

Ideally, given a pair of weights (u,v) that satisfies some A_p bump condition, we would be able to insert a factored pair between them that satisfies (essentially) the same bump condition. Such a result is true in varying degrees, depending on the bump condition on the weights. Our proof will be adapted from the proof of the Jones factorization theorem due to Coifman, Jones and Rubio de Francia [26] (see also Neugebauer [163]), so central to it is a version of the iteration algorithm. This requires the boundedness of the maximal operator; therefore, we will restrict ourselves to Young functions A and B such that (6.3) implies that $M : L^p(v) \to L^p(u)$ and $M : L^{p'}(u^{1-p'}) \to L^{p'}(v^{1-p'})$. It would be of interest to give a factorization theorem for the case when we have only one of these boundedness conditions on M, since, as we noted in Section 1.1, this corresponds to the class of weights conjectured to control weak type inequalities for singular integrals.

We begin with a result for arbitrary B_p bumps. We omit the proof since it is an immediate variation of the proof of Theorem 6.9 below.

Theorem 6.8. *Given p, $1 < p < \infty$, suppose that the pair of weights (u, v) satisfies*

$$\|u^{1/p}\|_{A,Q}\|v^{-1/p}\|_{B,Q} \le K < \infty,$$

where $\bar{A} \in B_{p'}$ and $\bar{B} \in B_p$. Then there exist weights w_1, w_2, such that the factored pair $(\tilde{u}, \tilde{v}) = \big(w_1(Mw_2)^{1-p}, (Mw_1)w_2^{1-p}\big)$ can be inserted between (u, v) and such that

$$\|\tilde{u}^{1/p}\|_{p,Q}\|\tilde{v}^{-1/p}\|_{p',Q} \le K < \infty.$$

Theorem 6.8 is implicit in Neugebauer [163] and essentially the same as results proved by Hernández [102] and Ruiz and Torrea [198]. The assumption that the pair (u, v) satisfies an A_p bump condition is stronger than necessary: it suffices to assume $M : L^p(v) \to L^p(u)$ and $M : L^{p'}(u^{1-p'}) \to L^{p'}(v^{1-p'})$. However, for consistency with our other results we give the hypotheses in terms of bump conditions.

To see the weakness of Theorem 6.8, note that we can restate the conclusion of Theorem 6.8 as $(\tilde{u}, \tilde{v}) \in A_p$. Hence, $M : L^p(\tilde{v}) \to L^{p,\infty}(\tilde{u})$ and $M : L^{p'}(\tilde{u}^{1-p'}) \to L^{p',\infty}(\tilde{v}^{1-p'})$. However, as we noted in Remark 6.6 above, the A_p condition is not sufficient for the corresponding strong type inequalities to hold. Thus we have lost critical information.

We are uncertain whether a stronger conclusion is true in Theorem 6.8 assuming only that the weights satisfy a B_p bump condition. To improve it we need to consider specific classes of Orlicz bumps. Because of their importance in the study of weighted norm inequalities, our principal result is for log bumps. Afterwards, we will consider exponential log bumps and power bumps.

Theorem 6.9. *Given p, $1 < p < \infty$, suppose that the pair of weights (u, v) satisfies*

$$\|u^{1/p}\|_{A,Q}\|v^{-1/p}\|_{B,Q} \le K < \infty, \tag{6.4}$$

where $A(t) = t^p \log(e+t)^{p-1+\delta}$, $B(t) = t^{p'} \log(e+t)^{p'-1+\delta}$, $\delta > 0$. Let $\tilde{A}(t) = t^p \log(e+t)^a$, $0 < a < \delta/p$, $\tilde{B}(t) = t^{p'} \log(e+t)^b$, $0 < b < \delta/p'$, $\tilde{\Phi}(t) = \tilde{A}(t^{1/p})$, and $\tilde{\Psi}(t) = \tilde{B}(t^{1/p'})$. Then there exist w_1, w_2 such that the factored pair

$$(\tilde{u}, \tilde{v}) = \big(w_1(M_{\tilde{\Psi}}w_2)^{1-p}, (M_{\tilde{\Phi}}w_1)w_2^{1-p}\big) \tag{6.5}$$

can be inserted between (u, v) and satisfies

$$\|\tilde{u}^{1/p}\|_{\tilde{A},Q}\|\tilde{v}^{-1/p}\|_{\tilde{B},Q} \le K < \infty. \tag{6.6}$$

Remark 6.10. As in Theorem 6.8, in Theorem 6.9 there is a loss of information in the factorization: we do not have, *a priori*, that M is bounded. However, if we make additional assumptions on the size of the exponent of the logarithm, we can recapture the desired boundedness conditions. For instance, if $A(t) = t^p \log(e+t)^{p^2-1+\delta}$ and $B(t) = t^{p'} \log(e+t)^{(p')^2-1+\delta}$, $\delta > 0$, then we can take $\tilde{A}(t) = t^p \log(e+t)^{p-1+\tilde{\delta}}$ and $\tilde{B}(t) = t^{p'} \log(e+t)^{p'-1+\tilde{\delta}}$, $0 < \tilde{\delta} < \min(\delta/p, \delta/p')$. Hence, $M : L^p(\tilde{v}) \to L^p(\tilde{u})$ and $M : L^{p'}(\tilde{u}^{1-p'}) \to L^{p'}(\tilde{v}^{1-p'})$.

If $p > 2$, then the exponent $p^2 - 1 + \delta$ is much larger than $p - 1 + \delta$; since this exponent is sharp for two-weight inequalities for the maximal operator and singular integrals (see Remark 5.15 and Theorem 9.21) we would like to improve Theorem 6.9. Ideally, we conjecture that for log bumps there is no significant loss of information in the factorization: i.e., we get an exponent of the form $p - 1 + \tilde{\delta}$, with $0 < \tilde{\delta} < \delta$. However, such a result seems currently out of reach. But based on early work on two-weight extrapolation and on our results for singular integrals in [47], we conjecture that we can improve Theorem 6.9 so that if we assume that the power on the logarithm in A is $2p - 1 + \delta$, then the power on the logarithm of $\tilde{A}$ will be $p - 1 + \tilde{\delta}$.

Proof of Theorem 6.9. We begin by defining the appropriate variant of the Rubio de Francia iteration algorithm. By Proposition 5.21 we have that

$$M_{\tilde{\Psi}} : L^p(v) \to L^p(u),$$
$$M_{\tilde{\Phi}} : L^{p'}(u^{1-p'}) \to L^{p'}(v^{1-p'}).$$

Suppose that $p \geq 2$ and let $\sigma = p/p' = p - 1 \geq 1$. (If $p < 2$, let $\sigma = p'/p$ and exchange the roles of $M_{\tilde{\Psi}}$ and $M_{\tilde{\Phi}}$ in the following argument.) Define the operator

$$Sf = u^{1/p} M_{\tilde{\Psi}}(fv^{-1/p}) + v^{-1/(p\sigma)} M_{\tilde{\Phi}}(f^\sigma u^{1/p})^{1/\sigma};$$

then S is bounded on L^p. Further, since $\sigma \geq 1$, S is sublinear. This follows from the triangle inequality for Orlicz norms and the fact that $\|f^\sigma u^{1/p}\|_{\tilde{\Phi},Q}^{1/\sigma} = \|fu^{1/\sigma p}\|_{\Theta,Q}$, where $\Theta(t) = \tilde{\Phi}(t^\sigma)$ is a Young function.

Therefore, we can define the iteration operator

$$\mathcal{R}h = \sum_{k=0}^{\infty} \frac{S^k h}{2^k \|S\|_{L^p}^k}.$$

Fix a non-negative function $h \in L^p$ and let $H = \mathcal{R}h$. Then $H \in L^p$ and $SH \leq 2\|S\|_{L^p} H$. In particular, we have that

$$M_{\tilde{\Phi}}(H^\sigma u^{1/p}) \leq CH^\sigma v^{1/p}, \qquad M_{\tilde{\Psi}}(Hv^{-1/p}) \leq CHu^{-1/p}.$$

Define the weights $w_1 = H^\sigma u^{1/p}$, $w_2 = Hv^{-1/p}$; then we have that

$$M_{\tilde{\Phi}} w_1 \leq C w_1 u^{-1/p} v^{1/p}, \qquad M_{\tilde{\Psi}} w_2 \leq C w_2 u^{-1/p} v^{1/p}.$$

Further, note that $w_1 w_2^{1-p} = u^{1/p} v^{1/p'}$. Therefore, it follows that

$$u = w_1 w_2^{1-p}(u/v)^{1/p'} \le C w_1 (M_{\tilde{\Psi}} w_2)^{1-p} \\ \le C(M_{\tilde{\Phi}} w_1) w_2^{1-p} \le C w_1 w_2^{1-p} (v/u)^{1/p} = Cv.$$

If we let

$$\tilde{u} = w_1 (M_{\tilde{\Psi}} w_2)^{1-p}, \qquad \tilde{v} = (M_{\tilde{\Phi}} w_1) w_2^{1-p};$$

then we have shown that we can insert the factored pair $(\tilde{u}, \tilde{v})$ between (u, v). Finally, note that by Theorem 6.4 this pair satisfies (6.6). This completes the proof. □

A stronger factorization theorem holds for pairs (u, v) that satisfy A_p bump conditions with exponential log bumps or power bumps. In these cases we can insert a factored pair $(\tilde{u}, \tilde{v})$ that satisfies essentially the same bump condition, so we have that $M : L^p(\tilde{v}) \to L^p(\tilde{u})$, and $M : L^{p'}(\tilde{u}^{1-p'}) \to L^{p'}(\tilde{v}^{1-p'})$. We omit the proof of Theorem 6.11 since it is nearly identical to the proof of Theorem 6.9. The only significant change is that one needs the analog of Proposition 5.21 for Orlicz maximal functions defined using exponential log bumps or power bumps; details of this are left to the reader.

Theorem 6.11. *Given p, $1 < p < \infty$, suppose that (u, v) satisfy*

$$\|u^{1/p}\|_{A,Q} \|v^{-1/p}\|_{B,Q} \le K < \infty, \tag{6.7}$$

where A is a p-Young function and B is a p'-Young function.

(a) *If $A(t) \approx t^p e^{\log(t^p)^r}$, $B(t) \approx t^{p'} e^{\log(t^{p'})^r}$, $0 < r \le 1/2$, then the factored pair $(\tilde{u}, \tilde{v})$ defined by (6.5) can be inserted between (u, v) and satisfies (6.6) with $\tilde{A}(t) \approx t^p e^{\log(t^p)^s}$, $\tilde{B}(t) \approx t^{p'} e^{\log(t^{p'})^s}$, whenever $0 < s < r$.*

(b) *If $A(t) = t^{rp}$, $B(t) = t^{rp'}$, $r > 1$, then the factored pair $(\tilde{u}, \tilde{v})$ defined by (6.5) can be inserted between (u, v) and satisfies (6.6) with $\tilde{A}(t) = t^{sp}$, and $\tilde{B}(t) = t^{sp'}$, whenever*

$$1 < s < \min\left(\frac{p'r}{r + p' - 1}, \frac{pr}{r + p - 1}\right).$$

Remark 6.12. For power bumps, since $r > 1$ we have that $1 < s < r$, but this condition further implies that s is bounded away from r.

It is clear from the proof of Theorem 6.9 (and so, by extension of Theorems 6.8 and 6.11) that in determining the bump conditions satisfied by $(\tilde{u}, \tilde{v})$, the bump on $\tilde{u}$ depends only on the bump on u and the bump on $\tilde{v}$ depends only on the bump on v. Therefore, we can combine these results to get a variety of factorization theorems for "mixed" bump conditions. We will not state a general result; details are left to the interested reader.

6.3 Inserting A_p weights

In [163], Neugebauer gave a necessary and sufficient condition on a pair of weights (u,v) for there to exist $w \in A_p$ such that $c_1 u \le w \le c_2 v$. As a corollary to our proofs of the factorization theorems we give a generalization of this result that allows us to insert A_∞ weights.

Theorem 6.13. *Given p, $1 < p < \infty$, suppose that (u,v) satisfy*

$$\|u^{1/p}\|_{A,Q}\|v^{-1/p}\|_{B,Q} \le K < \infty,$$

where A is a p-Young function and B is a p'-Young function.

(a) *If $A(t) = t^{rp}$, $r > 1$, and $\bar{B} \in B_p$, then there exists a weight w such that for every $q > p$, $w \in A_q$, and $c_1 u \le w \le c_2 v$.*

(b) *If $\bar{A} \in B_{p'}$ and $B(t) = t^{rp'}$, $r > 1$, then there exists a weight w such that for every $q < p$, $w^{1-p'} \in A_{q'}$, and $c_1 u \le w \le c_2 v$.*

(c) *If $A(t) = t^{rp}$, $B(t) = t^{rp'}$, $r > 1$, then there exists a weight w such that $w \in A_p$ and $c_1 u \le w \le c_2 v$.*

Proof. Theorem 6.13 is an immediate consequence of the Jones factorization theorem, Proposition 6.2, and the proofs of Theorems 6.8 and 6.11.

In Case (a) we have that $c_1 u \le \tilde{u} \le \tilde{v} \le c_2 v$, where

$$(\tilde{u},\tilde{v}) = \big(w_1(Mw_2)^{1-p}, (M_{\tilde{\Phi}}w_1)w_2^{1-p}\big),$$

and $\tilde{\Phi}(t) = t^s$ for some $s > 1$. If we let $w = M_{\tilde{\Phi}}w_1(Mw_2)^{1-p}$, then $c_1 u \le w \le c_2 v$. Further, by Proposition 6.2, $M_{\tilde{\Phi}}w_1 \in A_1$ and $(Mw_2)^\delta \in A_1$ for all $0 < \delta < 1$, so by the Jones factorization theorem, $w \in A_q$ for all $q > p$.

Cases (b) and (c) are proved in exactly the same fashion, with

$$(\tilde{u},\tilde{v}) = \big(w_1(M_{\tilde{\Psi}}w_2)^{1-p}, (Mw_1)w_2^{1-p}\big)$$

in Case (b) and

$$(\tilde{u},\tilde{v}) = \big(w_1(M_{\tilde{\Psi}}w_2)^{1-p}, (M_{\tilde{\Phi}}w_1)w_2^{1-p}\big)$$

in Case (c). □

Remark 6.14. There are several results for weights that satisfy power bump conditions whose proofs immediately extend to weights that satisfy exponential log bump conditions. As we will see in Chapter 9 below, we can use Theorem 6.13 to give very simple proofs of these results for power bumps. Therefore, it would be interesting to determine if it is also possible to insert A_∞ weights between weights that satisfy exponential log bumps.

6.4 Weights for fractional operators

In this section we prove factorization and reverse factorization theorems for the pairs of weights associated to the two-weight norm inequalities for fractional operators. More precisely, we will consider weights that satisfy conditions of the form

$$|Q|^{\alpha/n}\|u^{1/p}\|_{A,Q}\|v^{-1/p}\|_{B,Q} \leq K < \infty,$$

where A, B are Young functions and $0 < \alpha < n$. These weights generalize the A^{α}_{pq} weights discussed in Section 5.6, and we refer to them as fractional weights. Many of the proofs are nearly identical to those of the corresponding results in Sections 6.1 and 6.2, and so are omitted.

For simplicity we will only consider the fractional weights corresponding to diagonal inequalities from $L^p(v)$ into $L^p(u)$. Our results can be extended to the weights for off-diagonal inequalities; details are left to the reader. (See also Hernandez [102].)

Reverse factorization and factored weights

The most important difference between the results given in Section 6.1 and the analogous results for fractional weights is that the relationship between fractional A_1 weights and reverse factorization is more complicated. Given α, $0 < \alpha < n$, and a Young function Φ, the obvious definition for fractional A_1 weights is to take the pairs $(u, M_{\Phi,\alpha}u)$. These are a natural generalization of the weights associated with the weak $(1,1)$ inequality for M_α: when $\Phi(t) = t$, the pair $(u, M_\alpha u)$ governs the two-weight, weak $(1,1)$ inequality for M_α. (See (5.26).) We can then define the pairs $((M_{\Phi,\alpha}u)^{1-p}, u^{1-p})$ to be the corresponding dual fractional A_1 weights.

These weights are very closely connected to the one-weight A_p classes, just as the A_1 and dual A_1 weights are.

Proposition 6.15. *Given α, $0 < \alpha < n$, let Φ be a Young function such that $\Phi(t)/t^{n/\alpha}$ is quasi-decreasing and $\Phi(t)/t^{n/\alpha} \to 0$ as $t \to \infty$. Then given a weight u,*

(a) *$M_{\Phi,\alpha}u \in A_1$;*

(b) *if $1 < p < \infty$, $(M_{\Phi,\alpha}u)^{1-p} \in A_p \cap RH_\infty$.*

Proposition 6.15 follows at once from Theorem 5.43 and the Jones factorization theorem. Note that its conclusions are stronger than those in Proposition 6.2.

The fractional A_1 and dual A_1 weights immediately yield fractional factored weights: given a fractional A_p bump condition, we can construct a factored pair that satisfies it. The proof of the following result is identical to the proof of Theorem 6.4.

Theorem 6.16. *Given p, $1 < p < \infty$, and α, $0 < \alpha < n$, let A be a p-Young function and B a p'-Young function. Then given any weights w_1, w_2, the pair*

$$(\tilde{u}, \tilde{v}) = \big(w_1(M_{\Psi,\alpha}w_2)^{1-p}, (M_{\Phi,\alpha}w_1)w_2^{1-p}\big), \tag{6.8}$$

where $\Phi(t) = A(t^{1/p})$ and $\Psi(t) = B(t^{1/p'})$, satisfies

$$|Q|^{\alpha/n}\|\tilde{u}^{1/p}\|_{A,Q}\|\tilde{v}^{-1/p}\|_{B,Q} \le 1. \tag{6.9}$$

As an immediate corollary of Theorems 5.37 and 6.16 we get the factored weights that govern the two-weight norm inequalities for M_α.

Corollary 6.17. *Given p, $1 < p < \infty$, let $\bar{B} \in B_p$. For any α, $0 < \alpha < n$, and weights w_1, w_2, define the pair*

$$(\tilde{u}, \tilde{v}) = \big(w_1(M_{\Psi,\alpha}w_2)^{1-p}, (M_\alpha w_1)w_2^{1-p}\big),$$

where $\Psi(t) = B(t^{1/p'})$. Then M_α is bounded from $L^p(\tilde{v})$ into $L^p(\tilde{u})$.

Up to this point the results for fractional weights are exactly parallel to those for non-fractional weights. A key difference emerges when we consider the fractional A_1 weights that control the strong (p,p) inequalities for M_α. For $1 < p < n/\alpha$,

$$\int_{\mathbb{R}^n} M_\alpha f(x)^p u(x)\,dx \le C\int_{\mathbb{R}^n} |f(x)|^p M_{p\alpha}u(x)\,dx. \tag{6.10}$$

(This result is due to Sawyer [200]; also see Corollary 6.19 below.) In other words, the fractional A_1 weights depend on p. We cannot get these pairs from Corollary 6.17. We can for the Hardy-Littlewood maximal operator (i.e., when $\alpha = 0$) by taking $w_2 = 1$. We cannot do this in the fractional case since for any Young function Φ and any $\alpha > 0$, $M_{\Phi,\alpha}(1) = +\infty$.

Thus, we are led to consider the more general class of fractional A_1 weights consisting of the pairs $(u, M_{\Phi,p\alpha}u)$, $1 < p < n/\alpha$. Further, though M_α is not a linear operator, taking the formal dual of inequality (6.10) suggests that the dual fractional A_1 weights are of the form $((M_{\Psi,p'\alpha}u)^{1-p}, u^{1-p})$, $1 < p' < n/\alpha$. The proof of Theorem 6.16 is easily modified to show the following.

Theorem 6.18. *Given α, $0 < \alpha < n$, and p, $1 < p < n/\alpha$, let A be a p-Young function and define $\Phi(t) = A(t^{1/p})$. Then the pair of weights $(u, M_{\Phi,p\alpha}u)$ satisfies (6.9).*

Similarly, if $1 < p' < n/\alpha$, let B be a p'-Young function and define $\Psi(t) = B(t^{1/p'})$. Then the pair of weights $\big((M_{\Psi,p'\alpha}u)^{1-p}, u^{1-p}\big)$ also satisfies (6.9).

As a corollary to Theorems 5.37 and 6.18 we get the following. The result for dual fractional A_1 weights appears to be new.

Corollary 6.19. *Fix α, $0 < \alpha < n$ and p. Suppose one of the following is true:*

(a) $1 < p < n/\alpha$ *and* $(\tilde{u}, \tilde{v}) = (u, M_{p\alpha}u)$;

(b) $1 < p' < n/\alpha$, $\bar{B} \in B_p$, and $(\tilde{u}, \tilde{v}) = \big((M_{\Psi,p'\alpha}u)^{1-p}, u^{1-p}\big)$, where $\Psi(t) = B(t^{1/p'})$.

Then $M_\alpha : L^p(\tilde{v}) \to L^p(\tilde{u})$.

Remark 6.20. We can extend Theorems 6.16 and 6.18 to get factored weights of the form

$$(\tilde{u}, \tilde{v}) = \big(w_1(M_{\Psi,\beta}w_2)^{1-p}, (M_{\Phi,\gamma}w_1)w_2^{1-p}\big),$$

where $\gamma/p + \beta/p' = \alpha$. (If $\beta = 0$ or $\gamma = 0$ we take the associated maximal operator to be the non-fractional Orlicz maximal operator.) Details are left to the reader.

Factorization of weights

There are also factorization theorems for fractional weights: here we state the analogs of Theorems 6.8, 6.9 and 6.11. As was the case in Section 6.2, we get different results depending on the nature of the bump condition. For brevity we combine all our results into one theorem.

Theorem 6.21. *Given p, $1 < p < \infty$, and α, $0 < \alpha < n$, suppose that (u, v) satisfy*

$$|Q|^{\alpha/n}\|u^{1/p}\|_{A,Q}\|v^{-1/p}\|_{B,Q} \le K < \infty,$$

where A is a p-Young function and B is a p'-Young function.

(a) *If $\bar{A} \in B_{p'}$ and $\bar{B} \in B_p$, then there exist w_1, w_2, such that the factored pair $(\tilde{u}, \tilde{v}) = \big(w_1(M_\alpha w_2)^{1-p}, (M_\alpha w_1)w_2^{1-p}\big)$ satisfies $u \le c_1\tilde{u}$ and $\tilde{v} \le c_2 v$ and*

$$|Q|^{\alpha/n}\|\tilde{u}^{1/p}\|_{p,Q}\|\tilde{v}^{-1/p}\|_{p',Q} \le K < \infty.$$

In other words, $(\tilde{u}, \tilde{v}) \in A^\alpha_{pp}$, so $M_\alpha : L^p(\tilde{v}) \to L^{p,\infty}(\tilde{u})$ and $M_\alpha : L^{p'}(\tilde{u}^{1-p'}) \to L^{p',\infty}(\tilde{v}^{1-p'})$.

(b) *If $A(t) = t^p\log(e+t)^{p-1+\delta}$, $B(t) = t^{p'}\log(e+t)^{p'-1+\delta}$, $\delta > 0$, then there exist w_1, w_2 such that the factored pair $(\tilde{u}, \tilde{v}) = \big(w_1(M_{\tilde{\Psi},\alpha}w_2)^{1-p}, (M_{\tilde{\Phi},\alpha}w_1)w_2^{1-p}\big)$, where $\tilde{A}(t) = t^p\log(e+t)^a$, $0 < a < \delta/p$, $\tilde{B}(t) = t^{p'}(e+t)^b$, $0 < b < \delta/p'$, and $\tilde{\Phi}(t) = \tilde{A}(t^{1/p})$, $\tilde{\Psi}(t) = \tilde{B}(t^{1/p'})$, satisfies $u \le c_1\tilde{u}$, $\tilde{v} \le c_2 v$, and*

$$|Q|^{\alpha/n}\|\tilde{u}^{1/p}\|_{\tilde{A},Q}\|\tilde{v}^{-1/p}\|_{\tilde{B},Q} \le K < \infty.$$

As in Case (a), $(\tilde{u}, \tilde{v}) \in A^\alpha_{pp}$.

(c) *If $A(t) \approx t^p e^{\log(t^p)^r}$, $B(t) \approx t^{p'}e^{\log(t^{p'})^r}$, $0 < r \le 1/2$, then the factored pair $(\tilde{u}, \tilde{v}) = \big(w_1(M_{\tilde{\Psi},\alpha}w_2)^{1-p}, (M_{\tilde{\Phi},\alpha}w_1)w_2^{1-p}\big)$, where $\tilde{A}(t) \approx t^p e^{\log(t^p)^s}$, $\tilde{B}(t) \approx t^{p'}e^{\log(t^{p'})^s}$, $0 < s < r$, $\tilde{\Phi}(t) = \tilde{A}(t^{1/p})$, $\tilde{\Psi}(t) = \tilde{B}(t^{1/p'})$, satisfies $u \le c_1\tilde{u}$, $\tilde{v} \le c_2 v$, and*

$$|Q|^{\alpha/n}\|\tilde{u}^{1/p}\|_{\tilde{A},Q}\|\tilde{v}^{-1/p}\|_{\tilde{B},Q} \le K < \infty.$$

(d) *If* $A(t) = t^{rp}$, $B(t) = t^{rp'}$, $r > 1$, *then the factored pair*

$$(\tilde{u}, \tilde{v}) = \big(w_1(M_{\tilde{\Psi},\alpha}w_2)^{1-p}, (M_{\tilde{\Phi},\alpha}w_1)w_2^{1-p}\big),$$

where $\tilde{A}(t) = t^{sp}$, *and* $\tilde{B}(t) = t^{sp'}$, $1 < s < \min(\frac{pr}{r+p-1}, \frac{p'r}{r+p'-1})$, $\tilde{\Phi}(t) = \tilde{A}(t^{1/p})$, $\tilde{\Psi}(t) = \tilde{B}(t^{1/p'})$, *satisfies* $u \le c_1\tilde{u}$, $\tilde{v} \le c_2 v$, *and*

$$|Q|^{\alpha/n}\|\tilde{u}^{1/p}\|_{\tilde{A},Q}\|\tilde{v}^{-1/p}\|_{\tilde{B},Q} \le K < \infty.$$

In both Cases (c) *and* (d) *we have that* $M_\alpha : L^p(\tilde{v}) \to L^p(\tilde{u})$ *and* $M_\alpha : L^{p'}(\tilde{u}^{1-p'}) \to L^{p'}(\tilde{v}^{1-p'})$.

Remark 6.22. As in the non-fractional case, we can have mixed bump conditions. As before, we will not state a general theorem but will make use of specific conditions as needed.

Finally, note that one difference between Theorem 6.21 and the results in Section 6.2 is that here we do not have the inequality $\tilde{u} \le \tilde{v}$. We proved this for non-fractional weights using the fact that $u \le M_\Phi u$, but when $\alpha > 0$ it is not the case that $u \le M_{\Phi,\alpha}u$. As a consequence there is no analog of Theorem 6.13 in the fractional case.

Chapter 7

Two-Weight Extrapolation

In this chapter we give our generalization of the Rubio de Francia extrapolation theorem to the two-weight setting. We can summarize our results as follows: given an operator T, suppose that for some p_0, $1 < p_0 < \infty$, and any pair of weights (u_0, v_0) that satisfy the A_p bump condition

$$\|u_0^{1/p_0}\|_{A_0,Q}\|v_0^{-1/p_0}\|_{B_0,Q} \leq K_0 < \infty,$$

we have the norm inequality

$$\int_{\mathbb{R}^n} |Tf(x)|^{p_0} u_0(x)\, dx \leq C \int_{\mathbb{R}^n} |f(x)|^{p_0} v_0(x)\, dx. \tag{7.1}$$

Then for all p, $1 < p < \infty$, we deduce conditions on Young functions A and B such that if (u, v) satisfy

$$\|u^{1/p}\|_{A,Q}\|v^{-1/p_0}\|_{B,Q} \leq K < \infty,$$

then

$$\int_{\mathbb{R}^n} |Tf(x)|^{p} u(x)\, dx \leq C \int_{\mathbb{R}^n} |f(x)|^{p} v(x)\, dx.$$

The conditions we give on the Young functions A and B are stated in terms of A_0 and B_0, and they are such that the various specific classes of Orlicz bumps that we considered in Chapter 5—log bumps, exponential log bumps, power bumps—are preserved. For example, if (7.1) holds when the pair (u_0, v_0) satisfies a log bump condition, then it holds for all p and all weights that satisfy a log bump condition that depends on p.

Unfortunately, in many cases (particularly for log bumps which are important in applications) the new bump conditions are not gotten simply by rescaling A_0 and B_0 with the changes of variables $t \mapsto t^{p/p_0}$, $t \mapsto t^{p'/p_0'}$. The exact changes required depend on the kind of bump we begin with initially and whether $p > p_0$

or $p < p_0$. Further, we lose information: by extrapolation we can only deduce weighted norm inequalities for weights that satisfy stronger bump conditions than those we began with. This phenomenon is analogous to the loss of information in the factorization theorems in Chapter 6.

Our proofs of two-weight extrapolation use roughly the same techniques as the proofs in the one-weight case in Chapters 2 and 3, but there are significant differences as well. In particular, we divide our proofs into two cases: $p > p_0$ and $p < p_0$. In doing so we sacrifice one of the nice features of our proof in the one-weight setting; our proof is reminiscent of the one-weight extrapolation proof due to García-Cuerva [83]. We take this approach since the conditions we derive for A and B are simpler if we divide the proof into two cases. The bumps on u and and v behave differently, with the bump on u behaving worse when $p < p_0$, and the bump on v behaving worse when $p > p_0$. This behavior is easier to see and understand when we separate the proof into two cases.

The two case approach is not necessary, and in an appendix to this chapter we give a proof that does not require two cases. Nevertheless, this proof still differs significantly from the proof we give in the one-weight case in Chapter 3. We will describe how that proof can be adapted to the two-weight setting, and indicate the shortcomings which led us to adopt our main proofs.

Two-weight extrapolation has been considered by a number of authors. Segovia and Torrea [209, 210] considered the special case of weights u and $v = \omega u$, where ω is a non-negative, measurable function and and $u, v \in A_p$. The more general case was considered by Bloom [13], Neugebauer [164], Hernández [102], and Ruiz and Torrea [198]. However, the approach used by these authors was to assume that the initial inequality (7.1) holds for all pairs (u, v) such that $M : L^p(v) \to L^p(u)$. As we noted in Chapter 1, this hypothesis is too strong to be useful given the current state of knowledge. However, Neugebauer [164] used an approach similar to ours to consider extrapolation for weak type inequalities that satisfy the two-weight A_p condition.

This chapter is organized as follows. In Sections 7.1 and 7.2 we state and prove our main two-weight extrapolation theorem. The next three sections give variations of this result. In Section 7.3 we give a generalization of Neugebauer's weak type extrapolation theorem. In Section 7.4 we prove an extrapolation theorem for factored weights, and in Section 7.5 we give an extrapolation theorem for fractional weights. As we did in Section 6.4, we will restrict ourselves to (p, p) inequalities and not consider off-diagonal extrapolation. For this more general setting, see Hernández [102]. In Section 7.6 we give a proof of our main extrapolation theorem that does not treat two cases depending on the size of p. In Chapter 8 we will give further variations related to endpoint estimates, rescaling and the one-weight, A_∞ extrapolation theorem discussed in Chapter 3.

Remark 7.1. All of the results in this chapter hold if we replace all the cubes and maximal operators by dyadic cubes and the corresponding dyadic maximal operators.

7.1 Two-weight extrapolation

In this section we state our main two-weight extrapolation theorem. As we discussed in Chapter 2, our approach to extrapolation does not require us to work directly with an operator. Therefore, in the two-weight setting we will use a convention analogous to the one established in Chapter 3. Hereafter, $\mathcal{F}$ will denote a family of ordered pairs of non-negative, measurable functions (f,g) that are not identically zero. To avoid technical complications we will assume that $f,\, g \in \bigcap_{p>1} L^p$. Given $p>0$ and a pair of weights (u,v), if we say that

$$\int_{\mathbb{R}^n} f(x)^p\, u(x)\, dx \le C \int_{\mathbb{R}^n} g(x)^p\, v(x)\, dx, \qquad (f,g)\in\mathcal{F},$$

we mean that this inequality holds for all pairs $(f,g)\in\mathcal{F}$ such that $f\in L^p(u)$, and that the constant depends only upon p and the constant in the A_p bump condition. Since all such inequalities are trivial if the right-hand side is infinite, we will also assume that $g\in L^p(v)$. In Chapters 9 and 10 we discuss how to work with these assumptions in specific applications.

Theorem 7.2. *Given p_0, $1<p_0<\infty$, a p_0-Young function A_0, and a p_0'-Young function B_0, suppose that*

$$\int_{\mathbb{R}^n} f(x)^{p_0}\, u_0(x)\, dx \le C \int_{\mathbb{R}^n} g(x)^{p_0}\, v_0(x)\, dx, \qquad (f,g)\in\mathcal{F}, \tag{7.2}$$

for every pair of weights (u_0,v_0) that satisfy

$$\|u_0^{1/p_0}\|_{A_0,Q}\|v_0^{-1/p_0}\|_{B_0,Q} \le K_0 < \infty. \tag{7.3}$$

Then for all p, $1<p<\infty$,

$$\int_{\mathbb{R}^n} f(x)^{p}\, u(x)\, dx \le C \int_{\mathbb{R}^n} g(x)^{p}\, v(x)\, dx, \qquad (f,g)\in\mathcal{F}, \tag{7.4}$$

whenever (u,v) satisfy

$$\|u^{1/p}\|_{A,Q}\|v^{-1/p}\|_{B,Q} \le K < \infty, \tag{7.5}$$

where A is a p-Young function and B is a p'-Young function such that:

(a) *if $p>p_0$, then $\bar{A}\in B_{p'}$, $\bar{A}(t^{1/p'}) \preceq \bar{A}_0(t^{1/p_0'})$, and $B_0(t^{1/p_0'}) \preceq B(t^{1/p'})$;*

(b) *if $p<p_0$, then $\bar{B}\in B_p$, $A_0(t^{1/p_0}) \preceq A(t^{1/p})$, and $\bar{B}(t^{1/p}) \preceq \bar{B}_0(t^{1/p_0})$.*

Remark 7.3. When we say that a pair of weights satisfies a condition such as (7.3) or (7.5), we are assuming that this inequality is true for all cubes Q in $\mathbb{R}^n$.

Remark 7.4. When $p>p_0$, the condition $\bar{A}(t^{1/p'}) \preceq \bar{A}_0(t^{1/p_0'})$ is equivalent to $\bar{A}_0^{-1}(t)^{p_0'} \le c\bar{A}^{-1}(t)^{p'}$; in turn, by (5.4) this is equivalent to $\big(A^{-1}(t)/t\big)^{p'} \le c\big(A_0^{-1}(t)/t\big)^{p_0'}$. Similarly, when $p<p_0$, the condition $\bar{B}(t^{1/p}) \preceq \bar{B}_0(t^{1/p_0})$ is equivalent to both $(B^{-1}(t)/t)^p \le c(B_0^{-1}(t)/t)^{p_0}$ and $\bar{B}_0^{-1}(t)^{p_0} \le c\bar{B}^{-1}(t)^p$.

Extrapolation and families of Orlicz bumps

We will prove Theorem 7.2 in the next section. First, we will consider it from a different perspective. As stated, this result answers the following question: given a fixed pair A_0, B_0, of Young functions for which the bump condition (7.3) implies the weighted L^{p_0} inequality, for which pairs A, B does the bump condition imply that the weighted L^p norm inequality holds? A different but closely related question is: suppose the hypothesis is true for all pairs A_0, B_0 in some class $\mathcal{C}_{p_0}$; can we describe the class $\mathcal{C}_p$ of pairs A, B such that the conclusion holds? In particular, we are interested in determining the class $\mathcal{C}_p$ given that $\mathcal{C}_{p_0}$ is one of the families of Orlicz bumps introduced in Section 5.4. Ideally, each such class would be preserved under extrapolation, but this is not always the case.

We will consider this question in detail when $p > p_0$; by the symmetry in the statement of Theorem 7.2 the case $p < p_0$ is gotten by exchanging the roles of A and B. Below, for each family of Orlicz bumps we consider, we will fix the class $\mathcal{C}_{p_0}$ and a pair A_0, B_0 in it, and then determine the resulting pairs A, B. To do so we will simply define $B(t) = B_0(t^{p'/p_0'})$ and attempt to define A by $\bar{A}(t) = \bar{A}_0(t^{p'/p_0'})$. This is not always possible, so we will take $\bar{A}$ as large as possible subject to the condition $\bar{A}(t) \preceq \bar{A}_0(t^{p'/p_0'})$ and $\bar{A} \in B_{p'}$. We will then consider whether an arbitrary pair A, B (in the same family as the pair obtained above) can be gotten in this fashion. This will determine the class $\mathcal{C}_p$.

Our results, for both the case $p > p_0$ and the case $1 < p < p_0$, are summarized in Table 7.1 below.

No bump condition

Let $\mathcal{C}_{p_0}$ consist of the pair $A_0(t) = t^{p_0}$, $B_0(t) = t^{p_0'}$ (i.e., the two-weight A_{p_0} condition). Then we immediately have $B(t) = t^{p'}$ and (by the definition of the complementary function) we would like to take $\bar{A}(t) = t^{p'}$. However, $t^{p'}$ does not satisfy the $B_{p'}$ condition, so we cannot define A in this way. However, given any A such that $\bar{A} \in B_{p'}$, then $\bar{A}(t^{1/p'}) \preceq t = \bar{A}_0(t^{1/p_0'})$. Therefore, we can take $\mathcal{C}_p$ to consist of the pairs A, B such that $\bar{A} \in B_{p'}$ and $B(t) = t^{p'}$.

Remark 7.5. Equivalently, we see that the A_p class is not preserved by extrapolation using Theorem 7.2. However, see the extrapolation theorem of Neugebauer [164] (a special case of Theorem 7.12 below).

B_p bumps

Let $\mathcal{C}_{p_0}$ consist of all pairs A_0, B_0 such that $\bar{A}_0 \in B_{p_0'}$ and $\bar{B}_0 \in B_{p_0}$. Fix one such pair. By a change of variables in the definition of the B_p condition, $\bar{A} \in B_{p'}$ if and only if $\bar{A}_0 \in B_{p_0'}$. The same, however, is not necessarily true for B. If we further assume that $\bar{B}_0$ is doubling, then by Lemma 5.25, $\bar{B} \in B_p$. However, the converse is not true in general: given B such that $\bar{B} \in B_p$, then $\bar{B}_0(t) = \bar{B}(t^{p_0'/p'})$ need not be in B_{p_0}. This is shown by the log bumps discussed below; another example can be gotten from the log-log bumps—details are left to the reader. Therefore, in this case the class $\mathcal{C}_p$ consists of all pairs A, B such that $\bar{A} \in B_{p'}$ and $B(t) = B_0(t^{p'/p_0'})$, where $\bar{B}_0 \in B_{p_0}$.

Log bumps

We consider one particular family of log bumps that is important for applications; the general case can be treated in exactly the same way. Let $\mathcal{C}_{p_0}$ consist of the pair A_0, B_0 such that

$$A_0(t) = t^{p_0} \log(e+t)^{p_0-1+\delta_0}, \qquad B_0(t) = t^{p_0'} \log(e+t)^{p_0'-1+\delta_0},$$

where $\delta_0 > 0$ is arbitrary. We immediately have that

$$B(t) \approx t^{p'} \log(e+t)^{p_0'-1+\delta_0}.$$

From the formula for the complementary function of a log bump (see Section 5.4), we have that

$$\bar{A}(t) \approx \frac{t^{p'}}{\log(e+t)^{1+(p_0'-1)\delta_0}},$$

so

$$A(t) \approx t^p \log(e+t)^{p-1+\delta}, \qquad \delta = \frac{\delta_0(p-1)}{p_0-1}.$$

Given an arbitrary pair A, B,

$$A(t) = t^p \log(e+t)^{p-1+\delta}, \qquad B(t) = t^{p'} \log(e+t)^{p'-1+\delta}, \tag{7.6}$$

is it possible to get it in this fashion? Since we can choose $\delta_0 > 0$ as close to 0 as desired, given any value of δ we can choose δ_0 so that $\delta = \delta_0(p-1)/(p_0-1)$. Therefore, we can take A to be any log bump of this form. However, since $p_0' > p'$, we cannot get an arbitrary $\delta > 0$ for B: we have to take $\delta > p_0' - p'$. Therefore, the class $\mathcal{C}_p$ consists of all pairs A, B of the form

$$A(t) = t^p \log(e+t)^{p-1+\delta}, \qquad B(t) = t^{p'} \log(e+t)^{p_0'-1+\delta},$$

where $\delta > 0$ is arbitrary.

Remark 7.6. The fact that B_p bumps and log bumps are not perfectly preserved under extrapolation means that Theorem 7.2 does not necessarily yield sharp results. To see this, consider the Hardy-Littlewood maximal operator. By Theorem 5.14 we have that for all p, $1 < p < \infty$,

$$\int_{\mathbb{R}^n} Mf(x)^p u(x)\,dx \le C \int_{\mathbb{R}^n} |f(x)|^p v(x)\,dx,$$

provided

$$\|u^{1/p}\|_{p,Q} \|v^{-1/p}\|_{B,Q} \le K < \infty,$$

where $\bar{B} \in B_p$. Fix a value p_0; if we begin from this inequality and extrapolate to values $p < p_0$, then we recapture this result. However, if we extrapolate to $p > p_0$, we lose information: we have to take a larger Young function A since we need $\bar{A} \in B_{p'}$. Further, we cannot take B to be any B_p bump: for example, in the scale of log bumps we would have to take $B(t) = t^{p'} \log(e+t)^{p_0'-1+\delta}$.

In Chapters 9 and 10 we will consider the question of when two-weight extrapolation yields sharp results for specific operators.

Remark 7.7. Based on numerous examples (see Chapters 9 and 10 below) we originally conjectured that we could take the pair A, B in the general form given by (7.6). However, the condition on B (and the parallel condition on A when $p < p_0$) appear to be intrinsic to our proof. It is an open question whether Theorem 7.2 can be sharpened to yield these pairs. Such a result would have important consequences; see for example, the discussion of singular integrals in Chapter 9.

Exponential log bumps

Let $\mathcal{C}_{p_0}$ consist of all pairs A_0, B_0 such that for $t > 0$ large,

$$A_0(t) \approx t^{p_0} e^{\log(t^{p_0})^r}, \qquad B_0(t) \approx t^{p_0'} e^{\log(t^{p_0'})^r},$$

where $0 < r < 1/2$ is arbitrary. Then we immediately have that

$$B(t) \approx t^{p'} e^{\log(t^{p'})^r}.$$

Further, if we choose s such that $r < s < 1/2$, then we have that

$$\bar{A}_0(t) \approx t^{p_0'} e^{-(p_0'/p_0)\log(t^{p_0'})^r} \succeq t^{p_0'} e^{-(p'/p)\log(t^{p_0'})^s}$$

(see Section 5.4). Therefore, we can take

$$A(t) = t^p e^{\log(t^p)^s}.$$

Clearly, given an arbitrary pair A, B,

$$A(t) \approx t^p e^{\log(t^p)^s}, \qquad B(t) \approx t^{p'} e^{\log(t^{p'})^s}, \quad 0 < s < 1/2,$$

since we can take $0 < r < s$, we can get this pair in this fashion. Hence, we can take the class $\mathcal{C}_p$ to consist of all such pairs.

Power bumps

Let $\mathcal{C}_{p_0}$ consist of all pairs A_0, B_0 such that $A_0(t) = t^{rp_0}$, $B_0(t) = t^{rp_0'}$, $r > 1$. (In practice, we think of r as very close to 1.) Then we have that $B(t) = t^{rp'}$, and by the definition of the complementary function we can take $A(t) = t^{sp}$, where

$$s = \frac{1}{p}\left(\frac{(r\,p_0)'\,p'}{p_0'}\right)' > r > 1.$$

$\mathcal{C}_{p_0}$: Initial Bumps	$\mathcal{C}_p$: $p > p_0$	$\mathcal{C}_p$: $p < p_0$
$A_0(t) = t^{p_0}$ $B_0(t) = t^{p_0'}$	$\bar{A} \in B_{p'}$ $B(t) = t^{p'}$	$A(t) = t^p$ $\bar{B} \in B_p$
$\bar{A}_0 \in B_{p_0'}$ $\bar{B}_0 \in B_{p_0}$	$\bar{A} \in B_{p'}$ $B(t) = B_0(t^{p'/p_0'})$	$A(t) = A_0(t^{p/p_0})$ $\bar{B} \in B_p$
$A_0(t) = t^{p_0} \log(e+t)^{p_0-1+\delta_0}$ $B_0(t) = t^{p_0'} \log(e+t)^{p_0'-1+\delta_0}$	$A(t) = t^p \log(e+t)^{p-1+\delta}$ $B(t) = t^{p'} \log(e+t)^{p_0'-1+\delta}$	$A(t) = t^p \log(e+t)^{p_0-1+\delta_0}$ $B(t) = t^{p'} \log(e+t)^{p'-1+\delta}$
$A_0(t) \approx t^{p_0} e^{\log(t^{p_0})^r}$ $B_0(t) \approx t^{p_0'} e^{\log(t^{p_0'})^r}$ $0 < r < 1/2$	$A(t) \approx t^p e^{\log(t^p)^s}$ $B(t) \approx t^{p'} e^{\log(t^{p'})^s}$ $0 < s < 1/2$	$A(t) \approx t^p e^{\log(t^{p_0})^s}$ $B(t) \approx t^{p'} e^{\log(t^{p_0'})^s}$ $0 < s < 1/2$
$A_0(t) = t^{rp_0}$ $B_0(t) = t^{rp_0'}$ $r > 1$	$A(t) = t^{sp}$ $B(t) = t^{sp'}$ $s > 1$	$A(t) = t^{sp}$ $B(t) = t^{sp'}$ $s > 1$

Table 7.1: Bump conditions via extrapolation

Hence, $\bar{A} \in B_{p'}$. Since given any $s > 1$ we can find r, $1 < r < s$ such that this equality holds, it follows that we can realize any pair of power bumps in this fashion. Hence, we can take $\mathcal{C}_p$ to consist of all pairs A, B such that $A(t) = t^{sp}$, $B(t) = t^{sp'}$, $s > 1$.

Remark 7.8. As was the case for factorization results in Chapter 6, we could also consider "mixed" bump conditions where A_0 and B_0 are drawn from different families of Orlicz bumps. Since the bump on the left of an A_p bump condition is independent of the bump on the right, such results follow immediately from what we have done above; details are left to the interested reader.

7.2 Proof of two-weight extrapolation

In this section we prove Theorem 7.2. As we noted above the proof is divided into two cases, depending on whether $p > p_0$ or $p < p_0$. Before giving the proof, however, we first make a reduction we will use not only in this proof but in subsequent proofs as well.

As we stated in Chapter 5, whenever we assume that (u, v) satisfy (7.5), we also assume that u, $v^{-1} < \infty$ a.e. However, it is possible that $u = 0$ and/or $v = \infty$ on sets of positive measure. If this were the case, the proof below would have technical problems since u^{-1} and v appear in it. To avoid this we would like to assume that u and v are bounded and bounded away from 0. The following argument shows that we can always do this.

Given $0 < \epsilon < 1 < N$, let $u_N = \min(u, N)$, $u_{\epsilon,N} = \max(u_N, \epsilon)$, and define v_N and $v_{\epsilon,N}$ analogously. Then $u_{\epsilon,N}$, $v_{\epsilon,N}$ are bounded and bounded away from 0, and $(u_{\epsilon,N}, v_{\epsilon,N})$ satisfy (7.5) with constant $K + 2$:

$$\begin{aligned}
&\|u_{\epsilon,N}^{1/p}\|_{A,Q} \|v_{\epsilon,N}^{-1/p}\|_{B,Q} \\
&\qquad \le \|u_N^{1/p}\|_{A,Q} \|v_{\epsilon,N}^{-1/p}\|_{B,Q} + \|\epsilon^{1/p}\|_{A,Q} \|v_{\epsilon,N}^{-1/p}\|_{B,Q} \\
&\qquad \le \|u_N^{1/p}\|_{A,Q} \|v_N^{-1/p}\|_{B,Q} + 1 \\
&\qquad \le \|u_N^{1/p}\|_{A,Q} \|v^{-1/p}\|_{B,Q} + \|u_N^{1/p}\|_{A,Q} \|N^{-1/p}\|_{B,Q} + 1 \\
&\qquad \le \|u^{1/p}\|_{A,Q} \|v^{-1/p}\|_{B,Q} + 2 \\
&\qquad \le K + 2.
\end{aligned}$$

Therefore, if in our proofs we replace (u, v) by $(u_{\epsilon,N}, v_{\epsilon,N})$, the resulting constants will not depend on ϵ, N. In particular, we will be able to prove that (7.4) holds for the pair $(u_{\epsilon,N}, v_{\epsilon,N})$ with C independent of ϵ, N. Given this we can then derive (7.4) for the weights (u, v). Fix $(f, g) \in \mathcal{F}$; then by Fatou's lemma,

$$\int_{\mathbb{R}^n} f(x)^p\, u(x)\, dx$$
$$= \int_{\mathbb{R}^n} \lim_{\substack{\epsilon\to 0\\ N\to\infty}} f(x)^p\, u_{\epsilon,N}(x)\, dx \le \liminf_{\substack{\epsilon\to 0\\ N\to\infty}} \int_{\mathbb{R}^n} f(x)^p\, u_{\epsilon,N}(x)\, dx.$$

For every ϵ and N the integral on the right-hand side is finite: since $f \in \bigcap_{p>1} L^p$,

$$\int_{\mathbb{R}^n} f(x)^p\, u_{\epsilon,N}(x)\, dx \le \int_{\mathbb{R}^n} f(x)^p\, u(x)\, dx + \epsilon \int_{\mathbb{R}^n} f(x)^p\, dx < \infty.$$

Therefore, we can use (7.4) with the pair of weights $(u_{\epsilon,N}, v_{\epsilon,N})$ and the fact that $g \in \bigcap_{p>1} L^p$ to conclude that

$$\begin{aligned}\int_{\mathbb{R}^n} f(x)^p\, u(x)\, dx &\le \liminf_{\substack{\epsilon\to 0\\ N\to\infty}} \int_{\mathbb{R}^n} f(x)^p\, u_{\epsilon,N}(x)\, dx \\ &\le C \liminf_{\substack{\epsilon\to 0\\ N\to\infty}} \int_{\mathbb{R}^n} g(x)^p\, v_{\epsilon,N}(x)\, dx \\ &\le C \int_{\mathbb{R}^n} g(x)^p\, v(x)\, dx + C \liminf_{\epsilon\to 0} \epsilon \int_{\{v\le\epsilon\}} g(x)^p\, dx \\ &= C \int_{\mathbb{R}^n} g(x)^p\, v(x)\, dx.\end{aligned}$$

Proof of Theorem 7.2: $p > p_0$

Fix $(f, g) \in \mathcal{F}$ and fix (u, v) satisfying (7.5). Define

$$f_0 = \frac{f^{p-1}\, u}{\|f^{p-1}\, u\|_{L^{p'}(u^{1-p'})}} = \frac{f^{p-1}\, u}{\|f\|_{L^p(u)}^{p-1}}, \qquad F_0 = M f_0.$$

It is immediate that $\|f_0\|_{L^{p'}(u^{1-p'})} = 1$.

We claim that $\|F_0\|_{L^{p'}(v^{1-p'})} \le C$, where C depends only on u, v and p. By assumption $\bar{A} \in B_{p'}$; further, since B is a p'-Young function and the pair (u, v) satisfies (7.5), we have that

$$\|(v^{1-p'})^{1/p'}\|_{p',Q} \|(u^{1-p'})^{-1/p'}\|_{A,Q}$$
$$= \|v^{-1/p}\|_{p',Q} \|u^{1/p}\|_{A,Q} \le C \|u^{1/p}\|_{A,Q} \|v^{-1/p}\|_{B,Q} \le K.$$

Therefore, by Theorem 5.14, $M : L^{p'}(u^{1-p'}) \to L^{p'}(v^{1-p'})$, and so $\|F_0\|_{L^{p'}(v^{1-p'})} \le C\|f_0\|_{L^{p'}(u^{1-p'})} = C$.

Now let

$$s = \frac{(p/p_0)'}{p'} = \frac{p-1}{p-p_0} > 1, \qquad s' = \frac{p-1}{p_0-1},$$

and define a new pair of weights

$$(U_0, V_0) = (f_0^{1/s}\, u^{1/s'}, F_0^{1/s}\, v^{1/s'}).$$

Assume for the moment that the pair (U_0, V_0) satisfies (7.3) with constant independent of f_0. Then

$$\begin{aligned}\|f\|_{L^p(u)}^{p_0} &= \|f\|_{L^p(u)}^{-(p-p_0)} \int_{\mathbb{R}^n} f(x)^{p_0}\, f(x)^{p-p_0}\, u(x)^{1/s}\, u(x)^{1/s'}\, dx \\ &= \int_{\mathbb{R}^n} f(x)^{p_0}\, f_0(x)^{1/s}\, u(x)^{1/s'}\, dx = \int_{\mathbb{R}^n} f(x)^{p_0}\, U_0(x)\, dx.\end{aligned}$$

Since $f \in L^p(u)$, the right-hand term is finite, so we can apply (7.2) and Hölder's inequality with exponent $p/p_0 > 1$ to conclude

$$\begin{aligned}\|f\|_{L^p(u)}^{p_0} &= \int_{\mathbb{R}^n} f(x)^{p_0}\, U_0(x)\, dx \\ &\le C \int_{\mathbb{R}^n} g(x)^{p_0}\, V_0(x)\, dx \\ &= C \int_{\mathbb{R}^n} g(x)^{p_0}\, F_0(x)^{1/s} v(x)^{-1/s} v(x)\, dx \\ &\le C \|g\|_{L^p(v)}^{p_0} \left(\int_{\mathbb{R}^n} F_0(x)^{p'}\, v(x)^{1-p'}\, dx \right)^{\frac{1}{(p/p_0)'}} \\ &\le C \|g\|_{L^p(v)}^{p_0}.\end{aligned}$$

To complete the proof we need to prove that the pair (U_0, V_0) satisfies (7.3). We begin by proving that

$$\|h_1 h_2\|_{A_0,Q} \le C \|h_1\|_{sp_0,Q}\, \|h_2^{p_0'/p'}\|_{A,Q}^{p'/p_0'}. \tag{7.7}$$

This inequality is a consequence of the generalized Hölder's inequality (5.6). A short computation shows that $sp_0 = (p_0'/p')'$, so let $D(t) = t^{sp_0} = t^{(p_0'/p')'}$ and $C(t) = A(t^{p_0'/p'})$. Both of these are Young functions since $p_0' > p'$. By our assumption on A and A_0 (see Remark 7.4),

$$D^{-1}(t) C^{-1}(t) = t^{\frac{1}{(p_0'/p')'}} \cdot A^{-1}(t)^{\frac{1}{p_0'/p'}} \le c A_0^{-1}(t).$$

Therefore,

$$\|h_1 h_2\|_{A_0,Q} \le C \|h_1\|_{D,Q}\, \|h_2\|_{C,Q} = \|h_1\|_{sp_0,Q}\, \|h_2^{p_0'/p'}\|_{A,Q}^{p'/p_0'}.$$

Given inequality (7.7), we have that

$$\|U_0^{1/p_0}\|_{A_0,Q} = \|f_0^{1/(sp_0)}\, u^{1/(s'p_0)}\|_{A_0,Q}$$
$$\leq C\|f_0^{1/(sp_0)}\|_{sp_0,Q}\|(u^{1/(s'p_0)})^{p_0'/p'}\|_{A,Q}^{p'/p_0'} = \|f_0\|_{1,Q}^{1/(sp_0)}\|u^{1/p}\|_{A,Q}^{p'/p_0'}. \quad (7.8)$$

Since $f \in \bigcap_{p>1} L^p$ and since we may assume that u is bounded, $f_0 \in L^1_{\mathrm{loc}}$, so the right-hand side is finite. Further, we can argue as we did in the proof of the reverse factorization theorem (Theorem 6.4) and then rescale, using the fact that $B_0(t^{1/p_0'}) \preceq B(t^{1/p'})$, to get

$$\|V_0^{-1/p_0}\|_{B_0,Q} = \|Mf_0^{-1/(sp_0)}\, v^{-1/(s'p_0)}\|_{B_0,Q}$$
$$\leq C\|f_0\|_{1,Q}^{-1/(sp_0)}\|(v^{-p'/p})^{1/p_0'}\|_{B_0,Q} \leq C\|f_0\|_{1,Q}^{-1/(sp_0)}\|v^{-1/p}\|_{B,Q}^{p'/p_0'}.$$

Combining these two inequalities we see that since the pair (u,v) satisfies (7.5),

$$\left\|U_0^{1/p_0}\right\|_{A_0,Q}\left\|V_0^{-1/p_0}\right\|_{B_0,Q} \leq C\left\|u^{1/p}\right\|_{A,Q}^{p'/p_0'}\left\|v^{-1/p}\right\|_{B,Q}^{p'/p_0'} \leq K^{p'/p_0'}.$$

Thus the pair (U_0, V_0) satisfies (7.3) and our proof is complete. □

Remark 7.9. We can also prove Theorem 7.2 when $p > p_0$ using duality. Fix a non-negative function $f_0 \in L^{(p/p_0)'}(u)$ with norm 1 and define $F_0 = (M(f_0^s\, u)v^{-1})^{1/s}$. Then $\|F_0\|_{L^{(p/p_0)'}(v)} \leq C\,\|f_0\|_{L^{(p/p_0)'}(u)} = C$. The pair of weights $(U_0, V_0) = (f_0\, u, F_0\, v)$ satisfies (7.3) with constant independent of f_0. Hence,

$$\|f\|_{L^{p_0}(f_0 u)} = \|f\|_{L^{p_0}(U_0)} \leq C\|g\|_{L^{p_0}(V_0)} = \|g\|_{L^{p_0}(F_0 v)}$$
$$\leq \|g\|_{L^p(v)}\,\|F_0\|_{L^{(p/p_0)'}(v)}^{1/p_0} \leq C\|g\|_{L^p(v)}.$$

If we take the supremum over all such f_0 we get the desired result.

It is not difficult to see that these proofs are equivalent and that we can pass from the function f_0 and the weights U_0, V_0 in one proof to those in the other with the appropriate changes. Details are left to the interested reader.

Remark 7.10. We can adapt the proof of Theorem 7.2 when $p > p_0$ to get a somewhat more general result. If we replace the assumption $\bar{A}(t^{1/p'}) \preceq \bar{A}_0(t^{1/p_0'})$ with the assumption that there exists $C \in B_{p'}$ such that $A^{-1}(t)\,C^{-1}(t)^{1/sp_0} \leq cA_0^{-1}(t)$, then there exists a Young function Φ such that $M_\Phi : L^{p'}(u^{1-p'}) \to L^{p'}(v^{1-p'})$. We can then define F_0 as $M_\Phi f_0$ and adapt the above argument accordingly.

If we let $C = \bar{A}$, then this new condition immediately reduces to one that is equivalent to our original hypothesis. Furthermore, in practice the conditions we derive for specific families of Young functions are not significantly better than those we obtained above. Therefore, we leave further details to the interested reader. However, we will use this approach in Section 7.4 to adapt our proof to the case of factored weights, and in Chapter 8 we will use it to prove an endpoint result when $p_0 = 1$.

Proof of Theorem 7.2: $1 < p < p_0$

The proof of this case is broadly similar to the proof above when $p > p_0$ and is organized in the same way. Fix $(f, g) \in \mathcal{F}$ and fix (u, v) satisfying (7.5). Define

$$g_0 = \frac{g}{\|g\|_{L^p(v)}}, \qquad G_0 = \frac{f}{\|f\|_{L^p(u)}} + Mg_0.$$

It is immediate that $\|g_0\|_{L^p(v)} = 1$. (The first term in the definition of G_0 is included for technical reasons that will be clear below. Intuitively, think of G_0 as equal to Mg_0.)

We claim that $\|G_0\|_{L^p(u)} \le C$, where C depends only on u, v and p. Since A is a p-Young function and the pair (u, v) satisfies (7.5),

$$\|u^{1/p}\|_{p,Q}\|v^{-1/p}\|_{B,Q} \le C\|u^{1/p}\|_{A,Q}\|v^{-1/p}\|_{B,Q} \le K.$$

Therefore, since $\bar{B} \in B_p$, by Theorem 5.14, $M : L^p(v) \to L^p(u)$, and so

$$\|G_0\|_{L^p(u)} \le 1 + \|Mg_0\|_{L^p(u)} \le 1 + C\|g_0\|_{L^p(v)} = C.$$

Now define the new pair of weights

$$(U_0, V_0) = (G_0^{-(p_0-p)}u, g_0^{-(p_0-p)}v).$$

Suppose that the pair (U_0, V_0) satisfies (7.3) with constant independent of g_0. By Hölder's inequality with exponent $p_0/p > 1$,

$$\begin{aligned}\|f\|_{L^p(u)} &= \left(\int_{\mathbb{R}^n} f(x)^p\, G_0(x)^{-p/(p_0/p)'} G_0(x)^{p/(p_0/p)'} u(x)\, dx\right)^{1/p} \\ &\le \left(\int_{\mathbb{R}^n} f(x)^{p_0} G_0^{-p_0/(p_0/p)'} u(x)\, dx\right)^{1/p_0} \left(\int_{\mathbb{R}^n} G_0(x)^p u(x)\, dx\right)^{\frac{1}{p\,(p_0/p)'}} \\ &\le C\left(\int_{\mathbb{R}^n} f(x)^{p_0}\, G_0(x)^{p-p_0} u(x)\, dx\right)^{1/p_0}.\end{aligned}$$

By our definition of G_0, $f \le G_0\|f\|_{L^p(u)}$; hence,

$$\begin{aligned}\left(\int_{\mathbb{R}^n} f(x)^{p_0}\, U_0(x)\, dx\right)^{1/p_0} &= \left(\int_{\mathbb{R}^n} f(x)^{p_0}\, G_0(x)^{p-p_0} u(x)\, dx\right)^{1/p_0} \\ &\le \|f\|_{L^p(u)}\|G_0\|_{L^p(u)}^{p/p_0} \le C\|f\|_{L^p(u)} < \infty.\end{aligned}$$

This inequality and the fact that the pair the pair of weights (U_0, V_0) satisfies (7.3) allow us to use (7.2) to conclude that

$$\|f\|_{L^p(u)} \le C\left(\int_{\mathbb{R}^n} f(x)^{p_0}\, U_0(x)\, dx\right)^{1/p_0}$$

$$
\begin{aligned}
&\leq C\left(\int_{\mathbb{R}^n} g(x)^{p_0}\, V_0(x)\, dx\right)^{1/p_0} \\
&= C\left(\int_{\mathbb{R}^n} g(x)^{p_0}\, g_0(x)^{-(p_0-p)}\, v(x)\, dx\right)^{1/p_0} \\
&= C\|g\|_{L^p(v)}.
\end{aligned}
$$

To complete the proof we need to prove that the pair of weights (U_0, V_0) satisfies (7.3). Set

$$
s = \frac{p_0-1}{p_0-p} > 1, \qquad s' = \frac{p_0-1}{p-1}.
$$

We first prove that

$$
\|h_1 h_2\|_{B_0,Q} \leq C\|h_1\|_{sp_0',Q}\|h_2^{p_0/p}\|_{B,Q}^{p/p_0}. \tag{7.9}
$$

This estimate is a consequence of the generalized Hölder's inequality (5.6). Note that $sp_0' = (p_0/p)'$ and let $D(t) = t^{s\,p_0'} = t^{(p_0/p)'}$ and $C(t) = B(t^{p_0/p})$. Both of these are Young functions since $p_0 > p$. By our assumption on B and B_0 (see Remark 7.4),

$$
D^{-1}(t)\,C^{-1}(t) = t^{\frac{1}{(p_0/p)'}} \cdot B^{-1}(t)^{\frac{1}{p_0/p}} \leq cB_0^{-1}(t).
$$

Therefore,

$$
\|h_1 h_2\|_{B_0,Q} \leq C\|h_1\|_{D,Q}\,\|h_2\|_{C,Q} = \|h_1\|_{sp_0',Q}\,\|h_2^{p_0/p}\|_{B,Q}^{p/p_0}.
$$

If we argue as we did in the proof of the reverse factorization theorem (Theorem 6.4) and rescale using the fact that $A_0(t^{1/p_0}) \preceq A(t^{1/p})$, then

$$
\begin{aligned}
\|U_0^{1/p_0}\|_{A_0,Q} = \|G_0^{-\frac{p_0-p}{p_0}}\, u^{1/p_0}\|_{A_0,Q} &\leq \|(Mg_0)^{-\frac{p_0-p}{p_0}}\, u^{1/p_0}\|_{A_0,Q} \\
&\leq \|g_0\|_{1,Q}^{-\frac{p_0-p}{p_0}}\,\|u^{1/p_0}\|_{A_0,Q} \leq C\|g_0\|_{1,Q}^{-\frac{p_0-p}{p_0}}\,\|u^{1/p}\|_{A,Q}^{p/p_0}.
\end{aligned}
$$

Further, by (7.9) we have that

$$
\begin{aligned}
\|V_0^{-1/p_0}\|_{B_0,Q} &= \|g_0^{\frac{p_0-p}{p_0}}\, v^{-1/p_0}\|_{B_0,Q} \\
&\leq C\|g_0^{\frac{p_0-p}{p_0}}\|_{sp_0',Q}\,\|(v^{-1/p_0})^{p_0/p}\|_{B,Q}^{p/p_0} = \|g_0\|_{1,Q}^{\frac{p_0-p}{p_0}}\,\|v^{-1/p}\|_{B,Q}^{p/p_0}.
\end{aligned} \tag{7.10}
$$

Note that since $g \in \cap_{p>1} L^p$, $g_0 \in L^1_{\mathrm{loc}}$, so the terms on the right-hand side are finite.

Combining these two inequalities we see that since the pair (u, v) satisfies (7.5),

$$
\|U_0^{1/p_0}\|_{A_0,Q}\|V_0^{-1/p_0}\|_{B_0,Q} \leq C\|u^{1/p}\|_{A,Q}^{p/p_0}\|v^{-1/p}\|_{B,Q}^{p/p_0} \leq K^{p/p_0}.
$$

Thus the pair (U_0, V_0) satisfies (7.3) and our proof is complete. □

Remark 7.11. As was the case in the first part of the proof of Theorem 7.2 (see Remark 7.10) the proof when $p < p_0$ can be modified to get a somewhat more general result. If we replace the assumption $\bar{B}(t^{1/p}) \preceq \bar{B}_0(t^{1/p_0})$ with the assumption that there exists $C \in B_p$ such that $B^{-1}(t)\, C^{-1}(t)^{1/(p_0/p)'} \le cB_0^{-1}(t)$, then there exists a Young function Φ such that $M_\Phi : L^p(v) \to L^p(u)$. We can then define G_0 using M_Φ and modify the above argument accordingly.

If we let $C = \bar{B}$, then this new condition immediately reduces to one that is equivalent to our original hypothesis. And, as before, in practice this does not lead to significantly better conditions than those we derived above. Therefore, we leave further details to the reader.

7.3 Two-weight, weak type extrapolation

As we discussed in Chapter 2 (see also Chapter 9 below), one of the advantages of stating and proving our extrapolation theorems for pairs of functions is that we automatically get weak type inequalities. However, in the two-weight setting some information is lost in doing this. By proving an extrapolation result specifically for weak type inequalities we can get a sharper result.

Theorem 7.12. *Given p_0, $1 < p_0 < \infty$, a p_0-Young function A_0, and a p_0'-Young function B_0, suppose that*

$$\|f\|_{L^{p_0,\infty}(u_0)} \le C\|g\|_{L^{p_0}(v_0)}, \qquad (f,g) \in \mathcal{F}, \tag{7.11}$$

for every pair of weights (u_0, v_0) that satisfy

$$\|u_0^{1/p_0}\|_{A_0,Q}\|v_0^{-1/p_0}\|_{B_0,Q} \le K_0 < \infty. \tag{7.12}$$

Then, for all p, $1 < p < p_0$,

$$\|f\|_{L^{p,\infty}(u)} \le C\|g\|_{L^p(v)}, \qquad (f,g) \in \mathcal{F}, \tag{7.13}$$

whenever (u, v) satisfy

$$\|u^{1/p}\|_{A,Q}\|v^{-1/p}\|_{B,Q} \le K < \infty, \tag{7.14}$$

where A is a p-Young function and B is a p'-Young function such that $A_0(t^{1/p_0}) \preceq A(t^{1/p})$ and $\bar{B}(t^{1/p}) \preceq \bar{B}_0(t^{1/p_0})$.

Remark 7.13. The condition on B can be rewritten in equivalent ways as in Remark 7.4. Further, we want to emphasize that, unlike in Theorem 7.2 when $p < p_0$, here we do not assume that $\bar{B} \in B_p$. Thus, for instance, if we let $A_0(t) = t^{p_0}$ and $B_0(t) = t^{p_0'}$, then we can take $A(t) = t^p$ and $B(t) = t^{p'}$—that is, we may assume that $(u, v) \in A_p$. This particular case is the one treated by Neugebauer [164]. (A related, off-diagonal estimate was proved by Harboure, Macías and Segovia [97].)

We do not know if a result analogous to Theorem 7.12 is true when $p > p_0$ (even if we restrict ourselves to pairs $(u, v) \in A_p$) but we suspect that it is not. Neugebauer [164] proves a result in this direction but the inequalities he gets after extrapolation involve a weighted L^{p,p_0} norm that seems unnatural.

Theorem 7.12 is true when $p = 1$; we defer the statement and proof of this result to Section 8.3 below (Theorem 8.11).

Proof. The proof is very similar to the proof of Theorem 7.2 when $p < p_0$. Fix $(f, g) \in \mathcal{F}$ and fix (u, v) satisfying (7.14). Define the functions

$$g_0 = \frac{g}{\|g\|_{L^p(v)}}, \qquad G_0 = \frac{f}{\|f\|_{L^{p,\infty}(u)}} + Mg_0.$$

Clearly, $g_0 \in L^p(v)$ and $\|g_0\|_{L^p(v)} = 1$.

We claim that $\|G_0\|_{L^{p,\infty}(u)} \le C$. Since A is a p-Young function, B is a p'-Young function and the pair (u, v) satisfies (7.14), we have that

$$\|u^{1/p}\|_{p,Q}\|v^{-1/p}\|_{p',Q} \le C\|u^{1/p}\|_{A,Q}\|v^{-1/p}\|_{B,Q} \le K.$$

Hence, $(u, v) \in A_p$, so $M : L^p(v) \to L^{p,\infty}(u)$ and

$$\|G_0\|_{L^{p,\infty}(u)} \le 1 + \|Mg_0\|_{L^{p,\infty}(u)} \le 1 + C\|g_0\|_{L^p(v)} = C.$$

Now define the pair of weights

$$(U_0, V_0) = (G_0^{-(p_0-p)}\, u, g_0^{-(p_0-p)}\, v).$$

Given our hypotheses on A and B, we can argue exactly as we did in the proof of Theorem 7.2 to get that the pair (U_0, V_0) satisfies (7.12).

We can now estimate as follows: by Hölder's inequality in the scale of Lorentz spaces,

$$\begin{aligned}
\lambda\, u(\{x : f(x) > \lambda\})^{1/p} &= \lambda \left(\int_{\mathbb{R}^n} \chi_{\{x:f(x)>\lambda\}}\, G_0(x)^{p_0-p}\, U_0(x)\, dx \right)^{1/p} \\
&\le \lambda\, \|\chi_{\{x:f(x)>\lambda\}}\|_{L^{p_0/p,1}(U_0)}^{1/p}\, \|G_0^{p_0-p}\|_{L^{(p_0/p)',\infty}(U_0)}^{1/p} \\
&\le C\lambda\, U_0(\{x : f(x) > \lambda\})^{1/p_0}\, \|G_0\|_{L^{p_0,\infty}(U_0)}^{\frac{p_0-p}{p}} \\
&\le C\|f\|_{L^{p_0,\infty}(U_0)}\, \|G_0\|_{L^{p_0,\infty}(U_0)}^{\frac{p_0-p}{p}}.
\end{aligned}$$

We estimate the second term separately:

$$\begin{aligned}
\|G_0\|_{L^{p_0,\infty}(U_0)} &= \sup_{t>0} t\, U_0(\{x : G_0(x) > t\})^{1/p_0} \\
&= \sup_{t>0} t \left(\int_{\mathbb{R}^n} \chi_{\{x:G_0(x)>t\}} G_0(x)^{-(p_0-p)}\, u(x)\, dx \right)^{1/p_0}
\end{aligned}$$

$$\leq \sup_{t>0} t^{p/p_0} u(\{x : G_0(x) > t\})^{1/p_0} = \|G_0\|_{L^{p,\infty}(u)}^{p/p_0} \leq C.$$

Therefore, if we combine this with the fact that $f \leq \|f\|_{L^{p,\infty}(u)} G_0$, we get that

$$\|f\|_{L^{p_0,\infty}(U_0)} \leq \|f\|_{L^{p,\infty}(u)} \|G_0\|_{L^{p_0,\infty}(U_0)} \leq C \|f\|_{L^{p,\infty}(u)} < \infty.$$

Since the pair (U_0, V_0) satisfies (7.12), we can apply (7.11) to get the desired result:

$$\begin{aligned} \|f\|_{L^{p,\infty}(u)} &= \sup_{\lambda>0} \lambda\, u(\{x : f(x) > \lambda\})^{1/p} \\ &\leq \|f\|_{L^{p_0,\infty}(U_0)} \|G_0\|_{L^{p_0,\infty}(U_0)}^{\frac{p_0-p}{p}} \leq C\|g\|_{L^{p_0}(V_0)} = \|g\|_{L^p(v)}. \end{aligned}$$

□

7.4 Extrapolation for factored weights

In this section we prove a version of Theorem 7.2 for factored weights. Recall Definition 6.3: given Young functions Φ and Ψ and non-negative functions w_1 and w_2,

$$(\tilde{u}, \tilde{v}) = \big(w_1(M_\Psi w_2)^{1-p}, (M_\Phi w_1)w_2^{1-p}\big)$$

is a pair of factored weights. For all such pairs we implicitly assume that $M_\Phi w_1$, $M_\Psi w_2 < \infty$ a.e., so we also have that $w_1, w_2 \in L^1_{\mathrm{loc}}$.

Theorem 7.14. *Given p_0, $1 < p_0 < \infty$, a p_0-Young function A_0, and a p_0'-Young function B_0, suppose that*

$$\int_{\mathbb{R}^n} f(x)^{p_0} \tilde{u}_0(x)\, dx \leq C \int_{\mathbb{R}^n} g(x)^{p_0} \tilde{v}_0(x)\, dx, \qquad (f,g) \in \mathcal{F}, \tag{7.15}$$

for every pair of factored weights

$$(\tilde{u}_0, \tilde{v}_0) = \big(w_1(M_{\Psi_0} w_2)^{1-p_0}, (M_{\Phi_0} w_1)w_2^{1-p_0}\big), \tag{7.16}$$

where $\Phi_0(t) = A_0(t^{1/p_0})$, $\Psi_0(t) = B_0(t^{1/p_0'})$. Then for all p, $1 < p < \infty$,

$$\int_{\mathbb{R}^n} f(x)^p \tilde{u}(x)\, dx \leq C \int_{\mathbb{R}^n} g(x)^p \tilde{v}(x)\, dx, \qquad (f,g) \in \mathcal{F}, \tag{7.17}$$

whenever

$$(\tilde{u}, \tilde{v}) = \big(w_1(M_\Psi w_2)^{1-p}, (M_\Phi w_1)w_2^{1-p}\big), \tag{7.18}$$

where $\Phi(t) = A(t^{1/p})$, $\Psi(t) = B(t^{1/p'})$, and A is a p-Young function and B is a p'-Young function such that:

(a) *when $p > p_0$, $B_0(t^{1/p_0'}) \preceq B(t^{1/p'})$, and there exists $C \in B_{(p/p_0)'}$ such that $A^{-1}(t)C^{-1}(t)^{1/p_0} \leq cA_0^{-1}(t)$;*

(b) *when $p < p_0$, $A_0(t^{1/p_0}) \preceq A(t^{1/p})$ and there exists $C \in B_{(p'/p_0')'}$ such that $C^{-1}(t)^{1/p_0'} B^{-1}(t) \leq cB_0^{-1}(t)$.*

Remark 7.15. The hypotheses in (a) imply $A_0(t^{1/p_0}) \preceq A(t^{1/p})$. By the definition of the B_p condition, $C^{-1}(t) \geq ct^{1/(p/p_0)'}$, and since A_0 is a p_0-Young function, $A_0^{-1}(t) \leq Ct^{1/p_0}$. Hence,

$$A^{-1}(t) \leq C\,\frac{A_0^{-1}(t)}{C^{-1}(t)^{1/p_0}} \leq C\,A_0^{-1}(t)^{p_0/p}\,\frac{A_0^{-1}(t)^{1-p_0/p}}{t^{1/(p_0(p/p_0)')}} \leq C\,A_0^{-1}(t)^{p_0/p}.$$

A similar argument shows that given (b), $B_0(t^{1/p_0'}) \preceq B(t^{1/p'})$.

The scale of log bumps is essentially preserved by these hypotheses. For instance, suppose $p > p_0$, $A_0(t) = t^{p_0}\log(e+t)^{p_0-1+\delta_0}$ and $A(t) = t^p\log(e+t)^{p-1+\delta}$. Then the hypotheses of (a) hold with

$$C(t) = t^{(p/p_0)'}\log(e+t)^{-1-\epsilon}, \qquad \epsilon = \left(\frac{p}{p_0}\right)'\left(\frac{\delta p_0}{p} - \delta_0\right).$$

Thus for $\delta > 0$ sufficiently large (compared to δ_0), $C \in B_{(p/p_0)'}$.

Remark 7.16. We can recast the hypotheses of Theorem 7.14 to resemble those in Theorem 7.2. By Theorem 6.4 the weights $(\tilde{u}_0, \tilde{v}_0)$ defined by (7.16) satisfy $\|\tilde{u}_0^{1/p_0}\|_{A_0,Q}\|\tilde{v}_0^{-1/p_0}\|_{B_0,Q} \leq 1$, and the weights $(\tilde{u}, \tilde{v})$ defined by (7.18) satisfy $\|\tilde{u}^{1/p}\|_{A,Q}\|\tilde{v}^{-1/p}\|_{B,Q} \lesssim 1$.

Suppose further that $p > p_0$ and A is as in (a) of Theorem 7.2. Then $\bar{A} \in B_{p'}$, so $C(t) = \bar{A}(t^{(p/p_0)'/p'}) \in B_{(p/p_0)'}$. By Remark 7.4, $t^{1-p'/p_0'}A^{-1}(t)^{p'/p_0'} \leq A_0^{-1}(t)$. It follows from a short calculation that $A^{-1}(t)C^{-1}(t)^{1/p_0} \leq cA_0^{-1}(t)$. Similarly, if $p < p_0$ and B_0 satisfies the conditions in (b) of Theorem 7.2, we can define the Young function $C(t) = \bar{B}(t^{(p'/p_0')'/p}) \in B_{(p'/p_0')'}$ and get that $C^{-1}(t)^{1/p_0'}B^{-1}(t) \leq cB_0^{-1}(t)$.

Finally, the hypotheses of Theorem 7.14 should be compared to the more general hypotheses in Remarks 7.10 and 7.11.

Remark 7.17. It follows from the previous remark that the specific examples that we computed for Theorem 7.2 (see Table 7.1) also hold for Theorem 7.14; details are left to the reader.

Proof of Theorem 7.14: $p > p_0$

Fix a pair of weights $(\tilde{u}, \tilde{v})$ as in (7.18), fix $(f, g) \in \mathcal{F}$, and assume that $f \in L^p(\tilde{u})$. Let $r = (p/p_0)'$ and define a new pair of weights

$$(\tilde{U}, \tilde{V}) = \left(w_2\,(M_\Phi w_1)^{1-r}, (M_\Psi w_2)\,w_1^{1-r}\right).$$

We claim that $M_{\Phi_0} : L^r(\tilde{V}) \to L^r(\tilde{U})$. Let $\tilde{A}(t) = \Phi(t^{r'}) = A(t^{1/p_0})$ and $\tilde{B}(t) = \Psi(t^r)$. Since $t^r \preceq \tilde{B}(t)$, by Theorem 6.4 we have that

$$\|\tilde{U}^{1/r}\|_{r,Q}\|\tilde{V}^{-1/r}\|_{\tilde{A},Q} \leq K\,\|\tilde{U}^{1/r}\|_{\tilde{B},Q}\|\tilde{V}^{-1/r}\|_{\tilde{A},Q} \leq K.$$

Furthermore, $C \in B_r$, and since $A^{-1}(t)^{p_0}C^{-1}(t) \le cA_0^{-1}(t)^{p_0}$,

$$\tilde{A}^{-1}(t)C^{-1}(t) \le c\Phi_0^{-1}(t).$$

Hence, by Theorem 5.14 we get the desired estimate for M_{Φ_0}.

Now define

$$f_0 = \frac{f^{p/r}\, w_1\, (M_\Psi w_2)^{-p/r}}{\|f^{p/r}\, w_1\, (M_\Psi w_2)^{-p/r}\|_{L^r(\tilde{V})}} = \frac{f^{p/r}\, w_1\, (M_\Psi w_2)^{-p/r}}{\|f\|_{L^p(\tilde{u})}^{p/r}}.$$

Clearly, $\|f_0\|_{L^r(\tilde{V})} = 1$ and therefore, $\|M_{\Phi_0} f_0\|_{L^r(\tilde{U})} \le C$.

Because we are working with factored weights we will need to use an approximation argument to insure that we can apply (7.15). For each $N > 0$, let $Q_N = [-N, N]^n$ and $E_N = Q_N \cap \{x : f(x) \le N\}$, and define the pair of factored weights

$$(\tilde{U}_0^N, \tilde{V}_0^N) = \big(f_0\chi_{E_N}(M_{\Psi_0}w_2)^{1-p_0}, M_{\Phi_0}(f_0\chi_{E_N})w_2^{1-p_0}\big).$$

(If we did not need to use an approximation argument we would eliminate the set E_N and define the pair $(\tilde{U}_0, \tilde{V}_0)$.)

Since $B_0(t^{1/p_0'}) \preceq B(t^{1/p'})$, $(M_\Psi w_2)^{1-p_0} \le c(M_{\Psi_0}w_2)^{1-p_0}$. Then,

$$\begin{aligned}
&\|f\|_{L^p(\tilde{u})}^{p_0} \\
&\quad = \|f\|_{L^p(\tilde{u})}^{-p/r} \int_{\mathbb{R}^n} f(x)^{p_0} f(x)^{p/r} w_1(x) M_\Psi w_2(x)^{-p/r} M_\Psi w_2(x)^{1-p_0}\, dx \\
&\quad \le C \lim_{N\to\infty} \int_{\mathbb{R}^n} f(x)^{p_0}\, \tilde{U}_0^N(x)\, dx.
\end{aligned}$$

For each $N > 0$, the pair $(\tilde{U}_0^N, \tilde{V}_0^N)$ satisfies (7.16) and so (7.15) holds for this pair provided that the left-hand side is finite. If we assume this temporarily, then

$$\begin{aligned}
\int_{\mathbb{R}^n} f(x)^{p_0}\, \tilde{U}_0^N(x)\, dx &\le C \int_{\mathbb{R}^n} g(x)^{p_0}\, \tilde{V}_0^N(x)\, dx \\
&\le C \int_{\mathbb{R}^n} g(x)^{p_0}\, M_{\Phi_0} f_0(x) w_2(x)^{1-p_0}\, dx \\
&= C \int_{\mathbb{R}^n} g(x)^{p_0} \tilde{v}(x)^{p_0/p} M_{\Phi_0} f_0(x) \tilde{U}(x)^{1/r}\, dx \\
&\le C\|g\|_{L^p(\tilde{v})}^{p_0} \|M_{\Phi_0} f_0\|_{L^r(\tilde{U})} \\
&\le C\|g\|_{L^p(\tilde{v})}^{p_0}.
\end{aligned}$$

This completes the proof provided that for all N sufficiently large,

$$\int_{\mathbb{R}^n} f(x)^{p_0}\, \tilde{U}_0^N(x)\, dx < \infty.$$

Since w_2 is not identically zero, for N large enough and for every $x \in E_N$, $M_{\Psi_0} w_2(x) \geq \|w_2\|_{\Psi_0,Q_N} > 0$. Then by the definition of $\tilde{U}_0^N$ and f_0, and again using the fact that $(M_\Psi w_2)^{1-p_0} \leq c(M_{\Psi_0} w_2)^{1-p_0}$, we have that

$$\begin{aligned}\int_{\mathbb{R}^n} f(x)^{p_0}\, \tilde{U}_0^N(x)\, dx &\leq \|f\|_{L^p(\tilde{u})}^{-p/r} \int_{E_N} f(x)^p w_1(x) M_{\Psi_0} w_2(x)^{1-p}\, dx \\ &\leq \|f\|_{L^p(\tilde{u})}^{-p/r} \|w_2\|_{\Psi_0,Q_N}^{1-p} N^p w_1(Q_N) \\ &< \infty;\end{aligned}$$

the last inequality holds since $w_1 \in L^1_{\mathrm{loc}}$. □

Proof of Theorem 7.14: $1 < p < p_0$

The proof is similar to the proof of the previous case. Fix a pair of weights $(\tilde{u}, \tilde{v})$ as in (7.18) and fix $(f,g) \in \mathcal{F}$. Let

$$r = (p'/p_0')' = \frac{p\,(p_0 - 1)}{p_0 - p},$$

and define a new pair of weights

$$(\tilde{U}, \tilde{V}) = \big(w_1\,(M_\Psi w_2)^{1-r}, (M_\Phi w_1)\, w_2^{1-r}\big).$$

We claim that $M_{\Psi_0} : L^r(\tilde{V}) \to L^r(\tilde{U})$. Let $\tilde{A}(t) = \Phi(t^r)$ and let $\tilde{B}(t) = B(t^{1/p_0'}) = \Psi(t^{r'})$. Since $t^r \preceq \Phi(t^r) = \tilde{A}(t)$, by Theorem 6.4,

$$\|\tilde{U}^{1/r}\|_{r,Q} \|\tilde{V}^{-1/r}\|_{\tilde{B},Q} \leq C \|\tilde{U}^{1/r}\|_{\tilde{A},Q} \|\tilde{V}^{-1/r}\|_{\tilde{B},Q} \leq C.$$

Furthermore, $C \in B_r$, and since $B^{-1}(t)^{p_0'} C^{-1}(t) \leq c B_0^{-1}(t)^{p_0'}$,

$$\tilde{B}^{-1}(t) C^{-1}(t) \leq c \Psi_0^{-1}(t).$$

Hence, by Theorem 5.14 we get the desired estimate for M_{Ψ_0}.

Now define

$$g_0 = \frac{g^{p/r}\, w_2^{1-p/r}}{\|g^{p/r}\, w_2^{1-p/r}\|_{L^r(\tilde{V})}} = \frac{g^{p/r}\, w_2^{1-p/r}}{\|g\|_{L^p(\tilde{v})}^{p/r}}.$$

Clearly, $\|g_0\|_{L^r(\tilde{V})} = 1$ and therefore, $\|M_{\Psi_0} g_0\|_{L^r(\tilde{U})} \leq C$.

For each $N > 0$, let $Q_N = [-N,N]^n$ and $E_N = Q_N \cap \{x : f(x) \leq N\}$, and define the factored pairs

$$(\tilde{U}_0^N, \tilde{V}_0^N) = \big(w_1\, \chi_{E_N}\, (M_{\Psi_0} g_0)^{1-p_0}, M_{\Phi_0}(w_1\, \chi_{E_N})\, g_0^{1-p_0}\big).$$

Since $r/(p_0/p)' = p/p_0'$, by Hölder's inequality with exponent p_0/p we have that

$$\begin{aligned}
&\|f\|_{L^p(\tilde{u})} \\
&= \left(\int_{\mathbb{R}^n} f(x)^p M_{\Psi_0}g_0(x)^{-p/p_0'} M_{\Psi_0}g_0(x)^{p/p_0'} M_{\Psi}w_2(x)^{1-p} w_1(x)\,dx\right)^{1/p} \\
&\le \left(\int_{\mathbb{R}^n} f(x)^{p_0} M_{\Psi_0}g_0(x)^{1-p_0}\, w_1(x)\,dx\right)^{1/p_0} \\
&\qquad \times \left(\int_{\mathbb{R}^n} M_{\Psi_0}g_0(x)^r M_{\Psi}w_2(x)^{1-r}\, w_1(x)\,dx\right)^{1/(r\,p_0')} \\
&= \lim_{N\to\infty} \|f\|_{L^{p_0}(\tilde{U}_0^N)} \|M_{\Psi_0}g_0\|_{L^r(\tilde{U})}^{1/p_0'} \\
&\le C \lim_{N\to\infty} \|f\|_{L^{p_0}(\tilde{U}_0^N)}.
\end{aligned}$$

We claim that $\|f\|_{L^{p_0}(\tilde{U}_0^N)} < \infty$. By definition, f is bounded on E_N, so this is true if $\tilde{U}_0^N \in L^1$. But by the definition of $\mathcal{F}$, g is not identically zero, so for N sufficiently large and $x \in E_N$, $M_{\Psi_0}g_0(x) \ge \|g_0\|_{\Psi_0,Q_N} > 0$. Further, $w_1 \in L^1_{\text{loc}}$, so it follows as before that $\tilde{U}_0^N \in L^1$.

As $(\tilde{U}_0^N, \tilde{V}_0^N)$ is a pair of factored weights satisfying (7.16), for each N, we can apply (7.15) and the definition of g_0 to get

$$\begin{aligned}
\|f\|_{L^{p_0}(\tilde{U}_0^N)} &\le C\,\|g\|_{L^{p_0}(\tilde{V}_0^N)} \\
&\le C\left(\int_{\mathbb{R}^n} g(x)^{p_0}\, M_{\Phi_0}w_1(x)\, g_0(x)^{1-p_0}\,dx\right)^{1/p_0} = C\,\|g\|_{L^p(\tilde{v})},
\end{aligned}$$

where we have used that $A_0(t^{1/p_0}) \preceq A(t^{1/p})$ and so $M_{\Phi_0}w_1(x) \le C\,M_{\Phi}w_1(x)$. Combining these two inequalities we get the desired result. □

7.5 Extrapolation for fractional weights

In this section we prove extrapolation theorems for the weights associated with fractional operators,

$$|Q|^{\alpha/n}\|u^{1/p}\|_{A,Q}\|v^{-1/p}\|_{B,Q} \le K < \infty.$$

In particular, we give versions of Theorems 7.2 and 7.12. The statement and proof of the results are nearly identical to those given above, so we will only sketch the changes required. Complete details, as well as extensions similar to those in the remarks in the previous sections, are left to the interested reader.

Remark 7.18. We have not proved an extrapolation theorem for factored fractional weights analogous to Theorem 7.14. This is probably possible, though it may be complicated by the alternative approaches to defining factored weights discussed in Section 6.4. We leave this as an open problem.

Theorem 7.19. *Given α, $0 < \alpha < n$, and p_0, $1 < p_0 < \infty$, a p_0-Young function A_0, and a p_0'-Young function B_0, suppose that*

$$\int_{\mathbb{R}^n} f(x)^{p_0} u_0(x)\, dx \le C \int_{\mathbb{R}^n} g(x)^{p_0} v_0(x)\, dx, \qquad (f,g) \in \mathcal{F}, \tag{7.19}$$

for every pair of weights (u_0, v_0) that satisfy

$$|Q|^{\alpha/n} \|u_0^{1/p_0}\|_{A_0,Q} \|v_0^{-1/p_0}\|_{B_0,Q} \le K_0 < \infty. \tag{7.20}$$

Then for all p, $1 < p < \infty$,

$$\int_{\mathbb{R}^n} f(x)^{p} u(x)\, dx \le C \int_{\mathbb{R}^n} g(x)^{p} v(x)\, dx, \qquad (f,g) \in \mathcal{F}, \tag{7.21}$$

whenever (u, v) satisfy

$$|Q|^{\alpha/n} \|u^{1/p}\|_{A,Q} \|v^{-1/p}\|_{B,Q} \le K < \infty, \tag{7.22}$$

where A is a p-Young function and B is a p'-Young function such that:

(a) *when $p > p_0$, $\bar{A} \in B_{p'}$, $\bar{A}(t^{1/p'}) \preceq \bar{A}_0(t^{1/p_0'})$, and $B_0(t^{1/p_0'}) \preceq B(t^{1/p'})$;*

(b) *when $p < p_0$, $A_0(t^{1/p_0}) \preceq A(t^{1/p})$, $\bar{B} \in B_p$, and $\bar{B}(t^{1/p}) \preceq \bar{B}_0(t^{1/p_0})$.*

Theorem 7.20. *Given α, $0 < \alpha < n$, p_0, $1 < p_0 < \infty$, a p_0-Young function A_0, and a p_0'-Young function B_0, suppose that*

$$\|f\|_{L^{p_0,\infty}(u_0)} \le C \|g\|_{L^{p_0}(v_0)}, \qquad (f,g) \in \mathcal{F}, \tag{7.23}$$

for every pair of weights (u_0, v_0) that satisfy (7.20). Then, for all p, $1 < p < p_0$,

$$\|f\|_{L^{p,\infty}(u)} \le C \|g\|_{L^{p}(v)}, \qquad (f,g) \in \mathcal{F}, \tag{7.24}$$

whenever (u, v) satisfy (7.22). where A is a p-Young function and B is a p'-Young function such that $A_0(t^{1/p_0}) \preceq A(t^{1/p})$ and $\bar{B}(t^{1/p}) \preceq \bar{B}_0(t^{1/p_0})$.

Remark 7.21. Theorem 7.20 is true when $p = 1$, analogous to Theorem 8.11; we defer the statement and proof of this result to Section 8.4 below (Theorem 8.15).

Proof of Theorem 7.19. The proof is nearly identical to that of Theorem 7.2. When $p > p_0$, define f_0 as before and let $F_0 = M_\alpha f_0$. Then by Theorem 5.37 (in place of Theorem 5.14) we get that $M_\alpha : L^{p'}(u^{1-p'}) \to L^{p'}(v^{1-p'})$, and so $\|F_0\|_{L^{p'}(v^{1-p'})} \le C$. The proof then proceeds without change until we prove that the pair (U_0, V_0) satisfies (7.20). In this case we have that (7.8) holds and that

$$\|V_0^{-1/p_0}\|_{B_0,Q} \le C |Q|^{-\frac{\alpha}{n\, s\, p_0}} \|f_0\|_{1,Q}^{-1/sp_0} \|v^{-1/p}\|_{B,Q}^{p'/p_0'}.$$

Thus, since (u, v) satisfy (7.22), we have that

$$|Q|^{\alpha/n}\,\|U_0^{1/p_0}\|_{A_0,Q}\,\|V_0^{-1/p_0}\|_{B_0,Q} \\ \leq C|Q|^{\frac{\alpha}{n}-\frac{\alpha}{n\,s\,p_0}}\,\|u^{1/p}\|_{A,Q}^{p'/p_0'}\|v^{-1/p}\|_{B,Q}^{p'/p_0'} \leq K^{p'/p_0'}.$$

When $1 < p < p_0$, we define G_0 with M replaced by M_α. Then by Theorem 5.37 (in place of Theorem 5.14), $M_\alpha : L^p(v) \to L^p(u)$, so $\|G_0\|_{L^p(u)} \leq C$. The proof again proceeds without change until we prove that the pair (U_0, V_0) satisfies (7.20). In this case we have that (7.10) holds and

$$\|U_0^{1/p_0}\|_{A_0,Q} \leq C|Q|^{-\frac{\alpha\,(p_0-p)}{n\,p_0}}\,\|g_0\|_{1,Q}^{-\frac{p_0-p}{p_0}}\,\|u^{1/p}\|_{A,Q}^{p/p_0},$$

Thus, since (u,v) satisfy (7.22), we have that

$$|Q|^{\alpha/n}\|U_0^{1/p_0}\|_{A_0,Q}\,\|V_0^{-1/p_0}\|_{B_0,Q} \\ \leq C|Q|^{\frac{\alpha}{n}-\frac{\alpha\,(p_0-p)}{n\,p_0}}\,\|u^{1/p}\|_{A,Q}^{p/p_0}\|v^{-1/p}\|_{B,Q}^{p/p_0} \leq K^{p/p_0}. \qquad \square$$

Proof of Theorem 7.20. The proof is essentially the same as that of Theorem 7.12. We define G_0, with M_α in place of M. Then $(u,v) \in A_{pp}^\alpha$ (see Section 5.6), so $M_\alpha : L^p(v) \to L^{p,\infty}(u)$. Hence, $\|G_0\|_{L^{p,\infty}(u)} \leq C$. When $p < p_0$ we can show that (U_0, V_0) satisfy (7.20) just as we did in the proof of Theorem 7.19. Thereafter the proof continues without change. □

7.6 Appendix: A one case proof of extrapolation

In this section we give a version of Theorem 7.2 whose proof does not divide into two cases depending on whether $p > p_0$ or $p < p_0$. Before doing so, however, we want to briefly discuss why we have not directly adapted the proofs in Part I to the two-weight setting. For simplicity we will concentrate on the proof of Theorem 1.4 given in Chapter 2.

Given an operator T, suppose that for some p_0, $1 < p_0 < \infty$, we have that

$$\int_{\mathbb{R}^n} |Tf(x)|^{p_0} u_0(x)\,dx \leq C \int_{\mathbb{R}^n} |f(x)|^{p_0} v_0(x)\,dx \tag{7.25}$$

whenever

$$\|u_0^{1/p_0}\|_{A_0,Q}\|v_0^{-1/p_0}\|_{B_0,Q} \leq K_0 < \infty. \tag{7.26}$$

We want to prove that for all p,

$$\int_{\mathbb{R}^n} |Tf(x)|^{p} u(x)\,dx \leq C \int_{\mathbb{R}^n} |f(x)|^{p} v(x)\,dx$$

whenever

$$\|u^{1/p}\|_{A,Q}\|v^{-1/p}\|_{B,Q} \leq K < \infty.$$

What do we have to assume about A_0, B_0, A and B?

The first step is to find a replacement for the iteration algorithms. Let $\Phi(t) = A_0(t^{1/p_0})$ and $\Psi(t) = B_0(t^{1/p_0'})$, and replace $\mathcal{R}h$ by $M_\Phi h$ and $\mathcal{R}'h$ by $M_\Psi(hu)$. Then we can argue exactly as we did in the proof of Theorem 1.4 to get that there exists h, $\|h\|_{L^{p'}(u)} = 1$, such that

$$\|Tf\|_{L^p(u)} \le \left(\int_{\mathbb{R}^n} |Tf(x)|^{p_0} M_\Psi f(x)^{1-p_0} h(x)u(x)\,dx\right)^{1/p_0} \times \left(\int_{\mathbb{R}^n} M_\Psi f(x)\,h(x)u(x)\,dx\right)^{1/p_0'}.$$

By Theorem 6.4 the pair of factored weights

$$(\tilde{U_0}, \tilde{V_0}) = \left(hu\,(M_\Psi f)^{1-p_0}, M_\Phi(hu)\,|f|^{1-p_0}\right)$$

satisfies (7.26). Therefore, we can apply (7.25) to get

$$\|Tf\|_{L^p(u)} \le \left(\int_{\mathbb{R}^n} |f(x)| M_\Phi(hu)(x)\,dx\right)^{1/p_0} \left(\int_{\mathbb{R}^n} M_\Psi f(x)\,h(x)u(x)\,dx\right)^{1/p_0'}.$$

If we apply Hölder's inequality to each term, then we would be done if $M_\Psi : L^p(v) \to L^p(u)$ and $M_\Phi : L^{p'}(u^{1-p'}) \to L^{p'}(v^{1-p'})$. By Theorem 5.14, this is the case provided that there exist $C \in B_p$ and $D \in B_{p'}$ such that

$$B^{-1}(t)C^{-1}(t) \le c\Psi^{-1}(t), \quad A^{-1}(t)D^{-1}(t) \le c\Phi^{-1}(t).$$

To see that these conditions are stronger than those we derived in Theorem 7.2, we will consider the case when A_0 and A are log bumps. Let $A_0(t) = t^{p_0}\log(e+t)^{p_0-1+\delta_0}$, $\delta_0 > 0$. If we take $D(t) = t^{p'}\log(e+t)^{-1-\epsilon}$ for some $\epsilon > 0$, then $D \in B_{p'}$, and we can take $A(t) = t^p \log(e+t)^{p_0 p-1+\delta}$, where $\delta = \delta_0 p + \epsilon(p-1) > 0$. If we compare this to the results summarized in Table 7.1 we see that the exponent on the log term that we get from this proof is considerably larger unless we know *a priori* that we can take p_0 arbitrarily close to 1.

Remark 7.22. If we start with A_0 and B_0 exponential log bumps or power bumps, then this proof yields essentially the same conditions as Theorem 7.2. Details are left to the reader.

We now turn to our actual result. With the exception of the conditions imposed on the Young functions A and B, the statement is identical to that of Theorem 7.2, but for the convenience of the reader we restate the full result.

Theorem 7.23. *Given p_0, $1 < p_0 < \infty$, a p_0-Young function A_0, and a p_0'-Young function B_0, suppose that*

$$\int_{\mathbb{R}^n} f(x)^{p_0} u_0(x)\,dx \le C \int_{\mathbb{R}^n} g(x)^{p_0} v_0(x)\,dx, \qquad (f,g) \in \mathcal{F}, \tag{7.27}$$

for every pair of weights (u_0, v_0) that satisfy

$$\|u_0^{1/p_0}\|_{A_0,Q}\|v_0^{-1/p_0}\|_{B_0,Q} \le K_0 < \infty. \tag{7.28}$$

Then for all p, $1 < p < \infty$,

$$\int_{\mathbb{R}^n} f(x)^p\, u(x)\, dx \le C \int_{\mathbb{R}^n} g(x)^p\, v(x)\, dx, \qquad (f,g) \in \mathcal{F}, \tag{7.29}$$

whenever (u, v) satisfy

$$\|u^{1/p}\|_{A,Q}\|v^{-1/p}\|_{B,Q} \le K < \infty, \tag{7.30}$$

where A is a p-Young function and B is a p'-Young function such that $\bar{A} \in B_{p'}$, $\bar{B} \in B_p$, and there exists q, $1 < q < \min(p, p_0)$ such that

$$\Big[\Big(\frac{A^{-1}(t)}{t}\Big)^{p'} \cdot t\Big]^{q-1} \le c\Big[\Big(\frac{A_0^{-1}(t)}{t}\Big)^{p_0'} \cdot t\Big]^{p_0-1}, \tag{7.31}$$

$$\Big[\Big(\frac{B^{-1}(t)}{t}\Big)^{p} \cdot t\Big]^{p'-1} \le c\Big[\Big(\frac{B_0^{-1}(t)}{t}\Big)^{p_0} \cdot t\Big]^{q'-1}. \tag{7.32}$$

Remark 7.24. By inequality (5.4), (7.31) and (7.32) are equivalent to

$$\Big(\frac{\bar{A}_0^{-1}(t)^{p_0'}}{t}\Big)^{p_0-1} \le c\Big(\frac{\bar{A}^{-1}(t)^{p'}}{t}\Big)^{q-1},$$

$$\Big(\frac{\bar{B}_0^{-1}(t)^{p_0}}{t}\Big)^{q'-1} \le c\Big(\frac{\bar{B}^{-1}(t)^{p}}{t}\Big)^{p'-1}.$$

Remark 7.25. Though it is not immediate from the statement, the conditions on A and B in Theorem 7.23 yield the same conditions for specific families of Orlicz bumps as we got from Theorem 7.2. For brevity we will consider only the case of log bumps; details for the remaining cases are left to the reader.

Suppose that we can extrapolate from p_0 with $A_0(t) = t^{p_0}\log(e+t)^{p_0-1+\delta_0}$, where $\delta_0 > 0$ is arbitrary. We want to show that if $p > p_0$ then we can take $A(t) = t^p \log(e+t)^{p-1+\delta}$, where $\delta > 0$ is again arbitrary, and if $1 < p < p_0$ we must take $A(t) = t^p\log(e+t)^{p_0-1+\delta}$, $\delta > 0$. These are the conditions we deduced in Section 7.1; see Table 7.1.

Assume first that $p > p_0$ and $A(t) = t^p\log(e+t)^{p-1+\delta}$. A straightforward computation shows that

$$\Big[\Big(\frac{A^{-1}(t)}{t}\Big)^{p'} \cdot t\Big]^{q-1} \approx \log(e+t)^{-(p-1+\delta)(q-1)/(p-1)},$$

$$\Big[\Big(\frac{A_0^{-1}(t)}{t}\Big)^{p_0'} \cdot t\Big]^{p_0-1} \approx \log(e+t)^{-(p_0-1+\delta_0)}.$$

Therefore, we need $\delta > 0$ and q, $1 < q < \min(p, p_0)$, such that

$$p_0 - 1 + \delta_0 = (p - 1 + \delta)\frac{q-1}{p-1}.$$

Solving for q we get

$$q = (p-1)\frac{p_0 - 1 + \delta_0}{p - 1 + \delta} + 1; \tag{7.33}$$

if δ is such that

$$\delta > \delta_0 \frac{p-1}{p_0 - 1} > 0,$$

then we have that $1 < q < p_0 < p$. Given any $\delta > 0$, there exists $\delta_0 > 0$ such that this inequality holds. Since δ_0 may be chosen arbitrarily close to 0, we may take A with the desired exponent on the logarithm.

Now suppose that $1 < p < p_0$. We can repeat the above argument assuming that $A(t) = t^p \log(e+t)^a$, $a > 0$. Then solving for q we see that

$$q = a^{-1}(p-1)(p_0 - 1 + \delta_0) + 1.$$

Therefore, we will only have $q < p$ if $a > p_0 - 1 + \delta_0$; since δ_0 may be chosen arbitrarily, we see that for any $\delta > 0$ we can take $a = p_0 - 1 + \delta$.

The proof of Theorem 7.23 is similar to the proofs in Chapter 3; the biggest difference is that we use duality to pass to an L^q inequality—instead of an L^1 inequality—before using Hölder's inequality to get an L^{p_0} inequality. To check that the resulting pair of weights (U_0, V_0) satisfy (7.28) we need two variants of the generalized Hölder's inequality (5.6) which we give here as a lemma.

Lemma 7.26. *Let A_0, B_0, A, B be Young functions. Let $1 < p_0 < \infty$, $1 < q < \min(p, p_0)$ and $s = \frac{p-1}{p-q}$.*

(a) *If A, A_0, satisfy (7.31), then*

$$\|h_1 h_2\|_{A_0,Q} \le C\|h_1\|_{p_0 s, Q}\, \|h_2^{p_0 s'/p}\|_{A,Q}^{p/(p_0 s')}.$$

(b) *If B, B_0, satisfy (7.32), then*

$$\|h_1 h_2\|_{B_0,Q} \le C\|h_1\|_{(p_0/q)',Q}\, \|h_2^{p_0 s'/p}\|_{B,Q}^{p/(p_0 s')}.$$

Proof. To prove (a), let $\tilde{B}(t) = t^{p_0 s}$ and $C(t) = A(t^{p_0 s'/p})$. Since $1 < q < \min(p, p_0)$, $p_0 s > 1$ and $p_0 s'/p = (p_0/q)(q'/p') > 1$, so $\tilde{B}$ and C are Young functions. Then by assumption we have that

$$\tilde{B}^{-1}(t)\, C^{-1}(t) = t^{1/(p_0 s)} \cdot A^{-1}(t)^{p/(p_0 s')} \le c A_0^{-1}(t).$$

Therefore, by Lemma 5.2 we have that

$$\|h_1h_2\|_{A_0,Q} \le C\|h_1\|_{\tilde{B},Q}\,\|h_2\|_{C,Q} = C\|h_1\|_{p_0 s,Q}\,\|h_2^{p_0 s'/p}\|_{A,Q}^{p/(p_0 s')}.$$

The proof of (b) is almost the same. Let $\tilde{A}(t) = t^{(p_0/q)'}$ and $C(t) = B(t^{p_0 s'/p})$; these are Young functions as before. Hence,

$$\tilde{A}^{-1}(t)\,C^{-1}(t) = t^{1/(p_0/q)'}\cdot B^{-1}(t)^{p/(p_0 s')} \le cB_0^{-1}(t).$$

Therefore, by Lemma 5.2,

$$\|h_1h_2\|_{B_0,Q} \le C\|h_1\|_{\tilde{A},Q}\,\|h_2\|_{C,Q} = C\|h_1\|_{(p_0/q)',Q}\,\|h_2^{p_0 s'/p}\|_{B,Q}^{p/(p_0 s')}. \qquad \square$$

Proof of Theorem 7.23. Fix $(f,g)\in\mathcal{F}$ and fix (u,v) satisfying (7.30). Define the functions

$$f_0 = \frac{f^{p-1}\,u}{\|f^{p-1}\,u\|_{L^{p'}(u^{1-p'})}} = \frac{f^{p-1}\,u}{\|f\|_{L^p(u)}^{p-1}}, \qquad F_0 = Mf_0,$$

$$g_0 = \frac{g}{\|g\|_{L^p(v)}}, \qquad G_0 = \frac{f}{\|f\|_{L^p(u)}} + Mg_0.$$

We immediately have that $\|f_0\|_{L^{p'}(u^{1-p'})} = 1$ and $\|g_0\|_{L^p(v)} = 1$. Furthermore, arguing as in the proof of Theorem 7.2, since $\bar{A}\in B_{p'}$, (7.30) implies that $M : L^{p'}(u^{1-p'}) \to L^{p'}(v^{1-p'})$, and so $\|F_0\|_{L^{p'}(v^{1-p'})} \le C$. Similarly, since $\bar{B}\in B_p$, $M : L^p(v)\to L^p(u)$, so $\|G_0\|_{L^p(u)} \le C$.

Let

$$s = \frac{(p/q)'}{p'} = \frac{p-1}{p-q} > 1, \qquad s' = \frac{p-1}{q-1},$$

and define the new pair of weights

$$(U_0,V_0) = \big(G_0^{-(p_0-q)}\,f_0^{1/s}\,u^{1/s'},\, g_0^{-(p_0-q)}\,F_0^{1/s}\,v^{1/s'}\big).$$

Assume for the moment that (U_0,V_0) satisfy (7.28). Then by Hölder's inequality with exponent $p_0/q>1$ we have that

$$\begin{aligned}
&\|f\|_{L^p(u)}\\
&= \left(\int_{\mathbb{R}^n} f(x)^q\, f_0(x)^{1/s}\, u(x)^{1/s'}\,dx\right)^{1/q}\\
&= \left(\int_{\mathbb{R}^n} f(x)^q G_0(x)^{-q/(p_0/q)'} G_0(x)^{q/(p_0/q)'} f_0(x)^{1/s}\, u(x)^{1/s'}\,dx\right)^{1/q}\\
&\le \left(\int_{\mathbb{R}^n} f(x)^{p_0} U_0(x)\,dx\right)^{1/p_0} \left(\int_{\mathbb{R}^n} G_0(x)^q f_0(x)^{1/s} u(x)^{1/s'}\,dx\right)^{\frac{1}{q(p_0/q)'}}.
\end{aligned}$$

By Hölder's inequality with exponent $p/q > 1$,

$$\int_{\mathbb{R}^n} G_0(x)^q f_0(x)^{1/s} u(x)^{1/s'}\,dx \leq \|G_0\|^q_{L^p(u)} \|f_0\|^{p'/(p/q)'}_{L^{p'}(u^{1-p'})} \leq C. \tag{7.34}$$

Combining these two inequalities we get

$$\|f\|_{L^p(u)} \leq C\left(\int_{\mathbb{R}^n} f(x)^{p_0}\, U_0(x)\,dx\right)^{1/p_0}.$$

In order to use (7.27) we must show that $\|f\|_{L^{p_0}(U_0)} < \infty$. However, since $f \leq G_0\|f\|_{L^p(u)}$, by (7.34)

$$\begin{aligned}\|f\|^{p_0}_{L^{p_0}(U_0)} &= \int_{\mathbb{R}^n} f(x)^{p_0} G_0(x)^{-(p_0-q)} f_0(x)^{1/s}\, u(x)^{1/s'}\,dx \\ &\leq \|f\|^{p_0}_{L^p(u)} \int_{\mathbb{R}^n} G_0(x)^q f_0(x)^{1/s}\, u(x)^{1/s'}\,dx \leq C\|f\|^{p_0}_{L^p(u)} < \infty.\end{aligned}$$

Therefore, we can apply (7.27) and Hölder's inequality with exponent $p/q > 1$ to conclude that

$$\begin{aligned}\|f\|_{L^p(u)} &\leq C\left(\int_{\mathbb{R}^n} f(x)^{p_0} U_0(x)\,dx\right)^{1/p_0} \\ &\leq C\left(\int_{\mathbb{R}^n} g(x)^{p_0} V_0(x)\,dx\right)^{1/p_0} \\ &= \left(\int_{\mathbb{R}^n} g(x)^{p_0}\, g_0(x)^{-(p_0-q)} F_0(x)^{1/s} v(x)^{1/s'}\,dx\right)^{1/p_0} \\ &= \|g\|^{1-q/p_0}_{L^p(v)} \left(\int_{\mathbb{R}^n} g(x)^q F_0(x)^{1/s} v(x)^{1/s'}\,dx\right)^{1/p_0} \\ &\leq \|g\|_{L^p(v)}\, \|F_0\|^{1/(sp_0)}_{L^{p'}(v^{1-p'})} \\ &\leq C\|g\|_{L^p(v)}.\end{aligned}$$

To complete the proof we need to show that the pair (U_0, V_0) satisfies (7.28). By inequality (a) in Lemma 7.26 we have that

$$\begin{aligned}\|U_0^{1/p_0}\|_{A_0,Q} &= \|G_0^{-1+q/p_0} f_0^{1/(sp_0)} u^{1/(s'p_0)}\|_{A_0,Q} \\ &\leq \|(Mg_0)^{-1+q/p_0} f_0^{1/(sp_0)} u^{1/(s'p_0)}\|_{A_0,Q} \\ &\leq \|g_0\|^{-1+q/p_0}_{1,Q} \|f_0^{1/(sp_0)} u^{1/(s'p_0)}\|_{A_0,Q} \\ &\leq C\|g_0\|^{-1+q/p_0}_{1,Q} \|f_0\|^{1/(sp_0)}_{1,Q} \|u^{1/p}\|^{p/(s'p_0)}_{A,Q}.\end{aligned}$$

Similarly, from (b) in Lemma 7.26 we get

$$\|V_0^{-1/p_0}\|_{B_0,Q} = \|g_0^{1-q/p_0} F_0^{-1/(sp_0)} v^{-1/(s'p_0)}\|_{B_0,Q}$$

$$
\begin{aligned}
&= \|g_0^{1-q/p_0}(Mf_0)^{-1/(sp_0)}v^{-1/(s'p_0)}\|_{B_0,Q} \\
&\leq \|f_0\|_{1,Q}^{-1/sp_0}\|g_0^{1-q/p_0}v^{-1/(s'p_0)}\|_{B_0,Q} \\
&\leq C\|f_0\|_{1,Q}^{-1/(sp_0)}\|g_0\|_{1,Q}^{1-q/p_0}\|v^{-1/p}\|_{B,Q}^{p/(s'p_0)}.
\end{aligned}
$$

If we combine these two inequalities with the fact that the pair (u, v) satisfy (7.30) we get

$$
\|U_0^{1/p_0}\|_{A_0,Q}\|V_0^{-1/p_0}\|_{B_0,Q} \leq C\|u^{1/p}\|_{A,Q}^{p/(s'p_0)}\|v^{-1/p}\|_{B,Q}^{p/(s'p_0)} \leq K^{p/(s'p_0)}.
$$

This completes the proof. □

Chapter 8

Endpoint and A_∞ Extrapolation

In this chapter we consider further variations of the two-weight extrapolation theorem proved in Chapter 7. Our basic result is to show that we can extrapolate from the endpoint inequality

$$\int_{\mathbb{R}^n} f(x)u(x)\,dx \le C\int_{\mathbb{R}^n} g(x)M_\Phi u(x)\,dx, \tag{8.1}$$

where Φ is some Young function. Such inequalities are common in applications, particularly for the weak type inequalities—that is, for pairs (f,g) of the form $(\lambda\chi_{\{x:|Tf(x)|>\lambda\}},|f|)$. Extrapolation then allows us to immediately deduce weak (p,p) inequalities, $p>1$.

More generally, by rescaling we can replace (8.1) with

$$\int_{\mathbb{R}^n} f(x)^{p_0}u(x)\,dx \le C\int_{\mathbb{R}^n} g(x)^{p_0}M_\Phi u(x)\,dx, \tag{8.2}$$

for arbitrary $p_0>0$. This leads to a two-weight version of the A_∞ extrapolation theorem (Corollary 3.15). If $p>p_0$, $r=p/p_0$, and (u,v) satisfy

$$\|u^{1/r}\|_A\|v^{-1/r}\|_{r',Q} \le K < \infty, \tag{8.3}$$

where A is a r-Young function, then

$$\int_{\mathbb{R}^n} f(x)^p u(x)\,dx \le C\int_{\mathbb{R}^n} g(x)^p v(x)\,dx. \tag{8.4}$$

If we can take p_0 arbitrarily close to 0, then r can be arbitrarily large. In the one-weight case (8.3) is equivalent to $w\in A_r$, so in this case (8.4) holds if $w\in A_r$ for some r large—equivalently, if $w\in A_\infty$. However, as we noted in Chapter 5, this condition is not as flexible as the A_∞ condition, since a pair of weights (u,v) may satisfy (8.3) for some value of r but not for others. (Contrast this to the one-weight case, where if $w\in A_r$, then $w\in A_q$ for all $q>r$.)

Extrapolation arguments of this kind when $p_0 = 1$ were implicit in a number of results for specific operators, beginning with unpublished work of Muckenhoupt and Wheeden [147], and later in [53] and [33]. In [54], a weaker version of these endpoint results was proved for strong and weak (p,p) inequalities for pairs of functions of the form $\big(|Tf|, |Sf|\big)$ and for A_1 pairs of the form $(u, M^k u)$. It was noted in passing that the proofs worked for pairs of arbitrary functions, but this idea was not exploited to combine the weak type and strong type inequalities into a single result.

In certain cases, the converse of the endpoint extrapolation theorem is true: if (8.4) holds whenever the pair (u,v) satisfy (8.3), then (8.2) holds for some Young function Ψ_0. However, the converse is not perfect. For example, if we start with the pair (u_0, Mu_0) in (8.1) (i.e., $\Phi_0(t) = t$) and extrapolate "up" from the endpoint and then extrapolate "down" to recapture the endpoint result, we get pairs of the form $(u_0, M_{\Psi_0} u_0)$, where Ψ_0 is a Young function such that $t \preceq \Psi_0(t)$. This is another example of the loss of information which occurs in the two-weight case.

It is not generally possible to recapture endpoint inequalities in the one-weight case. For example, $M^2 = M \circ M$ is bounded on $L^p(w)$, $w \in A_p$, $1 < p < \infty$, but does not even satisfy the weak $(1,1)$ inequality on $L^1(w)$ if $w \in A_1$. However, as we discussed in Proposition 3.21 at the end of Section 3.3 above, there is a deep connection between endpoint results and extrapolation in the one-weight case. Rubio de Francia extrapolation (Theorem 3.9) is equivalent to the existence of $p_0 > 1$ such that for all p, $1 < p < p_0$, and all $w \in A_1$,

$$\int_{\mathbb{R}^n} f(x)^p w(x)\,dx \le C \int_{\mathbb{R}^n} g(x)^p w(x)\,dx, \qquad (f,g) \in \mathcal{F}. \tag{8.5}$$

Similarly, as we showed in Proposition 3.20, A_∞ extrapolation is equivalent to the existence of $p_0 > 0$ such that inequality (8.5) holds for $0 < p < p_0$ and for all $w \in A_1$. There are two-weight versions of both of these equivalences.

This chapter is organized as follows. In Section 8.1 we prove the endpoint extrapolation theorem and we compare it to another kind of extrapolation theorem proved in [54]. In Section 8.2 we consider three special cases of these results for A_1 pairs of the form (u, Mu). These are important in their own right in applications, and the ideas can be used to prove similar results for more general pairs. As a consequence we also get the two-weight analogs of the A_1 results discussed above. In Section 8.3 we state and prove the converse of the endpoint extrapolation theorem, and give explicit examples to show the loss of information. We also prove an endpoint result for weak type inequalities that extends Theorem 7.12. Finally, in Section 8.4 we briefly consider endpoint extrapolation for fractional operators.

Remark 8.1. All of the results in this chapter hold if we everywhere replace cubes and maximal operators by dyadic cubes and the corresponding dyadic maximal operators.

8.1 Endpoint extrapolation

In this section we state and prove the endpoint extrapolation theorem. In the Introduction we described two different results, but for simplicity we combine them into one theorem. However, the case $p_0 = 1$ figures prominently in the proof.

Theorem 8.2. *Given p_0, $0 < p_0 < \infty$, and a Young function Φ_0, suppose that for every weight u_0,*

$$\int_{\mathbb{R}^n} f(x)^{p_0}\, u_0(x)\, dx \le \int_{\mathbb{R}^n} g(x)^{p_0}\, M_{\Phi_0} u_0(x)\, dx, \qquad (f,g) \in \mathcal{F}. \tag{8.6}$$

Then, for all p, $p_0 < p < \infty$,

$$\int_{\mathbb{R}^n} f(x)^p u(x)\, dx \le C \int_{\mathbb{R}^n} g(x)^p v(x)\, dx, \qquad (f,g) \in \mathcal{F}, \tag{8.7}$$

whenever (u,v) satisfy

$$\|u^{1/r}\|_{A,Q} \|v^{-1/r}\|_{r',Q} \le K < \infty, \tag{8.8}$$

where $r = p/p_0$, A is an r-Young function, C is an r'-Young function such that $C \in B_{r'}$, and $A^{-1}(t)\, C^{-1}(t) \preceq \Phi_0^{-1}(t)$.

Remark 8.3. Since the A_1 pair $(u_0, M_{\Phi_0} u_0)$ can be thought of as a pair of factored weights, it is reasonable to ask if there is an endpoint result for factored weights similar to Theorem 7.14. There is, but it is the same as this result: the factored pairs gotten are precisely those pairs that by Theorem 6.4 satisfy (8.8).

Proof. We first prove this result for $p_0 = 1$ (so $r = p$) using the same approach as in the proof of Theorem 7.2 for $p > p_0$. Though many details are the same, we repeat them here for the convenience of the reader.

Fix $(f,g) \in \mathcal{F}$ such that $f \in L^p(u)$, and fix (u,v) satisfying (8.8). We define a new pair of weights $(U_0, M_{\Phi_0} U_0)$, where

$$U_0 = \frac{f^{p-1}\, u}{\|f^{p-1}\, u\|_{L^{p'}(u^{1-p'})}} = \frac{f^{p-1}\, u}{\|f\|_{L^p(u)}^{p-1}}.$$

It is immediate that $\|U_0\|_{L^{p'}(u^{1-p'})} = 1$.

We claim that $\|M_{\Phi_0} U_0\|_{L^{p'}(v^{1-p'})} \le C$, where C depends only on u, v and p. By assumption, $C \in B_{p'}$; further, (u,v) satisfy (8.8) with $r = p$, so

$$\|(v^{1-p'})^{1/p'}\|_{p',Q} \|(u^{1-p'})^{-1/p'}\|_{A,Q} = \|u^{1/p}\|_{A,Q} \|v^{-1/p}\|_{p',Q} \le K.$$

Therefore we have, by Theorem 5.14, $M_{\Phi_0} : L^{p'}(u^{1-p'}) \to L^{p'}(v^{1-p'})$ and so $\|M_{\Phi_0} U_0\|_{L^{p'}(v^{1-p'})} \le C \|U_0\|_{L^{p'}(u^{1-p'})} = C$.

Since $f \in L^p(u)$, the left-hand side of the following inequality is finite, so we can use (8.6) and Hölder's inequality to conclude that

$$\begin{aligned}\|f\|_{L^p(u)} &= \|f\|_{L^p(u)}^{-(p-1)} \int_{\mathbb{R}^n} f(x)f(x)^{p-1}u(x)\,dx \\ &= \int_{\mathbb{R}^n} f(x)U_0(x)\,dx \\ &\leq C \int_{\mathbb{R}^n} g(x)\,M_{\Phi_0}U_0(x)\,dx \\ &= C \int_{\mathbb{R}^n} g(x)v(x)^{1/p}\,M_{\Phi_0}U_0(x)v(x)^{-1/p}\,dx \\ &\leq C\|g\|_{L^p(v)}\|M_{\Phi_0}U_0\|_{L^{p'}(v^{1-p'})} \\ &\leq C\|g\|_{L^p(v)}.\end{aligned}$$

This completes the proof for the case $p_0 = 1$. The general case now follows by rescaling. Fix $p_0 > 0$ and the family $\mathcal{F}$ satisfying (8.6). Define a new family

$$\tilde{\mathcal{F}} = \{(\tilde{f}, \tilde{g}) = (f^{p_0}, g^{p_0}) : (f, g) \in \mathcal{F}\}.$$

Then we can rewrite (8.6) as

$$\int_{\mathbb{R}^n} \tilde{f}(x)u_0(x)\,dx \leq C \int_{\mathbb{R}^n} \tilde{g}(x)M_{\Phi_0}u_0(x)\,dx, \qquad (\tilde{f}, \tilde{g}) \in \tilde{\mathcal{F}}.$$

Therefore, by the above argument, we have that for all r, $1 < r < \infty$,

$$\int_{\mathbb{R}^n} \tilde{f}(x)^r u(x)\,dx \leq C \int_{\mathbb{R}^n} \tilde{g}(x)^r v(x)\,dx, \qquad (\tilde{f}, \tilde{g}) \in \tilde{\mathcal{F}},$$

whenever (u, v) satisfy (8.8). However, since $p = rp_0$, this is equivalent to (8.7). □

As we noted in the Introduction, a weaker version of Theorem 8.2 was proved in [54] using a duality argument which is essentially the same as the above proof. (See Remark 7.10.) There, the theorem was stated in terms of iterations of the maximal operator, M^k, $k \geq 1$: if (8.6) held for pairs $(u, M^{k_0}u)$, then (8.7) held for pairs of the form $(u, M^{k_p}u)$, where $k_p = [k_0 r] + 1$, $r = p/p_0$, and $[kr]$ is the largest integer less than or equal to $k_0 r$. Since $M^{k_p}u \approx M_\Phi u$, where $\Phi(t) = t\log(e+t)^{k_p - 1}$, by Theorem 6.4 the pair $(u, M^{k_p}u)$ satisfies (8.8).

In [54], a different extrapolation theorem was proved using norm inequalities for a weighted maximal operator: if (8.6) holds for pairs of weights of the form (w, Mw), then (8.7) holds for pairs of the form $(w, (Mw)^{p/p_0}w^{1-p/p_0})$. These pairs of weights are not comparable to those in Theorem 8.2: sometimes this result yields sharper estimates and other times it does not. To see this, we consider an example on $[0, \infty)$. Suppose that (8.6) holds for $p_0 = 1$ and $\Phi_0(t) = t$. Then by Theorem

6.4, pairs of the form $(u, M_\Phi u)$, $\Phi(t) = t\log(e+t)^{p-1+\delta}$, $\delta > 0$, satisfy (8.8), so (8.7) holds. Now let $w(x) = u(x) = \chi_{[0,1]} + x^{-1}\chi_{(1,\infty)}$. Then for x large,

$$Mw(x) \approx \frac{\log(x)}{x} \quad \text{and} \quad Mw(x)^p w(x)^{1-p} \approx \frac{\log(x)^p}{x},$$

but

$$M_\Phi u(x) \approx \frac{\log(x)^{p+\delta}}{x}.$$

Thus $(Mw)^p w^{1-p} \le M_\Phi u$, so the result in [54] yields a sharper inequality.

On the other hand, again by Theorem 6.4, dual A_1 pairs of the form $((Mu)^{1-p}, u^{1-p})$ satisfy (8.8). Define u as above but now let $w = (Mu)^{1-p}$. Then for x large,

$$w(x) = Mu(x)^{1-p} \approx \left(\frac{x}{\log(x)}\right)^{p-1},$$

and so $(Mw)^p w^{1-p} = +\infty$. Therefore, in this case Theorem 8.2 yields a sharper inequality.

8.2 Three special cases for the pairs (u, Mu)

In this section we consider Theorem 8.2 when inequality (8.6) holds for pairs of the form (u, Mu). Such inequalities appear in many applications; further, they provide a model for analyzing those situations where (8.6) holds with some larger Orlicz maximal function on the right-hand side. We will discuss three cases: (8.6) holds for $p_0 = 1$; for all p_0 such that $0 < p_0 < p_1$ for some $p_1 \le 1$; and for all p_0 such that $1 < p_0 < p_1$ for some $p_1 > 1$.

Case 1: $p_0 = 1$

Corollary 8.4. *Suppose that for all weights u_0,*

$$\int_{\mathbb{R}^n} f(x)u_0(x)\,dx \le C\int_{\mathbb{R}^n} g(x)Mu_0(x)\,dx, \qquad (f,g)\in\mathcal{F}. \tag{8.9}$$

Then for all $p > 1$,

$$\int_{\mathbb{R}^n} f(x)^p u(x)\,dx \le C\int_{\mathbb{R}^n} g(x)^p v(x)\,dx, \qquad (f,g)\in\mathcal{F}, \tag{8.10}$$

whenever (u, v) satisfy

$$\|u^{1/p}\|_{A,Q}\|v^{-1/p}\|_{p',Q} \le K < \infty, \tag{8.11}$$

where $\bar{A} \in B_{p'}$.

By Theorem 6.4, inequality (8.11) is satisfied by pairs of A_1 weights of the form $(u, M_\Phi u)$, where $\Phi(t) = A(t^{1/p})$. In particular, in the scale of log bumps we can take $\Phi(t) = t\log(e+t)^{p-1+\delta}$, $\delta > 0$. Similarly, we can take dual A_1 pairs of the form $((Mu)^{1-p}, u^{1-p})$.

Corollary 8.4 is not sharp. For example if we take

$$\mathcal{F} = \{(\lambda\chi_{\{x:Mf(x)>\lambda\}}, |f|), f \in L^1(Mu)\},$$

then (8.9) becomes the weak $(1,1)$ inequality for the Hardy-Littlewood maximal operator, and (8.10) becomes the weak (p,p) inequality. The necessary and sufficient condition for this inequality to hold is $(u,v) \in A_p$, which is strictly weaker than (8.11). Furthermore, pairs of the form (u, Mu) do not satisfy (8.11) but the strong (p,p) inequalities hold for these pairs by Theorems 5.14 and 6.4. On the other hand, the dual A_1 weights $((Mu)^{1-p}, u^{1-p})$ associated to (8.11) are the same as those associated to the two-weight A_p condition, so for weights of this form the inequality is sharp.

It is immediate from the proof why Corollary 8.4 is not sharp: the proof uses that the maximal operator satisfies a strong type inequality, so we lose information when considering weak-type inequalities for the maximal operator. In Chapters 9 and 10, however, we will see that for other, more singular operators this result does yield sharp inequalities.

Case 2: $0 < p_0 < p_1$

Corollary 8.5. *Suppose that for some $p_1 \le 1$ and for all p_0, $0 < p_0 < p_1$, and for all weights u_0,*

$$\int_{\mathbb{R}^n} f(x)^{p_0} u_0(x)\,dx \le C \int_{\mathbb{R}^n} g(x)^{p_0} Mu_0(x)\,dx, \quad (f,g) \in \mathcal{F}. \tag{8.12}$$

Then for all $p > 0$, (8.10) holds if for some $p_0 < \min(p_1, p)$ and $r = p/p_0 > p$, the pair (u,v) satisfies

$$\|u^{1/r}\|_{A,Q}\|v^{-1/r}\|_{r',Q} \le K < \infty, \tag{8.13}$$

where $\bar{A} \in B_{r'}$.

In the one-weight case, the reverse Hölder inequality implies that (8.13) is equivalent to $u \in A_r$ for some $r > 1$; that is, $u \in A_\infty$. Thus Corollary 8.5 is the two-weight analog of Proposition 3.20.

As we noted in Section 5.4 above, for an arbitrary A, if the pair (u,v) satisfies (8.13) for $r = p$, then it need not satisfy it for any value of $r > p$. However, if we restrict to the case of log bumps of the form $A(t) = t^p\log(e+t)^{p-1+\delta}$, $\delta > 0$, then it does hold for some $r > p$ (Proposition 5.20). This gives us a slightly sharper version of Corollary 8.5 that shows in this particular case assuming the endpoint inequality holds for either $p_0 = 1$ or $p_0 < 1$ yields the same conclusion.

Corollary 8.6. *Suppose that for all p_0, $0 < p_0 < 1$, inequality* (8.12) *holds. Then for all $p > 1$, if (u, v) satisfy* (8.11)*, where $A(t) = t^p \log(e+t)^{p-1+\delta}$, $\delta > 0$, then* (8.10) *holds. In particular, we can take A_1 weights of the form $(u, M_\Phi u)$, where $\Phi(t) = A(t^{1/p})$.*

Case 3: $1 < p_0 < p_1$

Corollary 8.7. *Suppose that for some $p_1 > 1$ and for all p_0, $1 < p_0 < p_1$,* (8.12) *holds. Then for all $p > 1$, if (u, v) satisfy*

$$\|u^{1/p}\|_{A,Q}\|v^{-1/p}\|_{B,Q} \le K < \infty, \tag{8.14}$$

where $\bar{A} \in B_{p'}$, $\bar{A}$ is doubling, and $B(t) = t^{sp'}$, $s > 1$, (8.10) *holds.*

Proof. Fix $p > 1$. By Hölder's inequality, we may assume that s is as close to 1 as desired. Therefore, after possibly changing s, we can find p_0, $1 < p_0 < \min(p, p_1)$, such that

$$\frac{p}{p_0} \ge \frac{p-1}{s} + 1.$$

By Proposition 5.24, if we define $r = p/p_0$, then $1 < r < p$, and so (u, v) satisfy

$$\|u^{1/r}\|_{C,Q}\|v^{-1/r}\|_{D,Q} \le K,$$

where $\bar{C} \in B_{r'}$ and $D(t) = t^{qr'}$, $q \ge 1$. Therefore, (applying Hölder's inequality if $q > 1$) the pair (u, v) satisfies the hypotheses of Theorem 8.2 and so (8.10) holds. □

In the one-weight case, the reverse Hölder inequality implies that (8.14) holds if $u \in A_p$. Thus Corollary 8.7 is the two-weight analog of Proposition 3.21.

In the proof of Corollary 8.7, since s may be small, we are using the hypothesis for values of p_0 very close to 1: that is, for r very close to p. But if we restrict ourselves to A_1 pairs, then this result is not optimal. For example, if $A(t) = t^p \log(e+t)^{p-1+\delta}$, then we get the A_1 pairs $(u, M_\Phi u)$, where $\Phi(t) = A(t^{1/p})$. Note that Φ does not depend on the value p_1. On the other hand, if we apply Theorem 8.2 directly and take values of p_0 very close to p_1, then we get the desired conclusion for A_1 pairs of the form $(u, M_\Psi u)$, where $\Psi(t) = t\log(e+t)^{p/p_0-1+\delta_0} = t\log(e+t)^{p/p_1-1+\delta}$, $\delta > 0$. As we will see in Chapter 10, such A_1 pairs are often the best possible.

8.3 The converse of endpoint extrapolation

In this section we prove a partial converse to Theorem 8.2, showing that endpoint estimates for A_1 pairs are a consequence of norm inequalities governed by suitable A_p bump conditions. To stress that we are looking for endpoint conditions, and to make comparison with Theorem 8.2 easier, we are going to reverse our usual notation of denoting weights and Young functions in the initial hypothesis with a subscript 0 and put this subscript on the weights and functions in the conclusion.

Theorem 8.8. *Given $p > 0$, suppose there exists $r > \max(p,1)$ such that if the weights (u,v) satisfy*

$$\|u^{1/r}\|_{A,Q}\|v^{-1/r}\|_{r',Q} \le K < \infty, \tag{8.15}$$

where A is some r-Young function, then

$$\int_{\mathbb{R}^n} f(x)^p u(x)\,dx \le C\int_{\mathbb{R}^n} g(x)^p v(x)\,dx, \quad (f,g)\in\mathcal{F}. \tag{8.16}$$

Then for all p_0, $p/r < p_0 < p$, and all weights u_0,

$$\int_{\mathbb{R}^n} f(x)^{p_0} u_0(x)\,dx \le C\int_{\mathbb{R}^n} g(x)^{p_0} M_{\Phi_0}u_0(x)\,dx, \quad (f,g)\in\mathcal{F}, \tag{8.17}$$

where $\Phi_0(t) = A(t^{1/r})$.

Proof. The proof follows from a rescaling argument. Define a new family

$$\tilde{\mathcal{F}} = \{(\tilde f,\tilde g) = (f^{p/r}, g^{p/r}) : (f,g)\in\mathcal{F}\}.$$

Then we can rewrite our hypothesis as: if (u,v) satisfy (8.15), then

$$\int_{\mathbb{R}^n} \tilde f(x)^r u(x)\,dx \le C\int_{\mathbb{R}^n} \tilde g(x)^r v(x)\,dx, \qquad (\tilde f,\tilde g)\in\tilde{\mathcal{F}}.$$

Fix r_0, $1 < r_0 < r$, and define A_0 by $A_0(t^{1/r_0}) = A(t^{1/r})$. Let B_0 be an r_0-Young function such that $\bar B_0 \in B_{r_0}$; then $\bar B_0(t^{1/r_0}) \preceq t$. Therefore, by Theorem 7.2, part (b), if the pair (u_0,v_0) satisfies

$$\|u_0^{1/r_0}\|_{A_0,Q}\|v_0^{-1/r_0}\|_{B_0,Q} \le K, \tag{8.18}$$

then

$$\int_{\mathbb{R}^n} \tilde f(x)^{r_0} u_0(x)\,dx \le C\int_{\mathbb{R}^n} \tilde g(x)^{r_0} v_0(x)\,dx, \qquad (\tilde f,\tilde g)\in\tilde{\mathcal{F}}. \tag{8.19}$$

In particular, by Theorem 6.4, (8.18) is satisfied by A_1 pairs of the form $(u_0, M_{\Phi_0}u_0)$, where $\Phi_0(t) = A_0(t^{1/r_0}) = A(t^{1/r})$. If we use this pair and let $p_0 = r_0 p/r$, then (8.19) is equivalent to (8.16), and the range of values for p_0 comes from the range for r_0. □

A priori, Theorem 8.8 may not yield the desired endpoint inequality. For example, given a fixed $p_0 > 0$, even if (8.16) holds for all $p > p_0 > 0$, the value of r associated to each p may be such that $p/r > p_0$. However, as the next two corollaries show, by making additional assumptions on the A_p bump condition we can reach the endpoint. In the first case, we have an A_∞-type condition that yields even more.

Corollary 8.9. *Let Φ_0 be a Young function. Given $p_1 > 0$, suppose that for all $p > p_1$ and for all $r > \max(p, 1)$ sufficiently large,* (8.16) *holds whenever (u, v) satisfy* (8.15) *with $A(t) = \Phi_0(t^r)$. Then for all p_0, $0 < p_0 \le p_1$ and all weights u_0,*

$$\int_{\mathbb{R}^n} f(x)^{p_0} u_0(x)\,dx \le C \int_{\mathbb{R}^n} g(x)^{p_0} M_{\Phi_0} u_0(x)\,dx, \quad (f, g) \in \mathcal{F}.$$

To prove this, fix p_0, $0 < p_0 \le p_1$, and take $p > p_1$ and $r > \max(p, 1)$ large enough that $p/r < p_0 < p$. Then we can apply Theorem 8.8.

In the second case we can reach the endpoint if we restrict ourselves to log bumps.

Corollary 8.10. *Given $p_0 \ge 1$, suppose that for all $p > p_0$,* (8.16) *holds whenever (u, v) satisfy* (8.15) *with $r = p$ and $A(t) = t^p \log(e + t)^{p-1+\delta}$, $\delta > 0$. Then* (8.17) *holds with $\Phi_0(t) = t \log(e + t)^{p_0-1+\epsilon}$, for any $\epsilon > 0$.*

Proof. Fix $\epsilon > 0$, and fix $p > p_0$ sufficiently close to p_0 so that there exists $\delta > 0$ such that $p - 1 + \delta = p_0 - 1 + \epsilon$. As in the proof of Proposition 5.20, (8.15) holds with A replaced by $\tilde{A} = t^r \log(e + t)^{r-1+\tilde{\delta}}$, where $r > p$ and $\tilde{\delta} > 0$ are such that $p - 1 + \delta = r - 1 + \tilde{\delta}$. Since $p/r < 1 \le p_0$, the desired conclusion follows from Theorem 8.8. □

Even though we can extrapolate to prove an endpoint result, we may not recapture the sharp result. For example, if we assume that

$$\int_{\mathbb{R}^n} f(x) u_0(x)\,dx \le C \int_{\mathbb{R}^n} g(x) M u_0(x)\,dx, \qquad (f, g) \in \mathcal{F}, \tag{8.20}$$

then by Corollary 8.4, the hypotheses of Corollary 8.10 hold $p_0 = 1$. However, by applying this result we do not recapture (8.20) but rather this inequality for the A_1 pairs $(u_0, M_{\Phi_0} u_0)$, where $\Phi_0(t) = t \log(e + t)^\epsilon$, $\epsilon > 0$.

We conclude this section with an extension of Theorem 7.12 that shows that an endpoint result can be gotten if we restrict ourselves to weak type inequalities.

Theorem 8.11. *Given p, $1 < p < \infty$, and a p-Young function A, suppose that*

$$\|f\|_{L^{p,\infty}(u)} \le C \|g\|_{L^p(v)}, \qquad (f, g) \in \mathcal{F}, \tag{8.21}$$

holds for every pair of weights (u, v) that satisfy

$$\|u^{1/p}\|_{A,Q} \|v^{-1/p}\|_{p',Q} \le K < \infty. \tag{8.22}$$

Then,

$$\|f\|_{L^{1,\infty}(u_0)} \le C \|g\|_{L^1(M_{\Phi_0} u_0)}, \qquad (f, g) \in \mathcal{F},$$

where $\Phi_0(t) = A(t^{1/p})$.

Proof. The proof is essentially the same as the proof of Theorem 7.12 and requires only minor changes that we sketch here. For brevity, let $v_0 = M_{\Phi_0}u_0$. Since $u_0 \le Mu_0 \le v_0$, by inequality (5.13) $M : L^1(v_0) \to L^{1,\infty}(u_0)$.

Next, we prove that the pair of weights

$$(U_0, V_0) = (G_0^{-(p-1)}\, u_0, g_0^{-(p-1)}\, v_0),$$

with

$$g_0 = \frac{g}{\|g\|_{L^1(v_0)}} \quad \text{and} \quad G_0 = \frac{f}{\|f\|_{L^{1,\infty}(u_0)}} + Mg_0,$$

satisfies (8.22). Since $\Phi_0(t) = A(t^{1/p})$, we have that

$$\begin{aligned}\|U_0^{1/p}\|_{A,Q} = \|G_0^{-1/p'}\, u_0^{1/p}\|_{A,Q} &\le \|Mg_0^{-1/p'}\, u_0^{1/p}\|_{A,Q} \\ &\le \|g_0\|_{1,Q}^{-1/p'}\, \|u_0^{1/p}\|_{A,Q} = \|g_0\|_{1,Q}^{-1/p'}\, \|u_0\|_{\Phi_0,Q}^{1/p}.\end{aligned}$$

Since the pair (u_0, v_0) satisfies $\|u_0\|_{\Phi_0,Q} \le M_{\Phi_0}u_0(x) = v_0(x)$, $x \in Q$, we have that

$$\begin{aligned}\|V_0^{-1/p}\|_{p',Q} &= \|g_0^{1/p'}\, v_0^{-1/p}\|_{p',Q} \\ &\le \|g_0^{1/p'}\|_{p',Q}\|u_0\|_{\Phi_0,Q}^{-1/p} = \|g_0\|_{1,Q}^{1/p'}\, \|u_0\|_{\Phi_0,Q}^{-1/p}.\end{aligned}$$

As we assumed in Chapter 7, $g \in \cap_{p>1} L^p$, so $g_0 \in L^1_{\mathrm{loc}}$. Therefore, $\|g_0\|_{1,Q} < \infty$. From these inequalities we see that the pair (U_0, V_0) satisfies (8.22). The rest of the proof is the same as the proof of Theorem 7.12 and we omit the details. □

8.4 Endpoint extrapolation for fractional operators

In this section we briefly consider endpoint extrapolation for fractional operators. We first prove the analog of Theorem 8.2 for $p_0 = 1$.

Theorem 8.12. *Given α, $0 < \alpha < n$, suppose that for every weight u_0,*

$$\int_{\mathbb{R}^n} f(x)u_0(x)\,dx \le C_0 \int g(x)M_{\Phi_0,\alpha}u_0(x)\,dx, \qquad (f, g) \in \mathcal{F}. \tag{8.23}$$

Then for all p, $1 < p < \infty$,

$$\int_{\mathbb{R}^n} f(x)^p u(x)\,dx \le C \int_{\mathbb{R}^n} g(x)^p v(x)\,dx,$$

whenever (u, v) satisfy

$$|Q|^{\alpha/n}\|u^{1/p}\|_{A,Q}\, \|v^{-1/p}\|_{p',Q} \le K < \infty, \tag{8.24}$$

where A is a p-Young function, C is a p'-Young functions such that $C \in B_{p'}$, and $A^{-1}(t)C^{-1}(t) \le \Phi_0^{-1}(t)$.

The proof of Theorem 8.12 is nearly identical to the proof of Theorem 8.2 when $p_0 = 1$, replacing M_{Φ_0} with $M_{\Phi_0,\alpha}$ and using Theorem 5.37 to get the boundedness of this operator.

There are two ways to generalize Theorem 8.12 to an endpoint result starting from an arbitrary $p_0 > 0$, but neither seems satisfactory. If we replace (f, g) by (f^{p_0}, g^{p_0}) in (8.23), then the rescaling argument in the proof of Theorem 8.2 goes through exactly as before. However, as we noted in Section 6.4, the A_1 pairs for fractional operators depend on p, so when $p_0 \neq 1$ the initial hypothesis is not appropriate. Instead, we should assume that (8.23) holds with $M_{\Phi_0,\alpha}$ replaced with $M_{\Phi_0,p_0\alpha}$. However, for the rescaling argument to work in this case we need to modify (8.24) and assume that the weights satisfy this condition with $|Q|^{\alpha/n}$ replaced by $|Q|^{\alpha p/p_0 n}$. Again, based on known results for fractional maximal operators (e.g., Theorem 5.37) this does not appear to be the correct condition.

A similar problem arises when we adapt the proof of Theorem 8.8 to the fractional case: we do not get the expected fractional A_1 pairs.

Theorem 8.13. *Suppose that for some α, $0 < \alpha < n$, and some $p > 0$, there exists $r > \max(p, 1)$, such that if the weights (u, v) satisfy*

$$|Q|^{\alpha/n}\|u^{1/r}\|_{A,Q}\|v^{-1/r}\|_{r',Q} \leq K < \infty, \tag{8.25}$$

where A is some r-Young function, then

$$\int_{\mathbb{R}^n} f(x)^p u(x)\,dx \leq C\int_{\mathbb{R}^n} g(x)^p v(x)\,dx, \quad (f, g) \in \mathcal{F}. \tag{8.26}$$

Then for all p_0, $p/r < p_0 < p$, and all weights u_0,

$$\int_{\mathbb{R}^n} f(x)^{p_0} u_0(x)\,dx \leq C\int_{\mathbb{R}^n} g(x)^{p_0} M_{\Phi_0,\alpha p_0 r/p} u_0(x)\,dx, \quad (f, g) \in \mathcal{F}, \tag{8.27}$$

where $\Phi_0(t) = A(t^{1/r})$.

The proof is exactly the same as the proof of Theorem 8.8, replacing Theorem 7.2 with Theorem 7.19, and Theorem 6.4 with Theorem 6.18.

Remark 8.14. The analogs of Corollaries 8.9 and 8.10 are also true; we leave their statement and proof to the interested reader.

We conclude this section with the analog of Theorem 8.11. In this case we do get the correct fractional A_1 pair.

Theorem 8.15. *Given p, $1 < p < \infty$, and a p-Young function A, suppose that* (8.21) *holds for every pair of weights (u,v) that satisfy*

$$|Q|^{\alpha/n}\|u^{1/p}\|_{A,Q}\|v^{-1/p}\|_{p',Q} \le K < \infty. \tag{8.28}$$

Then,

$$\|f\|_{L^{1,\infty}(u_0)} \le C\|g\|_{L^1(M_{\Phi_0,\alpha}u_0)}, \qquad (f,g) \in \mathcal{F},$$

where $\Phi_0(t) = A(t^{1/p})$.

Proof of Theorem 8.15. The proof is essentially the same as that of Theorem 8.11 and we sketch the changes. Let $v_0 = M_{\Phi_0,\alpha}u_0$; then, since $t \preceq \Phi_0(t)$, by (5.26), $M_\alpha : L^1(v_0) \to L^{1,\infty}(u_0)$. We can then define G_0 with M_α replacing M. Finally, the pair (U_0, V_0) satisfies (8.28) since

$$\|U_0^{1/p}\|_{A,Q} \le |Q|^{-\frac{\alpha(p-1)}{np}}\|g_0\|_{1,Q}^{-1/p'}\|u_0\|_{\Phi_0,Q}^{1/p}$$

and

$$\|V_0^{-1/p}\|_{p',Q} \le |Q|^{-\frac{\alpha}{np}}\|g_0\|_{1,Q}^{1/p'}\|u_0\|_{\Phi_0,Q}^{-1/p}.$$

The rest of the proof is the same as before. □

Chapter 9

Applications of Two-Weight Extrapolation

In this chapter and the next we apply the two-weight extrapolation theorems in Chapters 7 and 8 to the theory of two-weight norm inequalities. We are interested in proving inequalities of the form

$$\int_{\mathbb{R}^n} |Tf(x)|^p u(x)\,dx \le C \int_{\mathbb{R}^n} |f(x)|^p v(x)\,dx$$

and

$$u(\{x \in \mathbb{R}^n : |Tf(x)| > \lambda\}) \le \frac{C}{\lambda^p} \int_{\mathbb{R}^n} |f(x)|^p v(x)\,dx,$$

where the pair of weights (u, v) satisfies an A_p bump condition

$$\|u^{1/p}\|_{A,Q} \|v^{-1/p}\|_{B,Q} \le K < \infty,$$

for given Young functions A and B. We considered inequalities of this type in a number of papers [33, 47, 53, 55, 56, 137, 171, 172], and as we noted before, some of our extrapolation results were implicit in them.

However, despite the progress which has been made, the theory of two-weight norm inequalities is not as well established as the one-weight theory: after thirty years, fundamental questions remain open. Progress has been incremental, and seemingly modest improvements have required the development of sophisticated techniques. Two-weight extrapolation is another such technique. Our goal in this chapter is two-fold: to prove new results and to show that known results, often with difficult proofs, are easy consequences of extrapolation.

In this chapter we consider three operators: the sharp maximal function of Fefferman and Stein, Calderón-Zygmund singular integral operators, and fractional integral operators. Our goal for the sharp maximal function is to prove inequalities of the form

$$\int_{\mathbb{R}^n} |f(x)|^p u(x)\,dx \le C \int_{\mathbb{R}^n} M^\# f(x)^p v(x)\,dx$$

that generalize the Fefferman-Stein inequality [76]. As an application we develop a very general theory of two-weight Coifman-Fefferman inequalities of the form

$$\int_{\mathbb{R}^n} |Tf(x)|^p u(x)\,dx \le C \int_{\mathbb{R}^n} |Sf(x)|^p v(x)\,dx.$$

These are of interest in their own right, and the techniques we develop will be applied to specific operators.

For both singular and fractional integrals we first state very general conjectures for two-weight, strong and weak type inequalities. These conjectures are based on the early (and still open) conjectures of Muckenhoupt and Wheeden for the Hilbert transform discussed in Chapter 1. We then review the known theorems, in some cases giving new proofs, and prove new results. We want to stress again that even though some of these seem to be only small improvements over known results, they represent real progress. In particular, factored weights give an entirely new approach to two-weight norm inequalities.

Throughout this chapter, to apply the extrapolation theorems we have to define an appropriate family $\mathcal{F}$ of pairs of functions. For each operator and for weak and strong type inequalities these are different, but in general will be gotten from a family of functions, such as C_c^∞, that is dense in $L^p(u)$. Because of this, standard arguments will prove that the inequalities hold in general.

Remark 9.1. Further applications are given in Chapter 10, where we consider the dyadic square function and the vector-valued maximal operator.

9.1 The sharp maximal operator

In this section we consider the sharp maximal function of Fefferman and Stein [76]. Given a locally integrable function f and a cube Q, let

$$f_Q = \fint_Q f(y)\,dy,$$

and define the sharp maximal function of f by

$$M^\# f(x) = \sup_{Q \ni x} \fint_Q |f(y) - f_Q|\,dy. \tag{9.1}$$

To apply our extrapolation results, let the family $\mathcal{F}$ consist of the pairs $(f, M^\# f)$, where f is a non-negative function such that $f \in \cup_{p>1} L^p$. Note that for all $\lambda > 0$, $\{x \in \mathbb{R}^n : f(x) > \lambda\}$ has finite measure. For brevity we will write $f \in \mathcal{F}$ if the associated pair is contained in $\mathcal{F}$.

Our main result is a two-weight generalization of the A_∞ inequality relating f and $M^\# f$ in [76]. We first give a lemma that is a factored weight inequality of interest in its own right.

Lemma 9.2. *For all p_0, $0 < p_0 \leq 1$, and all $f \in \mathcal{F}$,*

$$\int_{\mathbb{R}^n} f(x)^{p_0} \tilde{u}(x)\, dx \leq C \int_{\mathbb{R}^n} M^\# f(x)^{p_0} \tilde{v}(x)\, dx, \tag{9.2}$$

where $(\tilde{u}, \tilde{v})$ is any pair of factored weights of the form

$$(w_1 (M_\Psi w_2)^{1-r}, (M w_1) w_2^{1-r}),$$

with $r \geq 1$ and Ψ any Young function. In particular, we can take the A_1 pair (u, Mu).

Remark 9.3. As a corollary to the proof we will see that we can replace $\tilde{v}$ on the right-hand side of (9.2) by the smaller weight $Mw_1(M_\Psi w_2)^{1-r}$.

Lemma 9.2 is a generalization of a result due to Lerner [125], who proved (9.2) for $p_0 = 1$, and for the A_1 pairs (u, Mu). Our proof is a modification of his and we postpone it to the end of this section.

Theorem 9.4. *Given p, $0 < p < \infty$, suppose there exist r, $r > 1$ and $r \geq p$, and $\bar{A} \in B_{r'}$, such that the pair (u, v) satisfies*

$$\|u^{1/r}\|_{A,Q} \|v^{-1/r}\|_{r',Q} \leq K < \infty. \tag{9.3}$$

Then for all $f \in \mathcal{F}$,

$$\int_{\mathbb{R}^n} f(x)^p u(x)\, dx \leq C \int_{\mathbb{R}^n} M^\# f(x)^p v(x)\, dx. \tag{9.4}$$

In particular (9.4) holds for A_1 pairs of the form $(u, M_\Phi u)$, where $\Phi(t) = A(t^{1/r})$, and dual A_1 pairs of the form $((Mu)^{1-r}, u^{1-r})$.

Given Lemma 9.2, Theorem 9.4 follows immediately if we apply Corollaries 8.4 and 8.5 (the latter with $p_1 = 1$) to (9.2) with the A_1 pair (u, Mu).

Remark 9.5. It is immediate that in the one-weight case we can take $w \in A_1$ in Lemma 9.2 and $w \in A_\infty$ in Theorem 9.4. This establishes the Fefferman-Stein inequality (3.32) discussed in Section 3.8, which is the main tool for proving one-weight Coifman-Fefferman inequalities. The two-weight case is discussed later in this section.

Remark 9.6. The techniques used to prove the main result in [17] can be modified to prove Theorem 9.4. Briefly, the proof there combines the decomposition argument used to prove Lemma 9.2 with another argument that, for a fixed cube Q, decomposes f into two functions whose supports are $2Q$ and $\mathbb{R}^n \setminus 2Q$. Details are left to the interested reader. Note that the above proof using extrapolation is considerably simpler.

If we restrict to the special case of log bumps, we can prove the analog of Theorem 9.4 for weighted $L^{p,\infty}$.

Theorem 9.7. *Given p, $0 < p < \infty$, suppose there exists r, $r > 1$ and $r \geq p$, such that the pair of weights (u, v) satisfies*

$$\|u^{1/r}\|_{A,Q}\|v^{-1/r}\|_{r',Q} \leq K < \infty, \tag{9.5}$$

where $A(t) = t^r \log(e+t)^{r-1+\delta}$, $\delta > 0$. Then for all $f \in \mathcal{F}$,

$$\|f\|_{L^{p,\infty}(u)} \leq C\|M^\# f\|_{L^{p,\infty}(v)}. \tag{9.6}$$

Corollary 9.8. *With the same hypotheses as Theorem 9.7, if $0 < p \leq 1$, then we can take the pair $(u, M_\Phi u)$, where $\Phi(t) = t\log(e+t)^\epsilon$, $\epsilon > 0$.*

Proof of Theorem 9.7. The proof uses an interpolation argument that first appeared in Carro, *et al.* [21].

Fix r and δ, and choose $r_1 > r$ and $\delta_1 > 0$ such that $r-1+\delta = r_1-1+\delta_1$. Then as we showed in Proposition 5.20,

$$\|u^{1/r_1}\|_{A_1,Q}\|v^{-1/r_1}\|_{r_1',Q} \leq K < \infty, \tag{9.7}$$

where $A_1(t) = t^{r_1}\log(e+t)^{r_1-1+\delta_1}$. Now fix

$$\epsilon = \frac{r_1 - r}{2}, \qquad q = \frac{p}{r+\epsilon} = \frac{p}{r_1-\epsilon};$$

then $q < \min(1, p)$ and

$$r = \frac{p}{q} - \epsilon, \qquad r_1 = \frac{p}{q} + \epsilon.$$

We now argue by duality. Since $p/q > 1$ we have that

$$\|f\|^q_{L^{p,\infty}(u)} = \|f^q\|_{L^{p/q,\infty}(u)} = \sup_h \int_{\mathbb{R}^n} f(x)^q u(x)h(x)\,dx,$$

where $\|h\|_{L^{(p/q)',1}(u)} = 1$. Fix such an h. Then by Lemma 9.2,

$$\begin{aligned}
\int_{\mathbb{R}^n} f(x)^q u(x)h(x)\,dx &\leq C\int_{\mathbb{R}^n} M^\# f(x)^q M(uh)(x)\,dx \\
&= C\int_{\mathbb{R}^n} M^\# f(x)^q M(uh)(x)v(x)^{-1}v(x)\,dx \\
&\leq C\|(M^\# f)^q\|_{L^{p/q,\infty}(v)}\|M(uh)v^{-1}\|_{L^{(p/q)',1}(v)} \\
&= C\|M^\# f\|^q_{L^{p,\infty}(v)}\|M(uh)v^{-1}\|_{L^{(p/q)',1}(v)}.
\end{aligned}$$

To complete the proof we need to show that

$$\|M(uh)v^{-1}\|_{L^{(p/q)',1}(v)} \leq C\|h\|_{L^{(p/q)',1}(u)}.$$

Since $r < p/q < r_1$, by interpolation in the scale of Lorentz spaces (see Stein and Weiss [218]) it will suffice to show that

$$\|M(uh)v^{-1}\|_{L^{r'}(v)} \leq C\|h\|_{L^{r'}(u)},$$
$$\|M(uh)v^{-1}\|_{L^{r'_1}(v)} \leq C\|h\|_{L^{r'_1}(u)},$$

or equivalently,

$$\|M(uh)\|_{L^{r'}(v^{1-r'})} \leq C\|hu\|_{L^{r'}(u^{1-r'})},$$
$$\|M(uh)\|_{L^{r'_1}(v^{1-r'_1})} \leq C\|hu\|_{L^{r'_1}(u^{1-r'_1})}.$$

By Theorem 5.14 these follow from (9.5) and (9.7). □

Proof of Corollary 9.8. Given p, $0 < p \leq 1$ and $\epsilon > 0$, choose $r > 1$ and $\delta > 0$ such that $r-1+\delta = \epsilon$. Then by Theorem 6.4, the pair $(u, M_\Phi u)$, with $\Phi(t) = t\log(e+t)^\epsilon$ satisfies (9.5). □

Fujii [81], via a very complicated decomposition argument, showed that (9.4) holds for all p, $0 < p < \infty$, if the pair (u, v) satisfies a different A_∞ type condition: there exist $0 < \alpha < 1$, $0 < \beta \leq 1$, and $C > 0$, such that, given any cube Q and $E \subset Q$ with $|E| < \alpha|Q|$,

$$u(E) \leq C\left(\frac{|E|}{|Q|}\right)^\beta v(Q \setminus E). \tag{9.8}$$

In the one-weight case, (9.8) is equivalent to the A_∞ condition. There is some overlap between this condition and (9.3), as there are pairs of weights that satisfy both, but there are also pairs that satisfy (9.3) but not (9.8). This is natural since (9.8) implies a stronger conclusion. However it is an open question whether (9.8) implies (9.3).

For example, consider pairs of the form

$$(u, v) = (w_1(Mw_2)^{1-r}, (M_\Phi w_1)w_2^{1-r}),$$

where $s, r > 1$ and $\Phi(t) = t^s$. By Theorem 6.4, this pair satisfies (9.3) with $A(t) = t^{sr}$. If we let $w = (M_\Phi w_1)(Mw_2)^{1-r}$, then by Proposition 6.2 and the reverse factorization property, $w \in A_q$, $q > r$. Further, $c_1 u \leq w \leq c_2 v$, which in turn immediately implies that (9.8) holds.

On the other hand, given an integer k, if $k > r - 1$, then by Theorem 6.4 the pair $(u, M_\Phi u)$, where $\Phi(t) = t\log(e+t)^k$, satisfies (9.3) with $A(t) = t^r\log(e+t)^{r-1+\delta}$ for some $\delta > 0$, but does not satisfy (9.8). Let

$$u(x) = \chi_{[0,1]} + \frac{1}{x}\chi_{(1,\infty)}.$$

Then for all $x > 0$ sufficiently large,

$$M_\Phi u(x) \approx M^{k+1}u(x) \approx \frac{\log(x)^{k+1}}{x}.$$

Fix $t > 0$ large, and let $E = [1, t]$ and $Q = [1, jt]$, $j > 1$. Then $u(E) = \log(t)$; on the other hand,

$$\left(\frac{|E|}{|Q|}\right)^\beta \int_{Q \setminus E} M_\Phi u(x)\, dx \approx \left(\frac{1}{j}\right)^\beta \left(\log(jt)^{k+2} - \log(t)^{k+2}\right).$$

Given any $\beta > 0$, the right-hand side tends to 0 as $j \to \infty$; it follows that (9.8) cannot hold.

Coifman-Fefferman type inequalities

As an application of the two-weight inequalities for the sharp maximal function, we give a general theory of two-weight Coifman-Fefferman type inequalities. As we discussed in Section 3.8, in the one-weight case these are norm inequalities of the form

$$\int_{\mathbb{R}^n} |Tf(x)|^p w(x)\, dx \le C \int_{\mathbb{R}^n} |Sf(x)|^p w(x)\, dx, \tag{9.9}$$

where S and T are operators (with S usually some kind of maximal operator), $0 < p < \infty$, and $w \in A_\infty$. Perhaps the most well-known example of such an inequality is for T a singular integral and S the Hardy-Littlewood maximal operator and is due to Coifman and Fefferman [24, 25]. However, there are two earlier examples in the literature. Fefferman and Stein [76] considered the case when T is the maximal operator and S is the sharp maximal operator; Burkholder and Gundy [17] considered square functions and maximal operators associated with martingales. Other well-known inequalities of this type relate fractional integrals and the fractional maximal operator (Muckenhoupt and Wheeden [154]), and square functions and the maximal operator (Gundy and Wheeden [94], Muckenhoupt and Wheeden [153]).

Originally, these results were proved using good-λ inequalities, an approach which cannot be extended to the two-weight setting. In [44] we introduced a new method that combined extrapolation theory and norm inequalities for the sharp maximal function. To state our results, we need a variant of the sharp maximal operator. For $0 < q \le 1$, let $M_q^\# f(x) = M^\#(|f|^q)(x)^{1/q}$. We say that a pair of operators S, T, satisfies a (pointwise) sharp function estimate if for all q, $0 < q < 1$, $M_q^\#(Tf)(x) \le C_q Sf(x)$ for all $f \in \mathcal{F}$, where $\mathcal{F} \subset \cup_{p>1} L^p$, and $x \in \mathbb{R}^n$. (We discuss examples of such pairs below.) Given such pairs of operators, we can use Theorems 9.4 and 9.7 to prove two-weight Coifman-Fefferman inequalities.

Theorem 9.9. *Let S and T be a pair of operators such that for all q, $0 < q < 1$, and all $f \in \mathcal{F}$, $M_q^\#(Tf)(x) \le C_q Sf(x)$. Fix p, $0 < p < \infty$. Suppose there exists $r > \max(p, 1)$ such that the pair (u, v) satisfies*

$$\|u^{1/r}\|_{A,Q}\|v^{-1/r}\|_{r',Q} \le K < \infty, \tag{9.10}$$

where $\bar{A} \in B_{r'}$. *Then*

$$\int_{\mathbb{R}^n} |Tf(x)|^p u(x)\,dx \le C \int_{\mathbb{R}^n} Sf(x)^p v(x)\,dx, \qquad f \in \mathcal{F}. \tag{9.11}$$

In particular we can take the A_1 *pair* $(u, M_\Phi u)$, *with* $\Phi(t) = A(t^{1/p})$.

In the scale of log bumps, if $p > 1$, *we can take* $r = p$ *and* $A(t) = t^p \log(e+t)^{p-1+\delta}$, $\delta > 0$. *If* $p = 1$, *we can take the* A_1 *pair* $(u, M_\Phi u)$, $\Phi(t) = t\log(e+t)^\epsilon$, $\epsilon > 0$. *If* $0 < p < 1$, *we can take the pair* (u, Mu).

Proof. Fix r and p, and take q, $0 < q < 1$, such that $r = p/q$. Then by Theorem 9.4 with $p = r$ we have that

$$\begin{aligned}\|Tf\|_{L^p(u)} = \||Tf|^q\|_{L^r(u)}^q \le C\|M^\#(|Tf|^q)\|_{L^r(v)}^q \\ = C\|M_q^\#(Tf)^q\|_{L^r(v)}^q \le C\|(Sf)^q\|_{L^r(v)}^q = C\|Sf\|_{L^p(v)}.\end{aligned}$$

In the scale of log-bumps, if $p > 1$, let $A(t) = t^p \log(e+t)^{p-1+\delta}$, $\delta > 0$. Then by Proposition 5.20, the pair (u, v) satisfies (9.10) for some $r > p$ with A replaced by $A_0(t) = t^r \log(e+t)^{r-1+\tilde{\delta}}$. Therefore, we can apply the above argument.

If $p = 1$, fix $\epsilon > 0$. Then there exist $r > 1$ and $\delta > 0$ such that $r - 1 + \delta = \epsilon$. Hence, if we let $A(t) = t^r \log(e+t)^{r-1+\delta}$, then by Theorem 6.4 the pair $(u, M_\Phi u)$ satisfies (9.10) and we can argue as before.

Finally, if $0 < p < 1$, fix q, $p < q < 1$. Then by Lemma 9.2 with $p_0 = p/q < 1$,

$$\int_{\mathbb{R}^n} |Tf(x)|^p u(x)\,dx \le C \int_{\mathbb{R}^n} M_q^\#(Tf)(x)^p Mu(x)\,dx \le C \int_{\mathbb{R}^n} Sf(x)^p Mu(x)\,dx. \qquad \square$$

Theorem 9.10. *Let* S *and* T *be a pair of operators such that for all* q, $0 < q < 1$, *and all* $f \in \mathcal{F}$, $M_q^\#(Tf)(x) \le C_q Sf(x)$. *Fix* p, $0 < p < \infty$, *and* r *such that* $r > 1$ *and* $r \ge p$. *If the pair* (u, v) *satisfies*

$$\|u^{1/r}\|_{A,Q}\|v^{-1/r}\|_{r',Q} \le K < \infty, \tag{9.12}$$

where $A(t) = t^r \log(e+t)^{r-1+\delta}$, $\delta > 0$, *then*

$$\|Tf\|_{L^{p,\infty}(u)} \le C\|Sf\|_{L^{p,\infty}(v)}, \qquad f \in \mathcal{F}. \tag{9.13}$$

If $0 < p \le 1$, *we can take the pair* $(u, M_\Phi u)$, $\Phi(t) = t\log(e+t)^\epsilon$, $\epsilon > 0$.

Proof. The proof of the first part of Theorem 9.10 is exactly the same as the proof of Theorem 9.9 for log bumps, replacing Theorem 9.4 with Theorem 9.7 and replacing the Lebesgue norm with the appropriate Lorentz norm. The proof of the second part for A_1 pairs is the same as the $p = 1$ case, but using Corollary 9.8. $\square$

Remark 9.11. In the one-weight case we can take $w \in A_\infty$ in Theorems 9.9 and 9.10. This yields the classical Coifman-Fefferman inequalities (9.9) without using extrapolation.

Remark 9.12. The C_p condition of Muckenhoupt [151], which is a more general condition implying one-weight Coifman-Fefferman inequalities, has a natural two-weight generalization:

$$\int_E u(x)\,dx \le C\left(\frac{|E|}{|Q|}\right)^{\epsilon} \int_{\mathbb{R}^n} M(\chi_Q)(x)^p v(x)\,dx, \tag{9.14}$$

where Q is any cube and $E \subset Q$. It is not known if this condition is sufficient for two-weight Coifman-Fefferman inequalities. However, in this connection, note that the condition (9.8) of Fujii discussed above implies (9.14) for all $p > 0$.

Remark 9.13. If we assume a stronger bump condition on the pair (u, v), then we can prove Theorems 9.9 and 9.10 without using the sharp function estimate. Suppose the operators S and T satisfy the one-weight inequality (9.9) for some $p > 0$; then by Corollary 3.15 it holds for all p, $0 < p < \infty$. Suppose that for some $p_0 > 1$ the pair (u, v) satisfies

$$\|u^{1/p_0}\|_{A,Q}\|v^{-1/p_0}\|_{B,Q} \le K < \infty,$$

with $A(t) = t^{sp_0}$, $s > 1$, and $\bar{B} \in B_{p_0}$. Then by Theorem 6.13 we can insert an A_∞ weight between u and v; hence, inequality (9.11) holds for all p. The Lorentz space inequality (9.13) is also true since (9.9) implies the one-weight Lorentz space inequality by A_∞ extrapolation.

In order to apply Theorems 9.9 and 9.10 we need pairs of operators that satisfy the sharp function estimate $M_q^{\#}(Tf)(x) \le C_q Sf(x)$. A number of such inequalities are known and we record them here.

(a) $M_q^{\#}(Tf)(x) \le CMf(x)$, $0 < q < 1$, where T is a Calderón-Zygmund singular integral. See Álvarez, Hounie and Pérez [4, 3]. They show that this inequality holds for other integral operators with more singular kernels.

(b) $M_q^{\#}(Mf)(x) \le CM^{\#}f(x)$, $0 < q < 1$; see [44].

(c) $M_q^{\#}(I_\alpha f)(x) \le CM_\alpha f(x)$, $0 < q \le 1$, where I_α, $0 < \alpha < n$, is the fractional integral. See Adams [1] for the case $q = 1$; when $q < 1$ the proof is essentially the same.

(d) $M_q^{\#}(g_\lambda^* f)(x) \le CMf(x)$, $0 < q < 1$, where g_λ^*, $\lambda > 2$, is the Littlewood-Paley square function. See [56].

(e) $M_q^{\#}(S_d f)(x) \le CM^d f(x)$, $0 < q < 1$, where M^d is the dyadic maximal function and S_d is the dyadic square function; this is proved in Section 10.1 below.

(f) $M_q^{\#}(\|Tf_j\|_{\ell^r})(x) \le CM(\|f_j\|_{\ell^r})(x)$, $0 < q < 1$, where T is a singular integral and $1 < r < \infty$; see Pérez and Trujillo-Gonzalez [178].

(g) $M_q^\#(\overline{N}_r f)(x) \leq CM(\|f_i\|_{\ell^r})(x)$, $0 < q < 1$, where $\overline{N}_r$, $1 < r < \infty$, is the vector-valued maximal operator associated with a smooth analog of the Hardy-Littlewood maximal operator. See Curbera *et al.* [57]. The smooth maximal operator is pointwise comparable to the Hardy-Littlewood maximal operator, so this inequality can be used to deduce Coifman-Fefferman type inequalities for $\overline{M}_r$.

Later in this chapter and in Chapter 10 we will give applications of these results to specific operators.

Remark 9.14. Commutators $[T, b]$, $b \in BMO$, are more singular and do not satisfy estimates of these types. However, there are analogous inequalities which can be used to deduce two-weight inequalities. For commutators of singular integrals see [173, 175]; for commutators of fractional integrals see [36].

Example 9.15. Theorem 9.9 lets us deduce weighted norm inequalities for the operator T. Here we illustrate this with a representative example for A_1 pairs. Suppose T satisfies the sharp function estimate $M_q^\#(Tf)(x) \leq CMf(x)$, $0 < q < 1$. Then for $1 < p < \infty$ and $\Phi(t) = t\log(e+t)^{p-1+\delta}$, the pair $(M_\Phi u, M(M_\Phi u))$ satisfies the bump condition (5.11). Hence, by Theorems 9.9 and 5.14, and by Example 5.30,

$$\begin{aligned}\int_{\mathbb{R}^n} |Tf(x)|^p u(x)\,dx &\leq C\int_{\mathbb{R}^n} Mf(x)^p M_\Phi u(x)\,dx \\ &\leq C\int_{\mathbb{R}^n} |f(x)|^p M(M_\Phi u)(x)\,dx \leq C\int_{\mathbb{R}^n} |f(x)|^p M_{\Phi_0} u(x)\,dx,\end{aligned}$$

where $\Phi_0(t) = t\log(e+t)^{p+\delta}$.

Example 9.16. The inequality in Example 9.15 need not be sharp. Assume T is self-adjoint. By Proposition 6.2, $(M_\Phi u)^{1-p'} \in A_\infty$. Therefore, by the one-weight Fefferman-Stein inequality and Corollary 6.5,

$$\begin{aligned}\int_{\mathbb{R}^n} |Tf(x)|^{p'} M_\Phi u(x)^{1-p'}\,dx &\leq C\int_{\mathbb{R}^n} M_q^\#(|Tf|)(x)^{p'} M_\Phi u(x)^{1-p'}\,dx \\ &\leq C\int_{\mathbb{R}^n} Mf(x)^{p'} M_\Phi u(x)^{1-p'}\,dx \\ &\leq C\int_{\mathbb{R}^n} |f(x)|^{p'} u(x)^{1-p'}\,dx.\end{aligned}$$

Hence, by duality,

$$\int_{\mathbb{R}^n} |Tf(x)|^p u(x)\,dx \leq C\int_{\mathbb{R}^n} |f(x)|^p M_\Phi u(x)\,dx.$$

Remark 9.17. This duality argument was first used in [172] to prove norm inequalities for singular integrals.

Proof of Lemma 9.2

Our proof is a modification of the original proof by Lerner [125]. First, by an approximation argument using the monotone convergence theorem, we may assume without loss of generality that w_1 is a bounded function of compact support and that $w_2 \geq \delta > 0$. In particular, $\tilde{u} \in L^\infty$. Similarly, since $M^\#(\min(f,N)) \leq 2M^\# f$, $N > 0$, (see Torchinsky [222, p. 218]) by an approximation argument we may also assume that $f \in L^\infty$.

Fix $a > 2^n$ and $m > 0$ such that $\|\tilde{u}\|_\infty \leq a^m$. For each $k \leq m$, let $\{Q_j^k\}$ be the Calderón-Zygmund cubes of $\tilde{u}$ at height a^k, and let $\Omega_k = \cup_j Q_j^k$. (See Proposition A.1 and Remark A.2.) For each k define the functions

$$b_k(x) = \sum_j (\tilde{u}(x) - \tilde{u}_{Q_j^k})\chi_{Q_j^k}(x),$$

$$g_k(x) = \tilde{u}(x) - b_k(x) = \begin{cases} \tilde{u}_{Q_j^k} & x \in Q_j^k \\ \tilde{u}(x) & x \in \mathbb{R}^n \setminus \Omega_k. \end{cases}$$

For all k we have $g_k(x) \leq 2^n a^k$ and $\|g_k\|_1 = \|\tilde{u}\|_1$.

Since the set Ω_m is empty, $b_m = 0$. Therefore, for every integer $l < 0$, we have the telescoping sequence

$$\tilde{u}(x) = \sum_{k=l}^{m-1} (b_k(x) - b_{k+1}(x)) + g_l(x).$$

Since $\tilde{u}_{Q_j^k} \leq 2^n a^k$, and since for each j and k,

$$\begin{aligned}(b_k(x) &- b_{k+1}(x))\chi_{Q_j^k}(x) \\ &= (\tilde{u}(x) - \tilde{u}_{Q_j^k})\chi_{Q_j^k}(x) - \sum_{Q_i^{k+1} \subset Q_j^k} (\tilde{u}(x) - \tilde{u}_{Q_i^{k+1}})\chi_{Q_i^{k+1}}(x), \end{aligned} \tag{9.15}$$

it follows immediately that for almost every x,

$$|b_k(x) - b_{k+1}(x)| \leq (1+a)\, 2^n a^k. \tag{9.16}$$

Further, by integrating (9.15) we see that

$$\int_{Q_j^k} b_k(x) - b_{k+1}(x)\, dx = 0. \tag{9.17}$$

We can now estimate as follows: for any $l < 0$,

$$\begin{aligned}\int_{\mathbb{R}^n} & f(x)^{p_0} \tilde{u}(x)\, dx \\ &= \sum_{k=l}^{m-1} \int_{\mathbb{R}^n} f(x)^{p_0} (b_k(x) - b_{k+1}(x))\, dx + \int_{\mathbb{R}^n} f(x)^{p_0} g_l(x)\, dx. \end{aligned} \tag{9.18}$$

The last term tends to 0 as $l \to -\infty$. To see this, fix $\epsilon > 0$. Then

$$\begin{aligned}\int_{\mathbb{R}^n} & f(x)^{p_0} g_l(x)\,dx \\ &= \int_{\{x:f(x)>\epsilon\}} f(x)^{p_0} g_l(x)\,dx + \int_{\{x:f(x)\leq\epsilon\}} f(x)^{p_0} g_l(x)\,dx \\ &\leq 2^n a^l \|f\|_\infty^{p_0} |\{x : f(x) > \epsilon\}| + \epsilon^{p_0} \|\tilde{u}\|_1.\end{aligned}$$

Since we can take ϵ arbitrarily close to 0, we get the desired limit.

Hence, if we pass to the limit in (9.18) we get

$$\int_{\mathbb{R}^n} f(x)^{p_0} \tilde{u}(x)\,dx = \sum_{k=-\infty}^{m-1} \int_{\mathbb{R}^n} f(x)^{p_0} (b_k(x) - b_{k+1}(x))\,dx. \tag{9.19}$$

To estimate the right-hand side, first note that $\big||a|^{p_0} - |b|^{p_0}\big| \leq |a-b|^{p_0}$ since $p_0 \leq 1$. Therefore, by (9.17) and (9.16),

$$\begin{aligned}\sum_{k=-\infty}^{m-1} & \int_{\mathbb{R}^n} f(x)^{p_0} (b_k(x) - b_{k+1}(x))\,dx \\ &= \sum_{k,j} \int_{Q_j^k} f(x)^{p_0} (b_k(x) - b_{k+1}(x))\,dx \\ &= \sum_{k,j} \int_{Q_j^k} (f(x)^{p_0} - f_{Q_j^k}^{p_0})(b_k(x) - b_{k+1}(x))\,dx \\ &\leq C \sum_{k,j} (1+a)\, 2^n\, a^k \int_{Q_j^k} \big|f(x)^{p_0} - f_{Q_j^k}^{p_0}\big|\,dx \\ &\leq C \sum_{k,j} \tilde{u}_{Q_j^k} \int_{Q_j^k} |f(x) - f_{Q_j^k}|^{p_0}\,dx.\end{aligned}$$

By Proposition 6.2, $W = (M_\Psi w_2)^{1-r} \in A_q \cap RH_\infty$, $q > r$. Hence, for any cube Q and $E \subset Q$,

$$\left(\frac{|E|}{|Q|}\right)^q \leq C \frac{W(E)}{W(Q)}. \tag{9.20}$$

(See Duoandikoetxea [68].) By Proposition A.1 there exists a collection of pairwise disjoint sets $E_j^k \subset Q_j^k$ and $\alpha > 0$ such that $|E_j^k| \geq \alpha |Q_j^k|$. Therefore, by the RH_∞ condition and (9.20),

$$\tilde{u}_{Q_j^k} \leq C \fint_{Q_j^k} w_1(x)\,dx \fint_{Q_j^k} M_\Psi w_2(x)^{1-r}\,dx$$

$$\leq C \fint_{Q_j^k} w_1(x)\,dx\,|Q_j^k|^{-1} \int_{E_j^k} M_\Psi w_2(x)^{1-r}\,dx.$$

Thus, since $p_0 \leq 1$, by Hölder's inequality,

$$\begin{aligned}
&\sum_{k,j} \tilde{u}_{Q_j^k} \int_{Q_j^k} |f(x) - f_{Q_j^k}|^{p_0}\,dx \\
&\quad \leq C \sum_{k,j} \fint_{Q_j^k} w_1(x)\,dx \int_{E_j^k} M_\Psi w_2(x)^{1-r}\,dx \fint_{Q_j^k} |f(x) - f_{Q_j^k}|^{p_0}\,dx \\
&\quad \leq C \sum_{k,j} \int_{E_j^k} M^\# f(x)^{p_0} M w_1(x) M_\Psi w_2(x)^{1-r}\,dx \\
&\quad \leq C \int_{\mathbb{R}^n} M^\# f(x)^{p_0} M w_1(x) M_\Psi w_2(x)^{1-r}\,dx \\
&\quad \leq C \int_{\mathbb{R}^n} M^\# f(x)^{p_0} \tilde{v}(x)\,dx.
\end{aligned}$$

This completes the proof of inequality (9.2). Note that the above inequality shows that we can replace $\tilde{v}$ by $Mw_1(M_\Psi w_2)^{1-r}$, which proves Remark 9.3.

9.2 Singular integral operators

In this section we consider Calderón-Zygmund singular integrals: operators defined by

$$Tf(x) = \text{p.v.} \int_{\mathbb{R}^n} K(x-y) f(y)\,dy,$$

where the kernel K satisfies

$$|K(x)| \leq \frac{C}{|x|^n} \qquad |\nabla K(x)| \leq \frac{C}{|x|^{n+1}}.$$

Our results also apply to non-convolution operators with kernels $K(x,y)$ that satisfy the appropriate decay estimates. For more information on all these operators, see [68, 88, 91, 92].

To apply our extrapolation results to singular integrals, we define two families of pairs of functions. For strong type inequalities, we will hereafter assume that we are applying our extrapolation results to the family $\mathcal{F}$ consisting of pairs of the form $(|Tf|, |f|)$, where $f \in C_c^\infty$. Similarly, for weak type inequalities, let $\mathcal{F}$ be the family of pairs of the form $(\lambda\chi_{\{x:|Tf(x)|>\lambda\}}, |f|)$, where $f \in C_c^\infty$ and $\lambda > 0$. (As before, we write $f \in \mathcal{F}$ to mean the corresponding pair is in $\mathcal{F}$.) By a standard density argument, if we can prove a weak or strong type two-weight norm inequality for T for $f \in \mathcal{F}$, then we can prove it for all f such that the right-hand side is finite.

The conjectures

As we discussed in Section 1.1, the two long-standing conjectures about two-weight norm inequalities for singular integrals are due to Muckenhoupt and Wheeden [147]. They conjectured that the Hilbert transform is bounded from $L^p(v)$ to $L^p(u)$ if the Hardy-Littlewood maximal operator satisfies

$$M : L^p(v) \to L^p(u),$$
$$M : L^{p'}(u^{1-p'}) \to L^{p'}(v^{1-p'}),$$

and that the Hilbert transform maps $L^p(v)$ into $L^{p,\infty}(u)$ if the maximal operator satisfies the second condition. (Though M is not linear, this condition is often referred to as the dual inequality.) By Theorem 5.14 we have sharp A_p bump conditions for the maximal operator to satisfy two-weight norm inequalities, and these bump conditions are the basis of our conjectures.

Conjecture 9.18. For each p, $1 < p < \infty$, there exists a constant C such that for all $f \in L^p(v)$,

$$\int_{\mathbb{R}^n} |Tf(x)|^p u(x)\, dx \le C \int_{\mathbb{R}^n} |f(x)|^p v(x)\, dx$$

whenever the pair (u, v) satisfies

$$\|u^{1/p}\|_{A,Q} \|v^{-1/p}\|_{B,Q} \le K < \infty, \tag{9.21}$$

where $\bar{A} \in B_{p'}$ and $\bar{B} \in B_p$.

Remark 9.19. A result related to this conjecture was proved by Treil, Volberg and Zheng [224] for the Hilbert transform. They replaced the A_p bump condition (9.21) with an "invariant" A_p condition using Möbius transforms. Their approach relies heavily on complex analysis.

Conjecture 9.20. For each p, $1 < p < \infty$, there exists a constant C such that for all $f \in L^p(v)$ and $\lambda > 0$,

$$u(\{x \in \mathbb{R}^n : |Tf(x)| > \lambda\}) \le \frac{C}{\lambda^p} \int_{\mathbb{R}^n} |f(x)|^p v(x)\, dx$$

whenever the pair (u, v) satisfies

$$\|u^{1/p}\|_{A,Q} \|v^{-1/p}\|_{p',Q} \le K < \infty, \tag{9.22}$$

where $\bar{A} \in B_{p'}$.

We have made progress in proving these conjectures but in their full generality they remain open. In this section we discuss the known results, and show how we can use extrapolation to either improve them or give proofs which are considerably simpler. We first consider the strong (p, p) inequality and then the weak (p, p) inequality. Finally we will consider the special case of norm inequalities for factored weights. This approach to norm inequalities for singular integrals is new, and we believe that it has great promise.

For some of our results, we will use the fact that Calderón-Zygmund singular integrals satisfy the sharp function estimate $M_q^{\#}(Tf)(x) \le C_q Mf(x)$, $0<q<1$. (See [3, 4] and Section 9.1.) This estimate is satisfied by more singular operators, such as certain kinds of pseudo-differential operators, and so the results we prove using it hold in greater generality. Our sharpest results, however, are proved using the specific structure of singular integrals.

Strong (p,p) inequalities

The best known results for strong (p,p) inequalities are a pair of theorems in the scale of log-bumps that were proved in [47] using the Calderón-Zygmund decomposition.

Theorem 9.21. [47] *Let T be a Calderón-Zygmund singular integral on $\mathbb{R}^n$, $n \ge 1$. Fix p, $n<p<\infty$, and assume that the pair of weights (u,v) satisfies*

$$\|u^{1/p}\|_{A,Q}\|v^{-1/p}\|_{B,Q} \le K < \infty, \tag{9.23}$$

where $A(t)=t^p\log(e+t)^{p-1+\delta}$, $\delta>0$, and $\bar{B}\in B_p$. Then

$$\int_{\mathbb{R}^n} |Tf(x)|^p u(x)\,dx \le C\int_{\mathbb{R}^n} |f(x)|^p v(x)\,dx. \tag{9.24}$$

Further, this inequality is sharp in the sense that there exist pairs of weights (u,v) such that (9.23) holds when $\delta=0$, but (9.24) does not hold for the Riesz transforms.

In the particular case when $n=1$ (e.g., for the Hilbert transform), we obtain the full range $1<p<\infty$. For $n\ge 2$, there is a gap when $1\le p\le n$. Since singular integrals are essentially self-adjoint (i.e., $T^*=-T$) this gap can be partly filled by duality. If $1<p\le n'$, $\bar{A}\in B_{p'}$ and $B(t)=t^{p'}\log(e+t)^{p'-1+\delta}$, then (9.24) holds. In addition, this gap could be completely filled by assuming a stronger hypothesis on the first bump.

Theorem 9.22. [47] *Let T be an operator that satisfies the sharp function estimate $M_q^{\#}(Tf)(x) \le CMf(x)$, $0<q<1$—e.g., T is a Calderón-Zygmund singular integral. If the pair of weights (u,v) satisfies (9.23) with $A(t)=t^p\log(e+t)^{2p-1+\delta}$, $\delta>0$, and $\bar{B}\in B_p$, then (9.24) holds.*

In the special case when both A and B are log bumps we can use two-weight extrapolation to improve both of these results. In this particular case the gap reduces to $n'\le p\le n$, and we will use Theorem 9.22 to partly fill it in.

Corollary 9.23. *Let T be a Calderón-Zygmund singular integral on $\mathbb{R}^n$, $n\ge 2$. Then Theorem 9.21 holds with $B(t)=t^{p'}\log(e+t)^{p'-1+\delta}$, $\delta>0$, for $p=n'$ and $p=n$. In particular, when $n=2$ we obtain the full range $1<p<\infty$.*

Corollary 9.24. *Let T be a Calderón-Zygmund singular integral on $\mathbb{R}^n$, $n \geq 3$. Assume either that $n/2 < p < n$, $A(t) = t^p \log(e+t)^{n-1+\delta}$ and $B(t) = t^{p'} \log(e+t)^{p'-1+\delta}$, $\delta > 0$, or that $n' < p < (n/2)'$, $A(t) = t^p \log(e+t)^{p-1+\delta}$ and $B(t) = t^{p'} \log(e+t)^{n-1+\delta}$ for some $\delta > 0$. Then* (9.24) *holds for pairs of weights that satisfy* (9.23).

While these two corollaries are only incremental improvements of Theorem 9.21, we want to stress that these results could not be obtained directly by the proof in [47]. Furthermore, neither corollary can be gotten by applying the earlier results on two-weight extrapolation discussed in Chapter 1. Finally, if our conjectured improvement to Theorem 7.2 is true (see Remark 7.7), then we would be able to extend Theorem 9.21 to the whole range $1 < p < \infty$, thereby establishing Conjecture 9.18 in the scale of log bumps.

Proof of Corollary 9.23. We first consider the case $p = n'$. Fix $\delta > 0$ and let $A(t) = t^p \log(e+t)^{p-1+\delta}$ and $B(t) = t^{p'} \log(e+t)^{p'-1+\delta}$. Define $p_0 = (n+\delta/2)' < n' = p$, $\delta_0 = \delta(p_0 - 1)/(p-1)$, $\delta_1 = \delta/2$, and let

$$A_0(t) = t^{p_0} \log(e+t)^{p_0-1+\delta_0},$$
$$B_0(t) = t^{p_0'} \log(e+t)^{p_0'-1+\delta_1}.$$

By Theorem 9.21 and the duality argument above, since $1 < p_0 < n'$, inequality (9.24) holds with p replaced by p_0 whenever (u, v) satisfy

$$\|u^{1/p_0}\|_{A_0,Q} \|v^{-1/p_0}\|_{B_0,Q} \leq K < \infty.$$

By our choice of p_0 and δ_0,

$$B(t) = t^{p'} \log(e+t)^{p'-1+\delta} = t^{p'} \log(e+t)^{p_0'-1+\delta_1}.$$

Therefore, since $p = n' > p_0$, by Theorem 7.2 applied in the scale of log bumps (see Table 7.1 and the discussion of log bumps in Section 7.1), if the pair (u, v) satisfies (9.23), then inequality (9.24) holds.

The case $p = n$ follows by duality. Alternatively we can modify the above argument, taking $p_0 = n + \delta/2$ and choosing A_0 and B_0 appropriately; details are left to the reader. □

Proof of Corollary 9.24. The proof is similar to the proof of Corollary 9.23. First suppose that $n/2 < p < n$ and let $p_0 = n$. We apply Theorem 7.2 with $p_0 = n$, $A_0(t) = t^{p_0} \log(e+t)^{p_0-1+\delta_0}$, $\delta_0 = \delta$ and $B_0(t) = t^{p_0'} \log(e+t)^{p_0'-1+\delta_1}$, $\delta_1 = \delta(p_0' - 1)/(p' - 1)$. (Here we use Corollary 9.23 to get the initial assumption (7.2).) Then again by Theorem 7.2 applied in the scale of log bumps we get that the desired estimates hold for the given Young functions A and B.

When $n' < p < (n/2)'$ the proof is essentially the same and the details are left to the reader. □

Remark 9.25. The particular case when $n = 2$ in Corollary 9.23 should be of use in proving a conjecture of Sarason for the composition of Toeplitz operators on Bergman spaces. See [47] for details and references.

Remark 9.26. From the proof we see that the conclusion of Corollary 9.24 holds for a larger range of p. However, we have only stated it for those values which improve Theorem 9.22. In particular, when $n = 3$ or $n = 4$, Corollary 9.24 improves on Theorem 9.22 in the full range $n' < p < n$. (When $n = 4$ and $p = 2$ the two results give the same exponent.)

Remark 9.27. In [47] we showed that versions of Theorems 9.21 and 9.22 held for vector-valued singular integrals. Given $f = \{f_i\}$, define $Tf = \{Tf_i\}$; then we prove inequalities of the form

$$\int_{\mathbb{R}^n} \|Tf(x)\|_{\ell^q}^p u(x)\,dx \le C \int_{\mathbb{R}^n} \|f(x)\|_{\ell^q}^p v(x)\,dx, \quad 1 < q < \infty.$$

Essentially the same proofs extend Corollaries 9.23 and 9.24 to this case. Precise statements and details are left to the reader.

A weaker version of Theorem 9.22 was proved in [56] assuming that A was a power bump. The proof used a very complicated decomposition argument reminiscent of a good-λ inequality. Here we show that this result is a direct consequence of two-weight extrapolation. We include it, though better results are known, as an illustration of the ways in which extrapolation can be applied.

For simplicity, we will state it for operators that are self-adjoint, but it will be clear from the proof that it is true for any operator (such as a singular integral) that is essentially self-adjoint.

Proposition 9.28. [56] *Fix p, $1 < p < \infty$, and let (u, v) be a pair of weights such that*

$$\|u^{1/p}\|_{A,Q}\|v^{-1/p}\|_{B,Q} \le K < \infty, \tag{9.25}$$

where $A(t) = t^{rp}$, $r > 1$, and $B(t) = t^{p'}\log(e+t)^{p'-1+\delta}$, $\delta > 0$. If T is a self-adjoint operator that satisfies the sharp function estimate $M_q^{\#}(Tf)(x) \le CMf(x)$, $0 < q < 1$, then

$$\int_{\mathbb{R}^n} |Tf(x)|^p u(x)\,dx \le C \int_{\mathbb{R}^n} |f(x)|^p v(x)\,dx. \tag{9.26}$$

Remark 9.29. In [56], Proposition 9.28 was proved only assuming that B is a B_p bump. The proof below does not extend to this case. Similarly, in [35] it was proved assuming that A is an exponential log bump. It is an open question whether this result can be proved using extrapolation. Similarly, it is an open question whether we can prove Theorem 9.22 itself using extrapolation. (Compare with Theorem 9.35 below.)

Proof. Our proof is modeled after the proof of Corollary 8.7; we will show that (9.26) is a consequence of Theorem 8.2. To do so, we first note that since T is self-adjoint, by duality it will suffice to prove (9.26) holds if we switch the log-bump and the power-bump in (9.25): that is, we will assume that

$$\|u^{1/p}\|_{A,Q}\|v^{-1/p}\|_{B,Q} \le K < \infty, \tag{9.27}$$

where $A(t) = t^p \log(e+t)^{p-1+\delta}$, $\delta > 0$, and $B(t) = t^{sp'}$, $s > 1$.

Fix $p_0 > 1$ such that

$$\frac{p}{p_0} = \frac{p-1}{s} + 1;$$

then $1 < p_0 < p$. Let $r = p/p_0$; then $s(1-p') = 1-r'$. By rescaling we can rewrite (9.27) as

$$\|u^{1/r}\|_{A_1,Q}\|v^{-1/r}\|_{r',Q} \le K^{p_0} < \infty,$$

where $A_1(t) \approx t^r \log(e+t)^{p-1+\delta}$. Let $\Phi_0(t) = t\log(e+t)^{p_0-1+\delta_0}$, where $\delta_0 > 0$ will be fixed below. Then by Theorem 9.38 below,

$$\int_{\mathbb{R}^n} |Tf(x)|^{p_0} u_0(x)\,dx \le C\int_{\mathbb{R}^n} |f(x)|^{p_0} M_{\Phi_0}u_0(x)\,dx;$$

therefore (9.26) will follow from Theorem 8.2 if we can show that there exists $C \in B_{r'}$ such that

$$A_1^{-1}(t)C^{-1}(t) \preceq \Phi_0^{-1}(t).$$

Fix $\delta_0 < \delta/r$, and fix $\eta > 0$ such that

$$\frac{\delta}{r} - \frac{\eta}{r'} = \delta_0.$$

Now define

$$C(t) = \frac{t^{r'}}{\log(e+t)^{1+\eta}} \in B_{r'};$$

Then

$$A_1^{-1}(t)C^{-1}(t) \approx \frac{t^{1/r}}{\log(e+t)^{(p-1+\delta)/r}}\, t^{1/r'}\log(e+t)^{(1+\eta)/r'} \approx \Phi_0^{-1}(t),$$

and so the above inequality holds. This completes the proof. □

As a final strong type inequality we state a two-weight generalization of the Coifman-Fefferman inequality relating singular integrals and maximal functions. It is an immediate consequence of Theorem 9.9.

Theorem 9.30. *Let T be an operator that satisfies the sharp function estimate $M_q^\#(Tf)(x) \le C_q Mf(x)$, $0 < q < 1$. Fix p, $0 < p < \infty$. Suppose there exists $r > \max(p,1)$ such that the pair (u,v) satisfies*

$$\|u^{1/r}\|_{A,Q}\|v^{-1/r}\|_{r',Q} \le K < \infty,$$

where $\bar{A} \in B_{r'}$. *Then*

$$\int_{\mathbb{R}^n} |Tf(x)|^p u(x)\, dx \le C \int_{\mathbb{R}^n} Mf(x)^p v(x)\, dx.$$

In the scale of log bumps, if $p > 1$ *we can take* $r = p$ *and* $A(t) = t^p \log(e+t)^{p-1+\delta}$, $\delta > 0$. *If* $p = 1$, *we can take the* A_1 *pair* $(u, M_\Phi u)$, $\Phi(t) = t\log(e+t)^\epsilon$, $\epsilon > 0$. *If* $0 < p < 1$, *we can take the pair* (u, Mu).

Weak (p,p) inequalities

Conjecture 9.20 would be an immediate consequence of Corollary 8.4 if we could prove that for all weights u and $\lambda > 0$,

$$u(\{x \in \mathbb{R}^n : |Tf(x)| > \lambda\}) \le \frac{C}{\lambda} \int_{\mathbb{R}^n} |f(x)| Mu(x)\, dx. \tag{9.28}$$

The truth of this inequality itself is an open question. As we noted in Section 1.1, this problem was first posed by Muckenhoupt and Wheeden for the Hilbert transform in the 1970's [147, 152]. The best known result was proved in [172], using Theorem 9.38 below and a Calderon-Zygmund decomposition argument. (A weaker result using power bumps was proved earlier by Cordoba and Fefferman [28].)

Theorem 9.31. [172] *Given a Young function* B *such that* $\bar{B} \in B_{q'}$ *for some* $q > 1$, *and* $\Phi(t) = B(t^{1/q})$, *then Calderón-Zygmund singular integrals satisfy the weak* $(1,1)$ *inequality*

$$u(\{x \in \mathbb{R}^n : |Tf(x)| > \lambda\}) \le \frac{C}{\lambda} \int_{\mathbb{R}^n} |f(x)| M_\Phi u(x)\, dx. \tag{9.29}$$

In particular, for any $\epsilon > 0$, *we can take* $\Phi(t) = t\log(e+t)^\epsilon$.

Remark 9.32. In [172], Theorem 9.31 is only stated for the particular case $\Phi(t) = t\log(e+t)^\epsilon$. However, the proof goes through for this more general case with essentially no change.

The best weak (p,p) result for singular integrals was proved in [53]. It proves Conjecture 9.20 in the scale of log bumps. The proof used an extrapolation argument that is a special case of Theorem 8.2.

Theorem 9.33. [53] *Fix* p, $1 < p < \infty$, *and a Young function* B *with* $\bar{B} \in B_{q'}$ *for some* $q > 1$, *and let* $\Phi(t) = B(t^{1/q})$. *Suppose* A *is a* p*-Young function,* C *is a* p'*-Young function with* $C \in B_{p'}$, *and* $A^{-1}(t)C^{-1}(t) \preceq \Phi^{-1}(t)$. *If the pair* (u, v) *satisfies*

$$\|u^{1/p}\|_{A,Q} \|v^{-1/p}\|_{p',Q} \le K < \infty, \tag{9.30}$$

then T *satisfies the weak* (p,p) *inequality*

$$u(\{x \in \mathbb{R}^n : |Tf(x)| > \lambda\}) \le \frac{C}{\lambda^p} \int_{\mathbb{R}^n} |f(x)|^p v(x)\, dx. \tag{9.31}$$

In particular, in the scale of log bumps we can take $A(t) = t^p \log(e+t)^{p-1+\delta}$, $\delta > 0$, *in* (9.30). *In this scale the inequality is sharp in the sense that there exists a pair of weights* (u, v) *that satisfy* (9.30) *with* $\delta = 0$ *but such that the weak* (p, p) *inequality does not hold for the Hilbert transform.*

Proof. Inequality (9.31) follows immediately from Theorems 8.2 and 9.31. When $A(t) = t^p \log(e+t)^{p-1+\delta}$, then arguing as in the proof of Proposition 5.21 we see that there exists $\epsilon > 0$ such that if $\Phi(t) = t\log(e+t)^\epsilon$, then there exists $C \in B_{p'}$ such that $A^{-1}(t)C^{-1}(t) \approx \Phi^{-1}(t)$. If we let $B(t) = \Phi(t^q)$, then $\bar{B} \in B_{q'}$ for all $q > 1$ such that $q - 1 < \epsilon$.

For the example showing that this result is sharp, see [53]. □

Remark 9.34. In the scale of log bumps, the weak (1,1) inequality in Theorem 9.31, the weak (p,p) inequality in Theorem 9.33, and the strong type (p,p) inequality for A_1 weights in Theorem 9.38 below are equivalent. We proved above that (9.29) implies (9.31), and the proof that (9.36) implies (9.29) is in [172]. That (9.31) implies (9.36) follows from Marcinkiewicz interpolation. To see this, fix $p > 1$ and $\delta > 0$. Then there exist $1 < p_0 < p < p_1$ and $\delta_0, \delta_1 > 0$ such that

$$p_0 - 1 + \delta_0 = p - 1 + \delta = p_1 - 1 + \delta_1.$$

Let $\Phi(t) = t\log(e+t)^{p-1+\delta}$. Then by Theorem 6.4, the pair $(u, M_\Phi u)$ satisfy (9.30) with p replaced by p_0 and p_1, so inequality (9.31) holds with p replaced by p_0 and p_1. Inequality (9.36) follows at once by interpolation.

A different and somewhat weaker version of Theorem 9.33 was proved in [55]. There it was assumed that the bump function A in (9.30) was a power bump, but inequality (9.31) was shown to hold for any operator T that satisfies the sharp function estimate $M_q^\#(Tf)(x) \le CMf(x)$, $0 < q < 1$. The proof was quite difficult since it relied on a careful decomposition argument to prove an inequality reminiscent of a good-λ inequality. Using extrapolation we give a very simple proof of a stronger result that is the weak type analog of Theorem 9.22.

Theorem 9.35. *Let* T *be an operator that satisfies the sharp function estimate* $M_q^\#(Tf)(x) \le CMf(x)$, $0 < q < 1$. *Then for all* $\lambda > 0$,

$$u(\{x \in \mathbb{R}^n : |Tf(x)| > \lambda\}) \le \frac{C}{\lambda} \int_{\mathbb{R}^n} |f(x)| M_\Psi u(x)\, dx, \tag{9.32}$$

where $\Psi(t) = t\log(e+t)^{1+\epsilon}$, $\epsilon > 0$.

Further, given $1 < p < \infty$, *if the pair* (u, v) *satisfies*

$$\|u^{1/p}\|_{A,Q} \|v^{-1/p}\|_{p',Q} \le K < \infty,$$

where $A(t) = t^p \log(e+t)^{2p-1+\delta}$, $\delta > 0$, *then*

$$u(\{x \in \mathbb{R}^n : |Tf(x)| > \lambda\}) \le \frac{C}{\lambda^p} \int_{\mathbb{R}^n} |f(x)|^p v(x)\, dx. \tag{9.33}$$

Remark 9.36. Obviously Theorem 9.35 is not sharp since better inequalities hold for Calderón-Zygmund singular integrals. However, it would be interesting to determine if there exists a more singular operator that satisfies the sharp function estimate for which this result is sharp.

Proof. Given the sharp function estimate, by Theorem 9.10 and the two-weight, weak $(1,1)$ inequality for the maximal operator (5.13),

$$\|Tf\|_{L^{1,\infty}(u)} \leq C\|Mf\|_{L^{1,\infty}(M_\Phi u)} \leq C\int_{\mathbb{R}^n} |f(x)|M(M_\Phi u)(x)\,dx, \tag{9.34}$$

where $\Phi(t) = t\log(e+t)^\epsilon$, $\epsilon > 0$. By Example 5.30, $M(M_\Phi u)(x) \leq CM_\Psi u(x)$. This proves (9.32).

To prove (9.33) we use an argument similar to the proof of Theorem 9.33. Fix $\delta > 0$, and choose $\nu > 0$ such that

$$\epsilon = \frac{\delta}{p} - \frac{\nu}{p'} > 0.$$

Let $C(t) = t^{p'}\log(e+t)^{-1-\nu} \in B_{p'}$; then $A^{-1}(t)C^{-1}(t) \approx \Psi^{-1}(t)$, where $\Psi(t) = t\log(e+t)^{1+\epsilon}$. Then by (9.32) and Theorem 8.2 with $p_0 = 1$ we get the desired weak (p,p) inequality. □

Remark 9.37. Though Theorem 9.35 is not sharp, the first inequality in (9.34) is sharp in the sense we cannot take $\epsilon = 0$ and $\Phi(t) = t$. Let $u = \chi_{(0,1)}$. Then for $x \geq 1$, $Mu(x) = x^{-1}$. If $f(x) = \log(x)^{-1}\chi_{(e,\infty)}$, then for $x \in (0,1)$,

$$Hf(x) = \int_e^\infty \frac{1}{x-y}\frac{1}{\log(y)}\,dy = -\infty.$$

Hence, $\|Hf\|_{L^{1,\infty}(u)} = \infty$. On the other hand, by L'Hôpital's rule, for $x > e$,

$$Mf(x) = \frac{1}{x-e}\int_e^x \frac{dy}{\log(y)} \approx \frac{1}{\log(x)}.$$

Thus, there exists $c < 1$ such that for $0 < t < 1$,

$$Mu(\{x : Mf(x) > t\}) \leq Mu(\{x > e : \log(x)^{-1} > ct\}) \leq \int_e^{e^{1/(ct)}} \frac{dx}{x} = \frac{1}{ct} - 1.$$

It follows that $\|Mf\|_{L^{1,\infty}(Mu)} \leq C < \infty$.

Inequalities for factored weights

Our first result for factored weights is actually for A_1 and dual A_1 pairs. The second part is stated for self-adjoint operators, but it is clear from the proof that it holds for essentially self-adjoint operators such as Calderón-Zygmund singular integrals. The proof itself is given in Example 9.16 above.

Theorem 9.38. *Let T be an operator that satisfies the sharp function inequality $M_q^{\#}(Tf)(x) \leq C_q Mf(x)$, $0 < q < 1$. For all p, $1 < p < \infty$, if $\bar{B} \in B_p$ and $\Psi(t) = B(t^{1/p'})$, then*

$$\int_{\mathbb{R}^n} |Tf(x)|^p M_\Psi u(x)^{1-p}\, dx \leq C \int_{\mathbb{R}^n} |f(x)|^p u(x)^{1-p}\, dx. \tag{9.35}$$

Suppose further that T is self adjoint; if $\bar{A} \in B_{p'}$ and $\Phi(t) = A(t^{1/p})$, then

$$\int_{\mathbb{R}^n} |Tf(x)|^p u(x)\, dx \leq C \int_{\mathbb{R}^n} |f(x)|^p M_\Phi u(x)\, dx. \tag{9.36}$$

In particular, we can take $A(t) = t^p \log(e+t)^{p-1+\delta}$, $\delta > 0$.

Theorem 9.38 was first proved for singular integrals in [172], where in the case of log bumps it was shown to be sharp: you cannot take $\Phi(t) = t\log(e+t)^{p-1}$ in (9.36).

By Theorem 6.4 the pairs $(u, M_\Phi u)$ and $((M_\Psi u)^{1-p}, u^{1-p})$ satisfy (9.21), and so we have proved Conjecture 9.18 for A_1 and dual A_1 pairs. However, we have not been able to use this fact to prove the conjecture in the case of factored weights in general. But we can improve Theorem 9.22, replacing the exponent $2p-1+\delta$ on the log bump with $p+\delta$.

Theorem 9.39. *Let T be an operator that satisfies the sharp function estimate $M_q^{\#}(Tf)(x) \leq CMf(x)$, $0 < q < 1$. Given p, $1 < p < \infty$, let $A(t) = t^p \log(e+t)^{p+\delta}$ and let B be a Young function such that $\bar{B} \in B_p$. Form the pair of factored weights*

$$(\tilde{u}, \tilde{v}) = (w_1(M_\Psi w_2)^{1-p}, (M_\Phi w_1)w_2^{1-p}),$$

where $\Phi(t) = A(t^{1/p})$ and $\Psi(t) = B(t^{1/p'})$, which satisfies the A_p bump condition

$$\|\tilde{u}^{1/p}\|_{A,Q}\|\tilde{v}^{-1/p}\|_{B,Q} \leq K < \infty.$$

Then T satisfies the strong (p,p) inequality

$$\int_{\mathbb{R}^n} |Tf(x)|^p \tilde{u}(x)\, dx \leq C \int_{\mathbb{R}^n} |f(x)|^p \tilde{v}(x)\, dx.$$

Proof. Let q be such that $0 < q < 1$; the exact value of q will be fixed below. By duality there exists non-negative $h \in L^{(p/q)'}$, $\|h\|_{L^{(p/q)'}} = 1$, such that

$$\left(\int_{\mathbb{R}^n} |Tf(x)|^p \tilde{u}(x)\, dx\right)^{q/p} = \int_{\mathbb{R}^n} |Tf(x)|^q w_1(x)^{q/p} M_\Psi w_2(x)^{-q/p'} h(x)\, dx.$$

By Lemma 9.2 (and Remark 9.3) and the sharp function estimate,

$$\int_{\mathbb{R}^n} |Tf(x)|^q w_1(x)^{q/p} M_\Psi w_2(x)^{-q/p'} h(x)\, dx$$

$$\leq C \int_{\mathbb{R}^n} M_q^{\#}(Tf)(x)^q M(w_1^{q/p}h)(x) M_\Psi w_2(x)^{-q/p'}\, dx$$
$$\leq C \int_{\mathbb{R}^n} Mf(x)^q M(w_1^{q/p}h)(x) M_\Psi w_2(x)^{-q/p'}\, dx.$$

Let $C(t) = t^{p/q}\log(e+t)^{p/q-1+\tilde{\delta}}$, where we choose q and $\tilde{\delta} > 0$ such that $p-1+\delta = p/q - 1 + \tilde{\delta}$. Then $\bar{C} \in B_{(p/q)'}$, and by Lemma 5.2 and rescaling,

$$M(w_1^{q/p}h) \leq cM_C(w_1^{q/p})M_{\bar{C}}h = c(M_{\Phi_0}w_1)^{q/p}M_{\bar{C}}h,$$

where $\Phi_0(t) = t\log(e+t)^{p-1+\delta}$. Combining this inequality with the two above, by Hölder's inequality, Theorem 5.13, Corollary 6.5, and Example 5.30 we have that

$$\left(\int_{\mathbb{R}^n} |Tf(x)|^p \tilde{u}(x)\, dx\right)^{q/p}$$
$$\leq C\left(\int_{\mathbb{R}^n} Mf(x)^p M_{\Phi_0}w_1(x) M_\Psi w_2(x)^{1-p}\, dx\right)^{q/p} \times \left(\int_{\mathbb{R}^n} M_{\bar{C}}h(x)^{(p/q)'}\, dx\right)^{1/(p/q)'}$$
$$\leq C\left(\int_{\mathbb{R}^n} |f(x)|^p M(M_{\Phi_0}w_1)(x) w_2(x)^{1-p}\, dx\right)^{q/p} \times \left(\int_{\mathbb{R}^n} |h(x)|^{(p/q)'}\, dx\right)^{1/(p/q)'}$$
$$\leq C\left(\int_{\mathbb{R}^n} |f(x)|^p M_\Phi w_1(x) w_2(x)^{1-p}\, dx\right)^{q/p}. \qquad \square$$

9.3 Fractional integral operators

In this section we consider the fractional integral operators: for each α, $0 < \alpha < n$, I_α is the convolution operator defined by

$$I_\alpha f(x) = \int_{\mathbb{R}^n} \frac{f(y)}{|x-y|^{n-a}}\, dy.$$

To apply our extrapolation results to fractional integrals, we define two families of pairs of functions. For strong type inequalities, we will hereafter assume that we are applying our extrapolation results to the family $\mathcal{F}$ consisting of pairs of the form $(|I_\alpha f|, |f|)$, where $f \in C_c^\infty$. Similarly, for weak type inequalities, let $\mathcal{F}$ be the family of pairs of the form $(\lambda\chi_{\{x:|I_\alpha f(x)|>\lambda\}}, |f|)$, where $f \in C_c^\infty$ and $\lambda > 0$. (As before, we write $f \in \mathcal{F}$ to mean the corresponding pair is in $\mathcal{F}$.) By a standard density argument, if we can prove a weak or strong type two-weight norm inequality for I_α for $f \in \mathcal{F}$, then we can prove it for all f such that the right-hand side is finite.

The conjectures

Formally, there are many similarities between singular and fractional integrals. In particular, just as the behavior of singular integrals is in some sense controlled by the Hardy-Littlewood maximal operator, the fractional integral operator is controlled by the fractional maximal operator M_α. It therefore seems natural to modify Conjectures 9.18 and 9.20 by substituting fractional A_p bump conditions that guarantee that M_α satisfies $M_\alpha : L^p(v) \to L^p(u)$ and/or the dual inequality $M_\alpha : L^{p'}(u^{1-p'}) \to L^{p'}(v^{1-p'})$. In fact, for the strong (p,p) inequality this does not yield a conjecture but a theorem first proved in [171].

Theorem 9.40. *Given p, $1 < p < \infty$, and α, $0 < \alpha < n$, suppose A and B are Young functions such that $\bar{A} \in B_{p'}$ and $\bar{B} \in B_p$. If the pair (u,v) satisfies*

$$|Q|^{\alpha/n}\|u^{1/p}\|_{A,Q}\|v^{-1/p}\|_{B,Q} \leq K < \infty, \tag{9.37}$$

then I_α satisfies the strong (p,p) inequality

$$\int_{\mathbb{R}^n} |I_\alpha f(x)|^p u(x)\,dx \leq C \int_{\mathbb{R}^n} |f(x)|^p v(x)\,dx.$$

We consider Theorem 9.40 to be strong evidence for the truth of Conjecture 9.18.

For the weak type inequality we make the following conjecture.

Conjecture 9.41. For each p, $1 < p < \infty$, there exists a constant C such that for all $f \in L^p(v)$ and $\lambda > 0$,

$$u(\{x \in \mathbb{R}^n : |I_\alpha f(x)| > \lambda\}) \leq \frac{C}{\lambda^p} \int_{\mathbb{R}^n} |f(x)|^p v(x)\,dx$$

whenever the pair (u,v) satisfies

$$|Q|^{\alpha/n}\|u^{1/p}\|_{A,Q}\|v^{-1/p}\|_{p',Q} \leq K < \infty, \tag{9.38}$$

where $\bar{A} \in B_{p'}$.

Below, we first consider the weak type inequality and then prove two results for factored weights.

Weak (p,p) inequalities

The best known weak (p,p) inequality for fractional integrals was proved in [55, 33], where it was assumed that A in (9.38) was a power bump. Conjecture 9.41 would follow immediately from Theorem 8.12 if the following inequality were true:

$$u(\{x \in \mathbb{R}^n : |I_\alpha f(x)| > \lambda\}) \leq \frac{C}{\lambda} \int_{\mathbb{R}^n} |f(x)| M_\alpha u(x)\,dx. \tag{9.39}$$

This estimate is analogous to the conjectured inequality (9.28) for singular integrals; however, Carro, *et al.* [21] gave a counter-example showing that (9.39) does not hold in general. This casts some doubt both on whether (9.28) and Conjecture 9.41 are true. However, using extrapolation we can prove a result in the scale of log bumps that is analogous to the weak type inequalities in Theorem 9.35. Our weak $(1,1)$ inequality ((9.40) below) was essentially proved in [21], and our proof is based on theirs.

Theorem 9.42. *Given α, $0<\alpha<n$, then for all $\lambda>0$,*

$$u(\{x\in\mathbb{R}^n : |I_\alpha f(x)|>\lambda\}) \le \frac{C}{\lambda}\int_{\mathbb{R}^n} |f(x)|M_{\Psi,\alpha}u(x)\,dx, \tag{9.40}$$

where $\Psi(t)=t\log(e+t)^{1+\epsilon}$. Further, given p, $1<p<\infty$, if the pair of weights (u,v) satisfies

$$|Q|^{\alpha/n}\|u^{1/p}\|_{A,Q}\|v^{-1/p}\|_{p',Q}\le K<\infty,$$

where $A(t)=t^p\log(e+t)^{2p-1+\delta}$, $\delta>0$, then

$$u(\{x\in\mathbb{R}^n : |I_\alpha f(x)|>\lambda\}) \le \frac{C}{\lambda^p}\int_{\mathbb{R}^n} |f(x)|^p v(x)\,dx. \tag{9.41}$$

Proof. The proof of Theorem 9.42 is very similar to the proof of Theorem 9.35; we will sketch the changes. Fractional integrals satisfy the sharp function inequality $M(I_\alpha f)(x)\le CM_\alpha f(x)$—see the list after Theorem 9.10. Therefore, by Theorem 9.10 and the two-weight, weak $(1,1)$ inequality for the fractional maximal operator (5.26),

$$\|I_\alpha f\|_{L^{1,\infty}(u)} \le C\|M_\alpha f\|_{L^{1,\infty}(M_\Phi u)} \le C\int_{\mathbb{R}^n} |f(x)|M_\alpha(M_\Phi u)(x)\,dx, \tag{9.42}$$

where $\Phi(t)=t\log(e+t)^\epsilon$, $\epsilon>0$. By Example 5.42, $M_\alpha(M_\Phi u)(x)\le CM_{\Psi,\alpha}u(x)$. This proves (9.40).

The proof of (9.41) is identical to that of the second half of Theorem 9.35, except that we use Theorem 8.12 instead of Theorem 8.2. □

Remark 9.43. There is a curious asymmetry between our results for singular and fractional integrals. For singular integrals we can prove our conjecture for the weak (p,p) inequality in the scale of log bumps, but can only prove the strong (p,p) inequality for all p if we replace the exponent $p-1+\delta$ on the log bump by $2p-1+\delta$. On the other hand, for the fractional integral we can prove the strong (p,p) inequality in the scale of log bumps, but for the weak (p,p) inequality we again must have an exponent $2p-1+\delta$ on the log bump.

The reason for this is that we rely on two different decomposition arguments. For singular integrals, the classical Calderón-Zygmund decomposition is used to prove a weak $(1,1)$ inequality (see [172]) which yields via extrapolation a weak (p,p) inequality. This argument does not work for fractional integrals since we cannot prove the corresponding weak $(1,1)$ inequality. On the other hand, we have a linearization argument for fractional integrals (see [171]) that yields strong (p,p) inequalities, and this technique does not work for singular integrals. We believe that some refinement of these two approaches will be needed to prove our conjectures.

Remark 9.44. In Remark 9.34 we showed that in the scale of log bumps various results for singular integrals were equivalent. This leads to the following conjecture for fractional integrals and another collection of inter-related results.

By analogy with Theorem 9.31, we conjecture that the weak $(1,1)$ inequality (9.41) is true if we replace M_α on the right-hand side by the larger fractional maximal operator $M_{\Phi,\alpha}$, $\Phi(t) = t\log(e+t)^\epsilon$, $\epsilon > 0$. If this were true, then arguing as in the proof of Theorem 9.33 we could prove Conjecture 9.41 in the special case of log bumps. Conversely, by Theorem 8.13 (or more precisely, by the corollaries alluded to in Remark 8.14) the weak (p,p) inequality implies the weak $(1,1)$ inequality. Finally, if the weak (p,p) inequality holds in the scale of log bumps, then by Theorem 6.18 it holds for pairs of the form $(u, M_{\Phi,p\alpha}u)$, $\Phi(t) = t\log(e+t)^{p-1+\delta}$, $\delta > 0$. Then by interpolation with change of measure (adapting an argument given in [32] for weights of the form $M_{p\alpha}u$), we get the strong type inequality (9.43) below. This estimate is the analog of Theorem 9.38 which is central to the proof of the weak $(1,1)$ inequality for singular integrals. However, as we noted above, we have been unable to extend this proof to the case of fractional integrals, so we cannot use the argument we used for singular integrals to prove these results.

Inequalities for factored weights

Our first result for factored weights is for fractional A_1 and dual A_1 pairs and is the analog of Theorem 9.38. It is a corollary of Theorem 9.40, but here we give a proof that does not require a decomposition argument and instead uses extrapolation.

Theorem 9.45. *Fix α, $0 < \alpha < n$. If $1 < p < n/\alpha$, and A is a p-Young function such that $\bar{A} \in B_{p'}$ and $A(t)/t^{n/\alpha}$ is quasi-decreasing and tends to 0 as $t \to \infty$, then*

$$\int_{\mathbb{R}^n} |I_\alpha f(x)|^p u(x)\,dx \le C \int_{\mathbb{R}^n} |f(x)|^p M_{\Phi,p\alpha}u(x)\,dx, \tag{9.43}$$

where $\Phi(t) = A(t^{1/p})$.

Similarly, if $1 < p' < n/\alpha$, $\bar{B} \in B_p$ and $B(t)/t^{n/\alpha}$ is quasi-decreasing and tends to 0 as $t \to \infty$, then

$$\int_{\mathbb{R}^n} |I_\alpha f(x)|^p M_{\Psi,p'\alpha}u(x)^{1-p}\,dx \le C \int_{\mathbb{R}^n} |f(x)|^p u(x)^{1-p}\,dx, \tag{9.44}$$

where $\Psi(t) = B(t^{1/p'})$.

Proof. Since I_α is self-adjoint, inequality (9.43) follows from (9.44) by duality. To prove this latter inequality we note that by Proposition 6.15, $M_{\Psi,p'\alpha}u \in A_1$, so

$M(M_{\Psi,p'\alpha}u) \approx M_{\Psi,p'\alpha}u$. Therefore, since I_α satisfies the sharp function inequality $M^\#(I_\alpha f)(x) \leq CM_\alpha f(x)$, by Theorem 9.4 and Corollary 6.19,

$$\begin{aligned}\int_{\mathbb{R}^n} &|I_\alpha f(x)|^p M_{\Psi,p'\alpha}u(x)^{1-p}\,dx\\ &\leq C\int_{\mathbb{R}^n} M_\alpha f(x)^p M(M_{\Psi,p'\alpha}u)(x)^{1-p}\,dx\\ &\leq C\int_{\mathbb{R}^n} M_\alpha f(x)^p M_{\Psi,p'\alpha}u(x)^{1-p}\,dx\\ &\leq C\int_{\mathbb{R}^n} |f(x)|^p u(x)^{1-p}\,dx.\end{aligned}$$

□

For factored weights we can improve Theorem 9.42. But, continuing the pattern noted in Remark 9.43, instead of being able to prove Conjecture 9.41 for factored weights we get a power of $p+\delta$ on the log bump, just as we did for the strong type inequality for singular integrals in Theorem 9.39.

Theorem 9.46. *Given α, $0<\alpha<n$, and p, $1<p<\infty$, let $A(t)=t^p\log(e+t)^{p+\delta}$, $\delta>0$, and $\Phi(t)=A(t^{1/p})$. Then the pair of factored weights*

$$(\tilde{u},\tilde{v}) = (w_1(M_\alpha w_2)^{1-p}, (M_{\Phi,\alpha}w_1)w_2^{1-p})$$

satisfies

$$|Q|^{\alpha/n}\|\tilde{u}^{1/p}\|_{A,Q}\|\tilde{v}^{-1/p}\|_{p',Q} \leq K < \infty,$$

and I_α satisfies the weak (p,p) inequality

$$\tilde{u}(\{x\in\mathbb{R}^n : |I_\alpha f(x)| > \lambda\}) \leq \frac{C}{\lambda^p}\int_{\mathbb{R}^n} |f(x)|^p \tilde{v}(x)\,dx.$$

Proof. The proof is very similar to the proof of Theorem 9.45. By Proposition 6.15, $M_\alpha w_2 \in A_1$, so $M(M_\alpha w_2) \approx M_\alpha w_2$. Let $A_0(t) = t^p\log(e+t)^{p-1+\delta}$, $\Phi_0(t) = A_0(t^{1/p})$, and define the weight

$$\tilde{u}_0 = M_{\Phi_0}w_1(M_\alpha w_2)^{1-p}.$$

Then the pair $(\tilde{u},\tilde{u}_0)$ satisfies (9.12) with $r=p$. Since we have the sharp function estimate $M_q^\#(I_\alpha f)(x) \leq CM_\alpha f(x)$ (see the list after Theorem 9.10), by Theorem 9.10

$$\|I_\alpha f\|_{L^{p,\infty}(\tilde{u})} \leq C\|M_\alpha f\|_{L^{p,\infty}(\tilde{u}_0)}.$$

By Example 5.42, $M_\alpha(M_{\Phi_0}w_1) \leq CM_{\Phi,\alpha}w_1$. Define the weight

$$\tilde{v}_0 = M_\alpha(M_{\Phi_0}w_1)w_2^{1-p} \leq C\tilde{v}.$$

By Theorem 6.16, the pair $(\tilde{u}_0,\tilde{v}_0)$ satisfies the A_{pp}^α condition (5.25), and so

$$\|M_\alpha f\|_{L^{p,\infty}(\tilde{u}_0)} \leq C\|f\|_{L^p(\tilde{v}_0)} \leq C\|f\|_{L^p(\tilde{v})}.$$

Combining these two inequalities we get the desired result. □

Chapter 10

Further Applications of Two-Weight Extrapolation

In this chapter we continue to apply the extrapolation theorems in Chapters 7 and 8 to the theory of two-weight norm inequalities. We consider two operators: the dyadic square function and the vector-valued maximal operator. There are two different approaches to these operators. On the one hand, they can be treated as vector-valued singular integrals, and so the natural conjectures for the weak and strong type inequalities would be the analogs of those for singular integrals in Section 9.2.

On the other hand, when considered from the perspective of extrapolation, a new approach emerges. When p_0 equals the critical index ($p_0 = 2$ for the dyadic square function, and $p_0 = q$ for the vector-valued maximal operator defined on ℓ^q, $1 < q < \infty$), it is much easier to prove weighted norm inequalities. We can then extrapolate up or down from this initial inequality, getting very different results in each case. Roughly speaking, for p below the critical index these operators behave like the Hardy-Littlewood maximal operator, and for p greater than the critical index they behave like singular integrals, but with the degree of singularity depending on the ratio p/p_0.

In the one-weight case either approach yields the same result—that $w \in A_p$ is a sufficient condition. However, in the two-weight case we get very different results, with the second approach yielding sharp inequalities for A_1 weights and the first approach yielding sharp inequalities for dual A_1 weights. By combining these two approaches we get new conjectures for both operators.

In this chapter we organize our results as we did in Chapter 9. We will state both kinds of conjectures for the weak and strong type inequalities, review known results and then prove new theorems. In particular, we have a large number of new results for factored weights. Additionally, we consider Coifman-Fefferman type inequalities for the dyadic square function, and prove the sharp function estimate stated above in Section 9.1.

Throughout this chapter, to apply the extrapolation theorems we have to define an appropriate family $\mathcal{F}$ of pairs of functions. For each operator and for weak and strong type inequalities these are different, but in general will be gotten from a family of functions, such as C_c^∞, that is dense in $L^p(u)$. Because of this, standard arguments will prove that the inequalities hold in general.

10.1 The dyadic square function

To define the dyadic square function, we first introduce some notation. Recall that $\mathcal{D}$ is the set of all dyadic cubes in $\mathbb{R}^n$. Given a dyadic cube Q, let $\widetilde{Q}$ denote its dyadic parent: that is, the unique dyadic cube such that $Q \subset \widetilde{Q}$ and $\ell(\widetilde{Q}) = 2\ell(Q)$. Throughout this section we will assume all cubes are dyadic unless otherwise specified.

Given a locally integrable function f, define the dyadic maximal function of f by

$$M^d f(x) = \sup_{\substack{Q\in\mathcal{D} \\ Q\ni x}} \fint_Q |f(y)|\,dy.$$

In the notation of Chapter 3, $M^d = M_{\mathcal{D}}$, but here we adopt the conventional notation. Given a locally integrable function f, define the dyadic square function of f by

$$S_d f(x) = \left(\sum_{Q\in\mathcal{D}} |f_{\widetilde{Q}} - f_Q|^2 \chi_Q(x) \right)^{1/2},$$

where

$$f_Q = \fint_Q f(y)\,dy.$$

The dyadic square function is the discrete analog of the area integral. It was first introduced by Paley [167] and more recently has been studied following the work of C. Fefferman [74]. For more information and references, see Wilson [235].

To apply our extrapolation results to the dyadic square function, we define two families of pairs of functions. For strong type inequalities, we will hereafter assume that we are applying our extrapolation results to the family $\mathcal{F}$ consisting of pairs of the form $(S_d f, |f|)$, where $f \in C_c^\infty$. Similarly, for weak type inequalities, let $\mathcal{F}$ be the family of pairs of the form $(\lambda\chi_{\{x:S_d f(x)>\lambda\}}, |f|)$, where $f \in C_c^\infty$ and $\lambda > 0$. (As before, we write $f \in \mathcal{F}$ to mean the corresponding pair is in $\mathcal{F}$.) By a standard density argument, if we can prove a weak or strong type two-weight norm inequality for S_d for $f \in \mathcal{F}$, then we can prove it for all f such that the right-hand side is finite.

The conjectures

If the dyadic square function is viewed as a vector-valued singular integral, using the ideas of Benedek, Calderón and Panzone [11] (also see [88, 196]), it is natural to make conjectures analogous to Conjectures 9.18 and 9.20 for singular integrals. For the strong type inequality we get the following.

Conjecture 10.1. For each p, $1 < p < \infty$, there exists a constant C such that for all $f \in L^p(v)$,

$$\int_{\mathbb{R}^n} S_d f(x)^p u(x)\, dx \leq C \int_{\mathbb{R}^n} |f(x)|^p v(x)\, dx \tag{10.1}$$

whenever the pair (u, v) satisfies

$$\|u^{1/p}\|_{A,Q} \|v^{-1/p}\|_{B,Q} \leq K < \infty, \tag{10.2}$$

where $\bar{A} \in B_{p'}$ and $\bar{B} \in B_p$, and $Q \in \mathcal{D}$.

This conjecture is true when A is a log bump; this was proved in [47] by modifying the proof of Theorem 9.21 to exploit the fact that the dyadic square function is more localized than a singular integral. For ease of reference below we record this fact as Theorem 10.2. The general case when $\bar{A}$ is a $B_{p'}$ bump remains open.

Theorem 10.2. *If* $A(t) = t^p \log(e+t)^{p-1+\delta}$, $\delta > 0$, $\bar{B} \in B_p$, *and if the pair* (u, v) *satisfies* (10.2), *then the dyadic square function satisfies the strong* (p, p) *inequality* (10.1).

In the case of weak type inequalities the corresponding conjecture is actually a theorem.

Theorem 10.3. *For each* p, $1 < p < \infty$, *there exists a constant* C *such that for all* $f \in L^p(v)$ *and* $\lambda > 0$,

$$u(\{x \in \mathbb{R}^n : S_d f(x) > \lambda\}) \leq \frac{C}{\lambda^p} \int_{\mathbb{R}^n} |f(x)|^p v(x)\, dx$$

whenever the pair (u, v) *satisfies*

$$\|u^{1/p}\|_{A,Q} \|v^{-1/p}\|_{p',Q} \leq K < \infty, \tag{10.3}$$

where $\bar{A} \in B_{p'}$ *and* $Q \in \mathcal{D}$.

The proof of this follows immediately from Corollary 8.4 and the weak $(1, 1)$ inequality

$$u(\{x \in \mathbb{R}^n : S_d f(x) > \lambda\}) \leq \frac{C}{\lambda} \int_{\mathbb{R}^n} |f(x)| M^d u(x)\, dx. \tag{10.4}$$

This was proved for the continuous square function by Chanillo and Wheeden [23]; Uchiyama [226] pointed out that their proof readily adapts to the dyadic square function.

Our second approach to the dyadic square function is motivated by attempts to prove norm inequalities using extrapolation. From this perspective $p = 2$ is a critical index, and the dyadic square function behaves significantly differently depending on the size of p: for $1 < p \leq 2$ it behaves like a maximal operator; for $p > 2$ it is more like a singular integral, but the "degree of singularity" depends on the ratio $p/2$. Based on our results below, we make the following conjectures.

Conjecture 10.4. For each p, $1 < p \leq 2$, there exists a constant C such that for all $f \in L^p(v)$,

$$\int_{\mathbb{R}^n} S_d f(x)^p u(x)\, dx \leq C \int_{\mathbb{R}^n} |f(x)|^p v(x)\, dx, \tag{10.5}$$

whenever the pair (u, v) satisfies

$$\|u^{1/p}\|_{p,Q} \|v^{-1/p}\|_{B,Q} \leq K < \infty, \tag{10.6}$$

where $\bar{B} \in B_p$ and $Q \in \mathcal{D}$.

For $2 < p < \infty$, inequality (10.5) holds whenever the pair (u, v) satisfies

$$\|u^{1/r}\|_{A,Q}^{1/2} \|v^{-1/p}\|_{B,Q} \leq K < \infty, \tag{10.7}$$

where $r = p/2$, $\bar{A} \in B_{r'}$, $\bar{B} \in B_p$, and $Q \in \mathcal{D}$.

Conjecture 10.5. For each p, $1 < p \leq 2$, there exists a constant C such that for all $f \in L^p(v)$,

$$u(\{x \in \mathbb{R}^n : S_d f(x) > \lambda\}) \leq \frac{C}{\lambda^p} \int_{\mathbb{R}^n} |f(x)|^p v(x)\, dx \tag{10.8}$$

whenever the pair (u, v) satisfies the two-weight dyadic A_p condition

$$\|u^{1/p}\|_{p,Q} \|v^{-1/p}\|_{p',Q} \leq K < \infty, \tag{10.9}$$

where $Q \in \mathcal{D}$.

For $2 < p < \infty$, inequality (10.8) holds whenever the pair (u, v) satisfies

$$\|u^{1/r}\|_{A,Q}^{1/2} \|v^{-1/p}\|_{p',Q} \leq K < \infty, \tag{10.10}$$

where $r = p/2$, $\bar{A} \in B_{r'}$, and $Q \in \mathcal{D}$.

In the case $1 < p \leq 2$ we have good evidence for both of these conjectures from our results for factored weights. The conjecture for the case $p > 2$ is more tentative since we cannot prove it even for factored weights. However it is true for A_1 and dual A_1 weights, and we have slightly weaker results for arbitrary factored weights. In the non-factored case we can only improve the condition on A from $\bar{A} \in B_{p'}$ to $\bar{A} \in B_{r'}$ at the cost of making the condition on B stronger—in essence B must become a power bump. We will make this more precise below in the statement of our results.

We will first consider strong type inequalities, and then our results for factored weights. This section is extensive since the proofs of several results do not use extrapolation but instead depend on more classical Calderón-Zygmund decomposition arguments and the properties of factored weights. We do not have a separate section on weak type inequalities because, with the exception of Theorem 10.3 above, we do not have any results in the non-factored case that are not trivial consequences of a strong type inequality. Finally, we consider Coifman-Fefferman type inequalities for the dyadic square function, building on earlier work of Wilson [230, 231, 232, 233, 234] and Chang, Wilson and Wolff [22].

Strong (p,p) inequalities

We first give two results related to Conjecture 10.1. By adapting the decomposition argument used in [56] to prove Proposition 9.28 we can prove the analogous result for the dyadic square function. (Details are left to the interested reader.)

Proposition 10.6. *Given p, $1<p<\infty$, and a pair of weights (u,v) such that for every $Q\in\mathcal{D}$,*

$$\|u^{1/p}\|_{A,Q}\|v^{-1/p}\|_{B,Q}\le K<\infty, \tag{10.11}$$

where $A(t)=t^{sp}$, $s>1$, B is a Young function such that $\bar{B}\in B_p$, and $Q\in\mathcal{D}$. Then there exists a constant C such that

$$\int_{\mathbb{R}^n} S_d f(x)^p u(x)\,dx\le C\int_{\mathbb{R}^n}|f(x)|^p v(x)\,dx. \tag{10.12}$$

We are unable to prove Proposition 10.6 using extrapolation and it would be of interest to find such a proof. We were able to prove Proposition 9.28 by using the fact that a singular integral is essentially self-adjoint: we proved a two-weight condition with a power bump on the right-hand term, and then by a duality argument we get a condition with a power bump on the left-hand term. If we make this condition with the switched bumps our hypothesis, we get the following result.

Proposition 10.7. *Given p, $1<p<\infty$, and a pair of weights (u,v) such that for every $Q\in\mathcal{D}$,* (10.11) *holds whenever A is a Young function such that $\bar{A}$ is doubling and $\bar{A}\in B_{p'}$, $B(t)=t^{sp'}$, $s>1$, and $Q\in\mathcal{D}$. Then there exists a constant C such that* (10.12) *holds.*

Proposition 10.7 follows immediately from Corollary 8.7 and an inequality due to Chanillo and Wheeden [23] and Uchiyama [226]: for $1<p_0\le 2$,

$$\int_{\mathbb{R}^n} S_d f(x)^{p_0} u(x)\,dx\le C\int_{\mathbb{R}^n}|f(x)|^{p_0} M^d u(x)\,dx. \tag{10.13}$$

Remark 10.8. We will prove a stronger version of inequality (10.13) below in Theorem 10.13.

We now consider results related to Conjecture 10.4. For $1 < p \le 2$ we have not been able to prove anything substantive for arbitrary pairs of weights. (For factored weights, see Theorem 10.13 below.) For $p > 2$ we have the following result.

Theorem 10.9. *Given p, $2 < p < \infty$, and a pair of weights (u, v) such that for every $Q \in \mathcal{D}$,*

$$\|u^{1/r}\|_{A,Q}\|v^{-1/r}\|_{r',Q} \le K < \infty, \tag{10.14}$$

where $r = p/2$ and A is a Young function such that $\bar{A} \in B_{r'}$. Then inequality (10.12) *holds.*

Theorem 10.9 follows immediately from Theorem 8.2 and inequality (10.13) with $p_0 = 2$.

Remark 10.10. The two conditions (10.14) and (10.2) are not immediately comparable: the bump on the left-hand term in (10.14) is better, but the bump on the right-hand term is considerably worse. To see this, consider the special case of log-bumps: $A(t) = t^r \log(e+t)^{r-1+\delta}$. If we let $A_0(t) = t^p \log(e+t)^{r-1+\delta}$, then by rescaling both terms we get

$$\|u^{1/r}\|_{A,Q}\|v^{-1/r}\|_{r',Q} = \|u^{1/p}\|^2_{A_0,Q}\|v^{-1/p}\|^2_{sp',Q},$$

where

$$s = \frac{p-1}{p/2-1}.$$

In (10.2) we can take the bump on the left to be a log bump of the form $A(t) = t^p \log(e+t)^{p-1+\delta}$, so for p large (10.14) requires a power on the log term approximately one half the size. On the other hand, we can take the bump on the right to be a B_p bump, as compared to a power bump above. Further, note that we cannot take an arbitrary power bump since s depends on p: $s > 2$ and $s \to \infty$ as p decreases to 2.

If we consider A_1 weights, then (10.14) yields a sharper condition than (10.2). Again if we restrict ourselves to log bumps, then in the latter we can take pairs of the form $(u, M^d_\Phi u)$, where $\Phi(t) = t\log(e+t)^{p-1+\delta}$, and in the former we can take pairs of the form $(u, M^d_{\Phi_0} u)$, where $\Phi_0(t) = t\log(e+t)^{p/2-1+\delta}$. Further, such pairs are sharp, as we will show in Example 10.11 below. (For the definition of M^d_Φ, see (5.7) and Remark 5.4.)

On the other hand, if we consider dual A_1 weights, then the reverse is true. In (10.14) we can take pairs of the form $((M^d u)^{1-r}, u^{1-r})$, and in (10.2) pairs of the form $((M^d_\Psi u)^{1-p}, u^{1-p})$, where $\Psi(t) = t\log(e+t)^{p'-1+\delta}$. To compare these, if we replace u by $u^{\frac{p-1}{r-1}}$ in the first pair, then we get

$$M^d(u^{\frac{p-1}{r-1}})^{r-1} = (M^d_s u)^{p-1} \ge (M^d_\Psi u)^{p-1};$$

thus, (10.2) yields a sharper condition.

Given the A_1 and dual A_1 weights above, it is natural to combine them to get a pair of factored weights:

$$(\tilde{u}, \tilde{v}) = \big(w_1(M^d_\Psi w_2)^{1-p}, (M^d_\Phi w_1)w_2^{1-p}\big),$$

where $\Phi(t) = t\log(e+t)^{p/2-1+\delta}$ and $\Psi(t) = t\log(e+t)^{p'-1+\delta}$. We conjecture that these pairs of weights are sufficient for the strong (p,p) inequality for $p \geq 2$. They satisfy the condition in Conjecture 10.4 when $p > 2$ and were the motivation for this conjecture.

In the one-weight case, by the Jones factorization theorem, a general result for A_p weights follows from inequalities for A_1 and dual A_1 weights by interpolation with change of measure. (See Sawyer [205] for a related interpolation argument.) However, this argument does not extend to the two-weight case, even for factored weights, since it depends on the reverse Hölder inequality. (See Theorems 10.14 and 10.16 for further evidence for this conjecture in the scale of factored weights.)

In the scale of A_1 weights defined using log bumps, Theorem 10.9 is sharp. The following example is adapted from one for the continuous square function due to Chanillo and Wheeden [23].

Example 10.11. Given $p > 2$, in dimension 1 the inequality

$$\int_{\mathbb{R}} S_d f(x)^p u(x)\,dx \leq C \int_{\mathbb{R}} |f(x)|^p M^d_\Phi u(x)\,dx$$

does not hold in general if $\Phi(t) = t\log(e+t)^{p/2-1}$.

Proof. Fix $p > 2$. For each $k \geq 1$ we will construct a pair of functions f_k and u_k on the real line such that

$$\sup_k \int_{\mathbb{R}} |f_k(x)|^p M^d_\Phi u_k(x)\,dx < \infty,$$

but

$$\lim_{k\to\infty} \int_{\mathbb{R}} S_d f_k(x)^p u_k(x)\,dx = \infty.$$

Fix $k \geq 1$ and define

$$u_k(x) = 2^{2k}\chi_{[0,2^{-2k})}(x).$$

Fix j, $1 \leq j \leq k$, and $x \in [2^{-2j}, 2^{-2j+1})$. We will first show that

$$M^d_\Phi u_k(x) \approx 2^{2j-1}(2k-2j+1)^{p/2-1}.$$

To see this, let

$$\varphi_\Phi(t) = \Phi^{-1}(t^{-1})^{-1} \approx t\log(e+1/t)^{p/2-1}.$$

Then given any interval P, a straightforward calculation using the definition shows that for any dyadic interval Q,

$$\|\chi_P\|_{\Phi,Q} = \varphi_\Phi\left(\frac{|P\cap Q|}{|Q|}\right).$$

Therefore, $M^d_\Phi(\chi_P)(x) = \varphi_\Phi(M^d(\chi_P)(x))$. Hence, if we let $P = [0, 2^{-2k})$ and let $x \in [2^{-2j}, 2^{-2j+1})$, then it is immediate that $M^d(\chi_P)(x) = 2^{2j-2k-1}$, and so

$$M^d_\Phi u_k(x) = 2^{2k} M^d_\Phi(\chi_P)(x) = 2^{2k}\varphi_\Phi(2^{2j-2k-1}) \approx 2^{2j-1}(2k-2j+1)^{p/2-1}.$$

Now define

$$f_k(x) = \sum_{j=1}^{k-1}(2k-2j+1)^{-1/2}\log(2k-2j+1)^{-1/2}\chi_{[2^{-2j},2^{-2j+1})}(x).$$

Then by a change of variables, since $p/2 > 1$,

$$\begin{aligned}
&\int_{\mathbb{R}} f_k(x)^p M^d_\Phi u_k(x)\,dx \\
&\qquad\approx \sum_{j=1}^{k-1}(2k-2j+1)^{-p/2}\log(2k-2j+1)^{-p/2}(2k-2j+1)^{p/2-1} \\
&\qquad= \sum_{j=2}^{k}(2j-1)^{-1}\log(2j-1)^{-p/2} \\
&\qquad\le C\sum_{j=2}^{\infty}(2j-1)^{-1}\log(2j-1)^{-p/2} \\
&\qquad< \infty.
\end{aligned}$$

On the other hand, for $1 \le i < k$, let $Q_i = [0, 2^{-2i+1})$. Then we have that $|(\widetilde{Q}_i \setminus Q_i)\cap \operatorname{supp}(f_k)| = 0$. Furthermore,

$$\begin{aligned}
(f_k)_{Q_i} = 2^{2i-1}&\sum_{j=i}^{k-1}(2k-2j+1)^{-1/2}\log(2k-2j+1)^{-1/2}2^{-2j} \\
&\ge 2^{-1}(2k-2i+1)^{-1/2}\log(2k-2i+1)^{-1/2}.
\end{aligned}$$

Therefore, if $x \in [0, 2^{-2k})$,

$$\begin{aligned}
S_d f_k(x)^2 &\ge \sum_{i=1}^{k-1}|(f_k)_{\widetilde{Q}_i} - (f_k)_{Q_i}|^2 \\
&= \frac{1}{4}\sum_{i=1}^{k-1}(f_k)^2_{Q_i}
\end{aligned}$$

$$\geq \frac{1}{16}\sum_{i=1}^{k-1}(2k-2i+1)^{-1}\log(2k-2i+1)^{-1}$$
$$= \frac{1}{16}\sum_{i=2}^{k}(2i-1)^{-1}\log(2i-1)^{-1}.$$

Hence,

$$\int_{\mathbb{R}} S_d f_k(x)^p u_k(x)\,dx \geq \left(\frac{1}{16}\sum_{i=2}^{k}(2i-1)^{-1}\log(2i-1)^{-1}\right)^{p/2}.$$

This sum diverges as $k \to \infty$, so our proof is complete. □

Inequalities for factored weights

Conjecture 10.4 for $p > 2$ holds for both A_1 and dual A_1 pairs. For A_1 weights this is an immediate consequence of Theorem 10.9; for dual A_1 weights it follows at once from Proposition 10.6. More generally, we have results for factored weights for this conjecture and for Conjecture 10.5 for all values of p. Because many of the proofs are lengthy, depending on Calderón-Zygmund decompositions and the properties of factored weights, we will first state all of our results and then give the proofs.

Our first result is a generalization of (10.13); the proof is adapted from the proof of this inequality (more precisely the analog for the continuous square function) due to Chanillo and Wheeden [23].

Theorem 10.12. *Let $\bar{B} \in B_2$ and $\Psi(t) = B(t^{1/2})$, and form the pair of factored weights*

$$(\tilde{u}, \tilde{v}) = (w_1(M^d_\Psi w_2)^{-1}, (M^d w_1)w_2^{-1}).$$

Then for every $Q \in \mathcal{D}$, the pair $(\tilde{u}, \tilde{v})$ satisfies

$$\|\tilde{u}^{1/2}\|_{2,Q}\|\tilde{v}^{-1/2}\|_{B,Q} \leq K < \infty,$$

and the dyadic square function satisfies the strong $(2,2)$ inequality

$$\int_{\mathbb{R}^n} S_d f(x)^2 \tilde{u}(x)\,dx \leq C\int_{\mathbb{R}^n} |f(x)|^2 \tilde{v}(x)\,dx.$$

Theorem 10.12 gives us Conjecture 10.4 for factored weights when $p = 2$. As a corollary, using the two-weight extrapolation theorem for factored weights, Theorem 7.14, we can also prove our conjecture for factored weights for $1 < p < 2$ in the scale of log bumps.

Theorem 10.13. *Given p, $1 < p < 2$, let $B(t) = t^{p'}\log(e+t)^{p'-1+\delta}$ and $\Psi(t) = B(t^{1/p'})$. Then the pair of factored weights*

$$(\tilde{u}, \tilde{v}) = (w_1(M^d_\Psi w_2)^{1-p}, (M^d w_1)w_2^{1-p})$$

satisfies, for every $Q \in \mathcal{D}$,

$$\|\tilde{u}^{1/p}\|_{p,Q}\|\tilde{v}^{-1/p}\|_{B,Q} \leq K < \infty,$$

and the dyadic square function satisfies the strong (p,p) inequality

$$\int_{\mathbb{R}^n} S_d f(x)^p \tilde{u}(x)\,dx \leq C \int_{\mathbb{R}^n} |f(x)|^p \tilde{v}(x)\,dx.$$

For $p > 2$ we cannot prove Conjecture 10.4 even in the scale of log bumps. However, we can prove two results that are only slightly worse. The first follows by extrapolation for factored weights.

Theorem 10.14. *Given $p > 2$, let $r = p/2$, $A(t) = t^p \log(e+t)^{r-1+\delta}$, and $B(t) = t^{p'} \log(e+t)^{1+\delta}$, $\delta > 0$. Let $\Phi(t) = A(t^{1/p})$ and $\Psi(t) = B(t^{1/p'})$ and form the pair of factored weights*

$$(\tilde{u}, \tilde{v}) = (w_1 (M_\Psi^d w_2)^{1-p}, (M_\Phi^d w_1) w_2^{1-p}).$$

Then the pair $(\tilde{u}, \tilde{v})$ satisfies, for every $Q \in \mathcal{D}$,

$$\|\tilde{u}^{1/p}\|_{A,Q}\|\tilde{v}^{-1/p}\|_{B,Q} \leq K < \infty,$$

and the dyadic square function satisfies the strong (p,p) inequality

$$\int_{\mathbb{R}^n} S_d f(x)^p \tilde{u}(x)\,dx \leq C \int_{\mathbb{R}^n} |f(x)|^p \tilde{v}(x)\,dx. \tag{10.15}$$

Remark 10.15. In the scale of log bumps, Conjecture 10.4 for $p > 2$ yields $B(t) = t^{p'} \log(e+t)^{p'-1+\delta}$. Since for $p > 2$, $1+\delta > p'-1+\delta$, we have that Theorem 10.14 is weaker. They agree asymptotically as p decreases to 2, but for p very large this result essentially increases the exponent on the logarithm by 1.

The second result that is slightly weaker than Conjecture 10.4 is gotten by a decomposition argument similar to the proof of Theorem 10.2, which in turn uses ideas from the proof of Lemma 9.2. An interesting feature of the proof is that we use the fact that $\tilde{u}$ is a factored weight to reduce the proof to a one-weight estimate that we prove using A_∞ extrapolation.

Theorem 10.16. *Given $p > 2$, let $r = p/2$, $A(t) = t^p \log(e+t)^{r+\delta}$, and $\bar{B} \in B_p$. Let $\Phi(t) = A(t^{1/p})$ and $\Psi(t) = B(t^{1/p'})$ and form the pair of factored weights*

$$(\tilde{u}, \tilde{v}) = (w_1 (M_\Psi^d w_2)^{1-p}, (M_\Phi^d w_1) w_2^{1-p}).$$

Then $(\tilde{u}, \tilde{v})$ satisfies, for every $Q \in \mathcal{D}$,

$$\|\tilde{u}^{1/p}\|_{A,Q}\|\tilde{v}^{-1/p}\|_{B,Q} \leq K < \infty,$$

and the dyadic square function satisfies the strong (p,p) inequality (10.15).

Remark 10.17. The exponent on the logarithm in the bump on the left-hand term in Theorem 10.16 is 1 greater than the power on the corresponding log bump given by Conjecture 10.4, but it is smaller than the log bump gotten in Theorem 10.2. This result is analogous to Theorem 9.39 for singular integrals and Theorem 9.46 for fractional integrals.

Our results for weak type inequalities are better than those given above for strong type inequalities. When $p < 2$ we can prove Conjecture 10.5 completely for factored weights; when $p > 2$ we can prove it for log-bumps. When p equals the critical index 2 we can almost prove it in the scale of log-bumps: we acquire an additional "fractional" logarithm in the bump on the left-hand term.

Theorem 10.18. *Given p, $1 < p < 2$, the pair of factored weights*

$$(\tilde{u}, \tilde{v}) = (w_1(M^d w_2)^{1-p}, (M^d w_1)w_2^{1-p})$$

satisfies the two-weight dyadic A_p condition (10.9), *and the dyadic square function satisfies the weak (p,p) inequality*

$$\tilde{u}(\{x \in \mathbb{R}^n : S_d f(x) > \lambda\}) \leq \frac{C}{\lambda^p} \int_{\mathbb{R}^n} |f(x)|^p \tilde{v}(x)\, dx. \tag{10.16}$$

Theorem 10.19. *Given $p \geq 2$, let $r = p/2$, $A(t) = t^p \log(e+t)^{r-1+\delta}$, $\delta > 0$, and $\Phi(t) = A(t^{1/p})$. Then the factored pair*

$$(\tilde{u}, \tilde{v}) = (w_1(M^d w_2)^{1-p}, (M_\Phi^d w_1)w_2^{1-p})$$

satisfies

$$\|\tilde{u}^{1/p}\|_{A,Q} \|\tilde{v}^{-1/p}\|_{p',Q} \leq K < \infty \tag{10.17}$$

for every $Q \in \mathcal{D}$, and the dyadic square function satisfies the weak (p,p) inequality (10.16).

Remark 10.20. When $p = 2$, $A(t) = t^2 \log(e+t)^\delta$, $\delta > 0$; our extrapolation techniques are not sharp enough to let us take $\delta = 0$ even for log bumps, and we are unable to modify the decomposition argument used to prove Theorem 10.12 to get weak type inequalities.

Remark 10.21. If Conjecture 10.5 were true for $p = 2$ for all factored weights, then it would also be true for $p > 2$: this would also follow from the two-weight extrapolation theorem for factored weights, Theorem 7.14.

We now give the proofs of the above theorems. We first give those proofs that depend on extrapolation; these proofs are straightforward and are all similar to one another. We then give those proofs that depend on a Calderón-Zygmund decomposition and properties of factored weights.

Proof of Theorems 10.13, 10.14 and 10.19

All three of these results are a consequence of Theorem 7.14, the two-weight extrapolation theorem for factored weights; the first two use Theorem 10.12 to start the extrapolation, and the last uses Theorem 10.18.

We first prove Theorem 10.13. Fix p, $1 < p < 2$, and let $A(t) = t^p$ and $B(t) = t^{p'} \log(e+t)^{p'-1+\delta}$, $\delta > 0$. Define $A_0(t) = t^2$, and $B_0(t) = t^2 \log(e+t)^{1+\delta_0}$, where we choose $\delta_0 > 0$ such that

$$\frac{\delta}{p'} - \frac{\delta_0}{2} > 0.$$

By Theorem 10.12, inequality (7.15) holds with $p_0 = 2$ for the pair $(\tilde{u}_0, \tilde{v}_0)$ defined by (7.16).

Further, $A_0(t^{1/2}) = A(t^{1/p})$. By our choice of δ_0, if we let

$$C(t) = \frac{t^{(p'/2)'}}{\log(e+t)^{1+\epsilon}},$$

where

$$\epsilon = \frac{\delta/p' - \delta_0/2}{1/p - 1/2} > 0,$$

then $C \in B_{(p'/2)'}$ and

$$\begin{aligned} C^{-1}(t)^{1/2} B^{-1}(t) &\approx t^{1/p-1/2} \log(e+t)^{(1+\epsilon)(1/p-1/2)} \frac{t^{1/p'}}{\log(e+t)^{1/p+\delta/p'}} \\ &= \frac{t^{1/2}}{\log(e+t)^{1/2+\delta_0/2}} \approx B_0^{-1}(t). \end{aligned}$$

The desired inequality now follows by Theorem 7.14.

The proof of Theorem 10.14 is essentially the same. Fix $p > 2$, $r = p/2$ and let $A(t) = t^p \log(e+t)^{r-1+\delta}$, $B(t) = t^{p'} \log(e+t)^{1+\delta}$, $\delta > 0$. Let $B_0(t) = t^2 \log(e+t)^{1+\delta} \approx B(t^{2/p'})$. Define $A_0(t) = t^2$ and let

$$C(t) = \frac{t^{(p/2)'}}{\log(e+t)^{1+\epsilon}}, \qquad \epsilon = \frac{\delta/p}{1/2 - 1/p} > 0.$$

Then $A^{-1}(t)C^{-1}(t)^{1/2} \leq cA_0^{-1}(t)$ and $C \in B_{(p/2)'}$, so we can apply Theorem 7.14 exactly as before.

Finally, we prove Theorem 10.19. Fix $p \geq 2$ and let $1 < p_0 < 2$; the exact value of p_0 will be fixed below. Let $A_0(t) = t^{p_0}$, $B_0(t) = t^{p_0'}$, $A(t) = t^p \log(e+t)^{p/2-1+\delta}$, and $B(t) = t^{p'}$. Then $B(t^{1/p'}) = B_0(t^{1/p_0'})$.

Now define

$$C(t) = \frac{t^{(p/p_0)'}}{\log(e+t)^{1+\epsilon}},$$

where

$$\frac{1}{p_0} = \frac{1}{2} + \frac{\delta}{2p}, \qquad \epsilon = \frac{\delta}{p+\delta-2}.$$

(Note that to get $p_0 > 1$, we need $\delta < p$; however, if a pair of weights satisfies (10.17) for some value of δ, it satisfies it for all smaller values, so we may assume $\delta < p$ without loss of generality.) Then

$$(1+\epsilon)\frac{1}{p_0} - \frac{\epsilon}{p} = \frac{1}{2} + \frac{\delta}{p},$$

and this is precisely what is needed to insure that $A^{-1}(t)C^{-1}(t)^{1/p_0} \leq cA_0^{-1}(t)$ and that $C \in B_{(p/p_0)'}$. Therefore, by Theorems 10.18 and 7.14 we get the desired inequality.

Proof of Theorem 10.12

By a straightforward approximation argument, we may assume without loss of generality that w_1 has compact support. For each integer k define

$$\Omega_k = \{x \in \mathbb{R}^n : M^d w_1(x) > 2^k\},$$

and

$$A_k = \{Q \in \mathcal{D} : 2^k < (w_1)_{\widetilde{Q}} \leq 2^{k+1}\}.$$

Then every cube Q such that $\tilde{u}(Q) > 0$ is contained in a unique A_k, and if $Q \in A_k$, then $Q \subset \widetilde{Q} \subset \Omega_k$. Further, $(w_1)_Q \leq 2^n (w_1)_{\widetilde{Q}} \leq 2^n \cdot 2^{k+1}$.

We now estimate as follows: by Proposition 6.2, $(M_\Psi^d w_2)^{-1} \in RH_\infty^d$: that is, it satisfies the RH_∞ condition on dyadic cubes. Hence,

$$\begin{aligned}
&\int_{\mathbb{R}^n} S_d f(x)^2 \tilde{u}(x)\, dx \\
&\qquad = \int_{\mathbb{R}^n} \sum_{Q \ni x} |f_{\widetilde{Q}} - f_Q|^2 \tilde{u}(x)\, dx \\
&\qquad = \sum_{Q \in \mathcal{D}} \tilde{u}_Q |f_{\widetilde{Q}} - f_Q|^2 |Q| \\
&\qquad = \sum_k \sum_{Q \in A_k} \tilde{u}_Q |f_{\widetilde{Q}} - f_Q|^2 |Q| \\
&\qquad \leq C \sum_k \sum_{Q \in A_k} ((M_\Psi^d w_2)^{-1})_Q (w_1)_Q |f_{\widetilde{Q}} - f_Q|^2 |Q|
\end{aligned}$$

$$
\begin{aligned}
&\le 2^{n+1}C\sum_k \sum_{Q\in A_k} 2^k((M_\Psi^d w_2)^{-1})_Q|f_{\tilde{Q}} - f_Q|^2|Q|;\\
&= 2^{n+1}C\sum_k \sum_{Q\in A_k} 2^k((M_\Psi^d w_2)^{-1})_Q|(f\chi_{\Omega_k})_{\tilde{Q}} - (f\chi_{\Omega_k})_Q|^2|Q|\\
&\le 2^{n+1}C\sum_k \sum_{Q\in \mathcal{D}} 2^k((M_\Psi^d w_2)^{-1})_Q|(f\chi_{\Omega_k})_{\tilde{Q}} - (f\chi_{\Omega_k})_Q|^2|Q|\\
&= 2^{n+1}C\sum_k 2^k \int_{\mathbb{R}^n} S_d(f\chi_{\Omega_k})(x)^2 M_\Psi^d w_2(x)^{-1}\,dx.
\end{aligned}
$$

By Theorem 6.4, the pair $((M_\Psi^d w_2)^{-1}, w_2^{-1})$ satisfies (10.2) with $p = 2$ and any p-Young function A, in particular with $A(t) = t^p \log(e+t)^{p-1+\delta}$. Hence, by Theorem 10.2,

$$
\begin{aligned}
\sum_k 2^k \int_{\mathbb{R}^n} & S_d(f\chi_{\Omega_k})(x)^2 M_\Psi^d w_2(x)^{-1}\,dx\\
&\le C\sum_k 2^k \int_{\Omega_k} |f(x)|^2 w_2(x)^{-1}\,dx\\
&\le C\int_{\mathbb{R}^n} |f(x)|^2 \sum_k 2^k \chi_{\Omega_k}(x) w_2(x)^{-1}\,dx.
\end{aligned}
$$

Fix $x \in \mathbb{R}^n$ with $M^d w_1(x) > 0$. Then there exists k_0 such that $2^{k_0} < M^d w_1(x) \le 2^{k_0+1}$. Therefore,

$$
\sum_k 2^k \chi_{\Omega_k}(x) w_2(x)^{-1} = \sum_{k\le k_0} 2^k w_2(x)^{-1} \le 2M^d w_1(x) w_2(x)^{-1} = 2\tilde{v}(x).
$$

This completes the proof.

Proof of Theorem 10.16

We begin with two lemmas. Hereafter, let $M^{d,\#}$ denote the dyadic sharp maximal operator, which is defined as in (9.1) except the supremum is taken over dyadic cubes. Similarly, let $M_Q^{d,\#}$ be the sharp function gotten when the supremum is taken over dyadic cubes contained in the dyadic cube Q. Finally, let A_∞^d denote the union of the dyadic A_p^d classes, $1 \le p < \infty$. (In passing, we note that these classes have properties for dyadic cubes similar to those described in Theorem 1.3.)

Lemma 10.22. *Given $w \in A_\infty^d$, then for $f \in L^1_{\text{loc}}$ and every $Q \in \mathcal{D}$,*

$$
\int_Q |f(x) - f_Q|\,w(x)\,dx \le C\int_Q M_Q^{d,\#} f(x)\,w(x)\,dx. \tag{10.18}
$$

Proof. The proof is very similar to the proof of Lemma 9.2; here we sketch the changes. Fix w, f and a dyadic cube Q. To adapt the proof of Lemma 9.2 we need to assume that f and w are bounded and that $w_Q = 1$. We will first show how to make this reduction.

First, suppose that we could prove inequality (10.18) for f bounded. Then given an arbitrary f, for $N > 0$ let $f_N = \max\big(\min(f, N), -N\big)$. Since for all real values of K,

$$M_Q^{d,\#}\big(\max(f,K)\big),\; M_Q^{d,\#}\big(\min(f,K)\big) \le 2M_Q^{d,\#} f,$$

(see Torchinsky [222, p. 218]), $M_Q^{d,\#} f_N \le 4M_Q^{d,\#} f$. On the other hand, since $|f_N| \le |f|$ and f_N converges pointwise to $f \in L^1_{\text{loc}}$, by the dominated convergence theorem, $(f_N)_Q \to f_Q$. Therefore, by Fatou's lemma and (10.18) for bounded functions,

$$\begin{aligned}\int_Q |f(x) - f_Q| w(x)\, dx &\le \liminf_{N\to\infty} \int_Q |f_N(x) - (f_N)_Q| w(x)\, dx \\ &\le C \liminf_{N\to\infty} \int_Q M_Q^{d,\#} f_N(x) w(x)\, dx \le 4C \int_Q M_Q^{d,\#} f(x) w(x)\, dx.\end{aligned}$$

Similarly, suppose that we could prove (10.18) assuming that w was bounded and $w_Q = 1$. For all $N > 0$, $\min(w, N) \in A^d_\infty$ with a constant independent of N (see [88]). Let

$$w_N = \frac{\min(w,N)\chi_Q}{(\min(w,N))_Q};$$

then w_N is bounded and $(w_N)_Q = 1$. The weight w_N is no longer in A^d_∞ but it is in $A^d_\infty(Q)$: i.e., $w \in A^d_p(Q)$ for some $p \ge 1$ and this is what we need to use Lemma A.4 below. Finally, if (10.18) holds with w replaced by w_N, then it follows immediately from the monotone convergence theorem that it holds for w.

Therefore, we may assume that f and w are bounded, $w_Q = 1$, $\operatorname{supp}(w) \subset Q$, and $w \in A^d_\infty(Q)$. Let $h(x) = |f(x) - f_Q|\,\chi_Q(x)$. Then $h \in \cup_{p>1} L^p$. Fix $a > 2^n$ and m such that $\|w\|_\infty \le a^m$. Then $m \ge 0$, and $m = 0$ if and only if $w = 1$ almost everywhere in Q. (Therefore, we may assume that $m \ge 1$, since the case $m = 0$ is trivial.)

The proof now goes through as in the proof of Lemma 9.2 with h in place of f and w in place of $\tilde{u}$ until (9.19). At that point, let $\{c_{Q^k_j}\}$ be a collection of constants whose values will be fixed below. Then

$$\begin{aligned}\int_Q |f(x) - f_Q|\, w(x)\, dx &= \sum_{k=-\infty}^{m-1} \int_{\mathbb{R}^n} h(x)\,(b_k(x) - b_{k+1}(x))\, dx \\ &= \sum_{k,j} \int_{Q^k_j} (h(x) - c_{Q^k_j})\,(b_k(x) - b_{k+1}(x))\, dx\end{aligned}$$

$$\leq C \sum_{k,j} (1+a)\, 2^n\, a^k \int_{Q_j^k} |h(x) - c_{Q_j^k}|\, dx$$
$$\leq C \Big(\sum_{k=-\infty}^{-1} \sum_j \cdots + \sum_{k=0}^{m-1} \sum_j \cdots \Big)$$
$$= C\,(I_1 + I_2).$$

Since $\operatorname{supp}(h) \subset Q$, then for each Q_j^k in each term, $Q \cap Q_j^k \neq \emptyset$.

We estimate I_1 and I_2 in turn. For each Q_j^k in I_1 let $c_{Q_j^k} = 0$. We have that $Q \subset Q_j^k$; for if $Q_j^k \subsetneq Q$, then by the maximality of Q_j^k it would follow that $1 = w_Q \leq a^k < 1$. Furthermore, since for each k the cubes $\{Q_j^k\}_j$ are pairwise disjoint and must intersect Q, this set contains a unique cube that we denote by Q^k. Hence, since $\operatorname{supp}(h) \subset Q$, $a > 1$ and $w_Q = 1$,

$$I_1 = \sum_{k=-\infty}^{-1} (1+a) 2^n a^k \int_{Q^k} |h(x)|\, dx \leq C \int_Q h(x)\, dx$$
$$= C \int_Q |f(x) - f_Q|\, dx \cdot w_Q \leq C \int_Q M_Q^{\#,d} f(x)\, w(x)\, dx.$$

To estimate I_2 let $c_{Q_j^k} = h_{Q_j^k}$. In this case we must have that $Q_j^k \subsetneq Q$: for if $Q \subset Q_j^k$, then the fact that $\operatorname{supp}(w) \subset Q$ would imply that

$$1 \leq a^k < w_{Q_j^k} = w(Q)\, |Q_j^k|^{-1} \leq w(Q)\, |Q|^{-1} = 1,$$

which would be a contradiction.

Let $\{E_j^k\}$ be the sets given in Proposition A.1. Then these sets are pairwise disjoint, $E_j^k \subset Q_j^k \subset Q$, and there exists $\alpha > 0$ such that $|Q_j^k| \leq \alpha |E_j^k|$. Since $w \in A_\infty^d(Q)$, by Lemma A.4 (which still holds in this case) there exists $\beta > 0$ such that $w(Q_j^k) \leq \beta w(E_j^k)$.

We can now repeat the argument in the proof of Lemma 9.2, getting

$$I_2 = \sum_{k=0}^{m-1} a^k \int_{Q_j^k} |h(x) - h_{Q_j^k}|\, dx$$
$$\leq C \sum_{k=0}^{m-1} w_{Q_j^k} \int_{Q_j^k} |h(x) - h_{Q_j^k}|\, dx$$
$$= C \sum_{k=0}^{m-1} w(Q_j^k) \fint_{Q_j^k} |h(x) - h_{Q_j^k}|\, dx$$
$$\leq C \sum_{k=0}^{m-1} w(E_j^k) \fint_{Q_j^k} |h(x) - h_{Q_j^k}|\, dx$$

$$\leq C \sum_{k=0}^{m-1} \int_{E_j^k} M_Q^{\#,d} h(x)\, w(x)\, dx$$
$$\leq C \int_Q M_Q^{\#,d} h(x)\, w(x)\, dx.$$

To complete the proof we will show that $M_Q^{\#,d}h(x) \leq 2\, M_Q^{\#,d} f(x)$ for every $x \in Q$. Fix $R \subset Q$, $R \in \mathcal{D}$; then for each $x \in R$,

$$|h(x) - h_R| \leq \fint_R \big||f(x) - f_Q| - |f(y) - f_Q|\big|\, dy$$
$$\leq \fint_R |f(x) - f(y)|\, dy \leq |f(x) - f_R| + \fint_R |f(y) - f_R|\, dy.$$

If we take the average over R, the desired inequality follows at once. □

Lemma 10.23. *Given $w \in A_\infty^d$, then for $f \in L^\infty$ and every $Q \in \mathcal{D}$,*

$$\int_Q S_d(f\chi_Q)(x)^2 w(x)\, dx \leq C \int_Q M_Q^d f(x)^2 w(x)\, dx.$$

Proof. Define the family

$$\mathcal{F} = (|f - f_Q|\chi_Q, (M_Q^{d,\#} f)\chi_Q), \quad f \in L^1_{\mathrm{loc}}.$$

Then we can apply A_∞ extrapolation, Corollary 3.15, starting from the strong $(1,1)$ inequality in Lemma 10.22 to get that for all p, $0 < p < \infty$, and all $w \in A_\infty^d$,

$$\int_Q |f(x) - f_Q|^p w(x)\, dx \leq C \int_Q M_Q^{d,\#} f(x)^p w(x)\, dx.$$

Hence, for all $f \in L^1_{\mathrm{loc}} \cap L^p(w, Q)$,

$$\int_Q |f(x)|^p w(x)\, dx \leq C \int_Q M^{d,\#} f(x)^p w(x)\, dx + C\Big(\fint_Q |f(x)|\, dx\Big)^p w(Q).$$

We want to apply this inequality when $p = 2/q$ and with f replaced by $S_d(f\chi_Q)^q$ for some q, $0 < q < 1$. However, S_d is weak $(1,1)$, and since $f \in L^\infty$, $f\chi_Q \in L^1$; therefore, by Kolmogorov's inequality, $S_d(f\chi_Q)^q \in L^1_{\mathrm{loc}}$. Similarly, since $w \in A_\infty^d$, $w \in RH_r^d$ for some $r > 1$. Because S_d is bounded on $L^{2r'/q}$ and $f\chi_Q \in L^{2r'/q}$, we have that

$$\int_Q S_d(f\chi_Q)(x)^{2/q} w(x)\, dx$$
$$\leq \Big(\int_Q S_d(f\chi_Q)(x)^{2r'/q}\, dx\Big)^{1/r'} \Big(\int_Q w(x)^r\, dx\Big)^{1/r} < \infty.$$

Therefore, we have that

$$\int_Q S_d(f\chi_Q)(x)^2 w(x)\,dx = \int_Q \left(S_d(f\chi_Q)(x)^q\right)^{2/q} w(x)\,dx$$
$$\leq C\int_Q M_q^{d,\#}(S_d(f\chi_Q))(x)^2 w(x)\,dx + C\Big(\fint_Q S_d(f\chi_Q)(x)^q\,dx\Big)^{2/q} w(Q).$$

We estimate each term separately. By the sharp function estimate in Theorem 10.24 below, for $0<q<1$,

$$\int_Q M_q^{d,\#}(S_d(f\chi_Q))(x)^2 w(x)\,dx$$
$$\leq C\int_Q M^d(f\chi_Q)(x)^2 w(x)\,dx = C\int_Q M_Q^d f(x)^2 w(x)\,dx.$$

To estimate the second term we use the fact that S_d is weak type $(1,1)$ (see [235]). Therefore, since $q<1$, by Kolmogorov's inequality,

$$\Big(\fint_Q (S_d(f\chi_Q))(x)^q\,dx\Big)^{2/q} w(Q)$$
$$\leq C\Big(\fint_Q |f(x)|\,dx\Big)^2 w(Q) \leq C\int_Q M_Q^d f(x)^2 w(x)\,dx.$$

This completes the proof. □

Proof of Theorem 10.16. Fix $p>2$ and f. Without loss of generality we may assume that f is non-negative and a bounded function of compact support. Fix the pair $(\tilde{u},\tilde{v})$; we may also assume that $w_1\in L^\infty$.

Let $r=p/2$; then by duality,

$$\|S_d f\|_{L^p(\tilde{u})}^2 = \|(S_d f)^2\|_{L^r(\tilde{u})} = \sup_h \int_{\mathbb{R}^n} S_d f(x)^2 \tilde{u}(x)^{2/p} h(x)\,dx,$$

where the supremum is taken over all h that are non-negative, bounded and have compact support and such that $\|h\|_{L^{r'}}=1$. Fix such an h; we will show that the integral on the right-hand side can be bounded by a constant independent of h. Let $\bar{w}_1 = w_1^{2/p}h$ and $\bar{w}_2 = (M_\Psi^d w_2)^{(1-p)2/p}$. Since $p>2$, by the dyadic version of Proposition 6.2, $\bar{w}_2\in A_p^d\cap RH_\infty^d$. Furthermore, $\bar{w}_1$ is bounded and has compact support, so $\bar{w}_1\in L^1(\bar{w}_2)$.

Since $\bar{w}_2\in A_\infty^d$, by Lemma A.4 it is dyadic doubling. Fix $a>[\bar{w}_2]_D$ (the doubling constant of $\bar{w}_2$) and $m>0$ such that $\|\bar{w}_1\|_{L^\infty}\leq a^m$. For each $k\leq m$, let $\{Q_j^k\}$ be the Calderón-Zygmund cubes of $\bar{w}_1$ with respect to $\bar{w}_2$ at height a^k. (See Proposition A.5.) Let $\Omega_k=\cup_j Q_j^k$. Define

$$W_{j,k} = \frac{1}{\bar{w}_2(Q_j^k)}\int_{Q_j^k} \bar{w}_1(x)\bar{w}_2(x)\,dx;$$

then $a^k < W_{j,k} \leq [\bar{w}_2]_D a^k$. Now define

$$b_k(x) = \sum_j (\bar{w}_1(x) - W_{j,k})\chi_{Q_j^k}(x),$$

$$g_k(x) = \bar{w}_1(x) - b_k(x) = \begin{cases} W_{j,k} & x \in Q_j^k, \\ \bar{w}_1(x) & x \in \mathbb{R}^n \setminus \Omega_k. \end{cases}$$

It follows from this definition that $g_k(x) \leq [\bar{w}_2]_D a^k$ and $\|g_k\|_{L^1(\bar{w}_2)} = \|\bar{w}_1\|_{L^1(\bar{w}_2)}$. Since $\Omega_m = \emptyset$, we set $b_m = 0$, and so for every $l < 0$,

$$\bar{w}_1(x) = \sum_{k=l}^{m-1} (b_k(x) - b_{k+1}(x)) + g_l(x).$$

As in the proof of Lemma 9.2,

$$|b_k(x) - b_{k+1}(x)| \leq (1+a)[\bar{w}_2]_D a^k$$

and

$$\int_{Q_j^k} (b_k(x) - b_{k+1}(x))\bar{w}_2(x)\,dx = 0.$$

We now estimate as follows:

$$\begin{aligned}\int_{\mathbb{R}^n} S_d f(x)^2 \tilde{u}(x)^{2/p} h(x)\,dx &= \int_{\mathbb{R}^n} S_d f(x)^2 \bar{w}_1(x)\bar{w}_2(x)\,dx \\ &= \sum_{k=l}^{m-1} \int_{\mathbb{R}^n} S_d f(x)^2 (b_k(x) - b_{k+1}(x))\bar{w}_2(x)\,dx \\ &\qquad + \int_{\mathbb{R}^n} S_d f(x)^2 g_l(x)\bar{w}_2(x)\,dx.\end{aligned}$$

We claim that the last term goes to 0 as $l \to -\infty$. By Hölder's inequality,

$$\begin{aligned}&\int_{\mathbb{R}^n} S_d f(x)^2 g_l(x)\bar{w}_2(x)\,dx \\ &\qquad \leq \left(\int_{\mathbb{R}^n} S_d f(x)^p \bar{w}_2(x)\,dx\right)^{1/r} \left(\int_{\mathbb{R}^n} g_l(x)^{r'} \bar{w}_2(x)\,dx\right)^{1/r'} = I_1 \times I_2.\end{aligned}$$

Since $\bar{w}_2 \in A_p^d$, S_d is bounded on $L^p(\bar{w}_2)$. (See Buckley [16]; this is also a corollary to Theorem 10.2.) Hence,

$$I_1 \leq C\left(\int f(x)^p \bar{w}_2(x)\,dx\right)^{1/r} \leq C\|f\|_{L^\infty}^2 \bar{w}_2(\mathrm{supp}(f))^{1/r} < \infty.$$

On the other hand,

$$I_2 \le Ca^{l/r} \left(\int_{\mathbb{R}^n} g_l(x)\bar{w}_2(x)\,dx \right)^{1/r'} = Ca^{l/r}\|\bar{w}_1\|_{L^1(\bar{w}_2)},$$

and so $I_2 \to 0$ as $l \to -\infty$.

Therefore, in the limit, with constants $c_{j,k}$ to be chosen below,

$$\begin{aligned}
\int_{\mathbb{R}^n} & S_d f(x)^2 \tilde{u}(x)^{2/p} h(x)\,dx \\
&= \sum_{k=-\infty}^{m-1} \int_{\mathbb{R}^n} S_d f(x)^2 (b_k(x) - b_{k+1}(x))\bar{w}_2(x)\,dx \\
&= \sum_{k,j} \int_{Q_j^k} S_d f(x)^2 (b_k(x) - b_{k+1}(x))\bar{w}_2(x)\,dx \\
&= \sum_{k,j} \int_{Q_j^k} [S_d f(x)^2 - c_{j,k}](b_k(x) - b_{k+1}(x))\bar{w}_2(x)\,dx \\
&\le C \sum_{k,j} a^k \int_{Q_j^k} |S_d f(x)^2 - c_{j,k}|\bar{w}_2(x)\,dx.
\end{aligned}$$

For $x \in Q_j^k$,

$$S_d f(x)^2 = \sum_{x \in Q \subsetneq Q_j^k} |(f\chi_{Q_j^k})_{\tilde{Q}} - (f\chi_{Q_j^k})_Q|^2 + \sum_{Q_j^k \subseteq Q} |f_{\tilde{Q}} - f_Q|^2. \tag{10.19}$$

The second term on the right is constant; set $c_{j,k}$ equal to this value. The first term is dominated by $S_d(f\chi_{Q_j^k})(x)^2$. Hence, by Lemma 10.23,

$$\begin{aligned}
\sum_{k,j} a^k & \int_{Q_j^k} |S_d f(x)^2 - c_{j,k}|\bar{w}_2(x)\,dx \\
&\le \sum_{j,k} a^k \int_{Q_j^k} S_d(f\chi_{Q_j^k})(x)^2 \bar{w}_2(x)\,dx \\
&\le C \sum_{j,k} a^k \int_{Q_j^k} M^d f(x)^2 \bar{w}_2(x)\,dx \\
&\le C \sum_{j,k} W_{j,k} \int_{Q_j^k} M^d f(x)^2 \bar{w}_2(x)\,dx.
\end{aligned}$$

Since $\bar{w}_2 \in RH_\infty^d$,

$$W_{j,k} = \frac{1}{\bar{w}_2(Q_j^k)} \int_{Q_j^k} \bar{w}_1(x)\bar{w}_2(x)\,dx \le C \fint_{Q_j^k} w_1(x)^{2/p} h(x)\,dx.$$

Let $\Phi_0(t) = t\log(e+t)^{r-1+\delta}$ and let $C(t) = \Phi_0(t^r)$. Then $\bar{C} \in B_{r'}$. Therefore, by the generalized Hölder's inequality and rescaling,

$$\begin{aligned}
&\sum_{j,k} W_{j,k} \int_{Q_j^k} M^d f(x)^2 \bar{w}_2(x)\,dx \\
&\qquad \le C \sum_{j,k} \left(\fint_{Q_j^k} w_1(x)^{2/p} h(x)\,dx \right) \left(\fint_{Q_j^k} M^d f(x)^2 \bar{w}_2(x)\,dx \right) |Q_j^k| \\
&\qquad \le C \sum_{j,k} \|w_1^{2/p}\|_{C,Q} \|h\|_{\bar{C},Q} \left(\fint_{Q_j^k} M^d f(x)^2 \bar{w}_2(x)\,dx \right) |Q_j^k| \\
&\qquad = C \sum_{j,k} \|w_1\|_{\Phi_0,Q}^{2/p} \|h\|_{\bar{C},Q} \left(\fint_{Q_j^k} M^d f(x)^2 \bar{w}_2(x)\,dx \right) |Q_j^k|.
\end{aligned}$$

By our choice of a (see Proposition A.5) there exists a collection $\{E_j^k\}$ of pairwise disjoint sets and $\alpha > 0$ such that $\bar{w}_2(Q_j^k) \le \alpha \bar{w}_2(E_j^k)$. Because $\bar{w}_2 \in A_\infty^d$, by Lemma A.4 there exists $\beta > 0$ such that $|Q_j^k| \le \beta |E_j^k|$. Therefore,

$$\begin{aligned}
&\sum_{j,k} \|w_1\|_{\Phi_0,Q}^{2/p} \|h\|_{\bar{C},Q} \left(\fint_{Q_j^k} M^d f(x)^2 \bar{w}_2(x)\,dx \right) |Q_j^k| \\
&\qquad \le C \sum_{j,k} \|w_1\|_{\Phi_0,Q}^{2/p} \|h\|_{\bar{C},Q} \left(\fint_{Q_j^k} M^d f(x)^2 \bar{w}_2(x)\,dx \right) |E_j^k| \\
&\qquad \le C \sum_{j,k} \fint_{Q_j^k} M^d f(x)^2 M_{\Phi_0}^d w_1(x)^{2/p} \bar{w}_2(x)\,dx \int_{E_j^k} M_{\bar{C}} h(x)\,dx \\
&\qquad \le C \sum_{j,k} \int_{E_j^k} M^d\big(M^d(f)^2 M_{\Phi_0}^d(w_1)^{2/p} \bar{w}_2\big)(x) M_{\bar{C}}^d h(x)\,dx \\
&\qquad \le C \int_{\mathbb{R}^n} M^d\big(M^d(f)^2 M_{\Phi_0}^d(w_1)^{2/p} \bar{w}_2\big)(x) M_{\bar{C}}^d h(x)\,dx \\
&\qquad \le C \left(\int_{\mathbb{R}^n} M^d\big(M^d(f)^2 M_{\Phi_0}^d(w_1)^{2/p} \bar{w}_2\big)(x)^r\,dx \right)^{1/r} \\
&\qquad\qquad \times \left(\int_{\mathbb{R}^n} M_{\bar{C}}^d h(x)^{r'}\,dx \right)^{1/r'} \\
&\qquad = J_1 \times J_2.
\end{aligned}$$

Since $\bar{C} \in B_{r'}$, by Theorem 5.13,

$$J_2 \le C \left(\int_{\mathbb{R}^n} h(x)^{r'}\,dx \right)^{1/r'} = C.$$

Similarly, by Corollary 6.5 (which is also true in the dyadic case) and Example 5.30,

$$\begin{aligned}J_1 &\le C\left(\int_{\mathbb{R}^n} M^d f(x)^p M^d_{\Phi_0} w_1(x) M^d_\Psi w_2(x)^{1-p}\,dx\right)^{1/r}\\&\le C\left(\int_{\mathbb{R}^n} f(x)^p M^d(M^d_{\Phi_0} w_1)(x) w_2(x)^{1-p}\,dx\right)^{1/r}\\&\le C\left(\int_{\mathbb{R}^n} f(x)^p M^d_\Phi w_1(x) w_2(x)^{1-p}\,dx\right)^{2/p}.\end{aligned}$$

This completes the proof. □

Proof of Theorem 10.18

Our proof uses a Calderón-Zygmund decomposition and is similar to the proof of (10.4) in [23].

Fix f; by a standard argument we may assume without loss of generality that f is non-negative, bounded and has compact support. Fix $\lambda > 0$, and let

$$\Omega = \{x \in \mathbb{R}^n : M^d f(x) > \lambda\}.$$

By Proposition A.1 there exists a collection $\{Q_j\}$ of maximal, disjoint dyadic cubes $\{Q_j\}$ such that $\Omega = \cup_j Q_j$ and $f_{Q_j} > \lambda$. Define the functions

$$g(x) = \begin{cases} f_{Q_j}, & x \in Q_j,\\ f(x) & x \in \mathbb{R}^n \setminus \Omega,\end{cases}$$

$$b_j(x) = \big(f(x) - f_{Q_j}\big)\chi_{Q_j}(x),$$

and let $b = \sum_j b_j$. Then $f = g + b$, $g(x) \le 2^n\lambda$ for almost every $x \in \mathbb{R}^n$, and $(b_j)_Q = 0$ for any $Q \in \mathcal{D}$ such that $Q_j \subset Q$.

We now estimate as follows:

$$\begin{aligned}\tilde{u}(\{x \in \mathbb{R}^n : S_d f(x) > \lambda\}) &\le \tilde{u}(\{x \in \mathbb{R}^n \setminus \Omega : S_d g(x) > \lambda/2\})\\&\quad + \tilde{u}(\{x \in \mathbb{R}^n \setminus \Omega : S_d b(x) > \lambda/2\})\\&\quad + \tilde{u}(\Omega)\\&= I_1 + I_2 + I_3.\end{aligned}$$

We estimate each term separately. Since $(\tilde{u}, \tilde{v})$ is in A_p^d, that is, it satisfies (10.9),

$$\begin{aligned}I_3 &= \sum_j \tilde{u}(Q_j)\\&\le \frac{1}{\lambda^p}\sum_j \tilde{u}(Q_j)\left(\frac{1}{|Q_j|}\int_{Q_j} f(x)\,dx\right)^p\end{aligned}$$

$$\leq \frac{1}{\lambda^p} \sum_j \tilde{u}(Q_j) \left(\frac{1}{|Q_j|} \int_{Q_j} f(x)^p \tilde{v}(x)\, dx \right) \|\tilde{v}^{-1/p}\|^p_{p',Q_j}$$

$$\leq \frac{C}{\lambda^p} \int_{\mathbb{R}^n} f(x)^p \tilde{v}(x)\, dx.$$

To estimate the second term, fix $x \in \mathbb{R}^n \setminus \Omega$. Then

$$S_d b(x) = \left(\sum_{x \in Q \subset \mathcal{D}} |b_{\tilde{Q}} - b_Q|^2 \right)^{1/2};$$

since $\operatorname{supp}(b) = \Omega$, the only possible non-zero terms in this sum come from cubes that contain x and such that $\tilde{Q} \cap \Omega \neq \emptyset$. But then $Q_j \subsetneq \tilde{Q}$ for one or more cubes Q_j. Since $b = \sum_j b_j$, by the definition of b_j it follows that $b_{\tilde{Q}} = b_Q = 0$, so $S_d b(x) = 0$. Therefore, $I_2 = 0$.

We now estimate I_1. By Chebyshev's inequality,

$$I_1 \leq \frac{C}{\lambda^2} \int_{\mathbb{R}^n} S_d g(x)^2 w_1(x) \chi_{\mathbb{R}^n \setminus \Omega}(x) M^d w_2(x)^{1-p}\, dx.$$

Let $\bar{w}_1 = w_1 \chi_{\mathbb{R}^n \setminus \Omega}$ and $\bar{w}_2 = (M^d w_2)^{p-1}$. Since $1 < p < 2$, by Proposition 6.2, $\bar{w}_2 \in A_1^d$, so by the reverse Hölder inequality, $\bar{w}_2 \approx M^d_\Psi \bar{w}_2$ a.e., where $\Psi(t) = t^s$ for some $s > 1$. Let $B(t) = \Psi(t^2)$; then $\bar{B} \in B_2$. Therefore, by Theorem 10.12 and the definition of g,

$$\begin{aligned}
&\frac{1}{\lambda^2} \int_{\mathbb{R}^n} S_d g(x)^2 w_1(x) \chi_{\mathbb{R}^n \setminus \Omega}(x) M^d w_2(x)^{1-p}\, dx \\
&\qquad \leq \frac{C}{\lambda^2} \int_{\mathbb{R}^n} S_d g(x)^2 \bar{w}_1(x) M^d_\Psi \bar{w}_2(x)^{-1}\, dx \\
&\qquad \leq \frac{C}{\lambda^2} \int_{\mathbb{R}^n} g(x)^2 M^d \bar{w}_1(x) \bar{w}_2(x)^{-1}\, dx \\
&\qquad \leq \frac{C}{\lambda^p} \int_{\mathbb{R}^n} g(x)^p M^d \bar{w}_1(x) M^d w_2(x)^{1-p}\, dx \\
&\qquad = \frac{C}{\lambda^p} \int_{\mathbb{R}^n \setminus \Omega} f(x)^p M^d \bar{w}_1(x) M^d w_2(x)^{1-p}\, dx \\
&\qquad\qquad + \frac{C}{\lambda^p} \sum_j (f_{Q_j})^p \int_{Q_j} M^d \bar{w}_1(x) M^d w_2(x)^{1-p}\, dx \\
&\qquad = J_1 + J_2.
\end{aligned}$$

Since $\bar{w}_1 \leq w_1$ and $w_2 \leq M^d w_2$ almost everywhere, we immediately have that

$$J_1 \leq \frac{C}{\lambda^p} \int_{\mathbb{R}^n \setminus \Omega} f(x)^p M^d w_1(x) w_2(x)^{1-p}\, dx.$$

To estimate the second term, by Lemma 5.28 and Remark 5.29, for $x \in Q_j$, $M^d\bar{w}_1(x) = M^d(w_1\chi_{\mathbb{R}^n\setminus\Omega})(x)$ is constant. Therefore, by Hölder's inequality,

$$\begin{aligned} J_2 &\leq \frac{C}{\lambda^p}\sum_j \fint_{Q_j} f(x)^p w_2(x)^{1-p}\,dx \\ &\qquad\qquad \times \|w_2\|_{1,Q_j}^{p-1}\int_{Q_j} M^d\bar{w}_1(x)M^d w_2(x)^{1-p}\,dx \\ &\leq \frac{C}{\lambda^p}\sum_j \fint_{Q_j} f(x)^p M^d\bar{w}_1(x)w_2(x)^{1-p}\,dx \\ &\qquad\qquad \times \int_{Q_j} M^d w_2(x)^{p-1}M^d w_2(x)^{1-p}\,dx \\ &\leq \frac{C}{\lambda^p}\sum_j \int_{Q_j} f(x)^p M^d w_1(x)w_2(x)^{1-p}\,dx \\ &\leq \frac{C}{\lambda^p}\int_{\mathbb{R}^n} f(x)^p M^d w_1(x)w_2(x)^{1-p}\,dx. \end{aligned}$$

This completes the proof.

Coifman-Fefferman inequalities

In this section we briefly consider Coifman-Fefferman inequalities relating the dyadic square function and the dyadic maximal operator. We begin by proving the sharp function inequality mentioned above in Section 9.1.

Proposition 10.24. *For all q, $0 < q < 1$, there exists C_q such that*

$$M_q^{d,\#}(S_d f)(x) \leq C_q M^d f(x). \tag{10.20}$$

Proof. Fix $x \in \mathbb{R}^n$; then it will suffice to show that given any $Q_0 \in \mathcal{D}$, $x \in Q_0$, there exists a constant C_q independent of Q_0 and a constant c depending on f and Q_0 such that

$$\fint_{Q_0} |S_d f(x)^q - c^q|\,dx \leq C_q M^d f(x)^q.$$

We argue as we did in the proof of Theorem 10.16 at (10.19). Let

$$c = \Bigg(\sum_{\substack{Q\in\mathcal{D}\\ Q_0\subseteq Q}} |f_{\widetilde{Q}} - f_Q|^2\Bigg)^{1/2};$$

then, since $0 < q < 1$,

$$\fint_{Q_0} |S_d f(x)^q - c^q|\,dx \leq \fint_{Q_0} |S_d f(x)^2 - c^2|^{q/2}\,dx$$

$$= \fint_{Q_0} \Bigg(\sum_{\substack{x\in Q\in\mathcal{D} \\ Q\subsetneq Q_0}} |f_{\tilde{Q}} - f_Q|^2 \Bigg)^{q/2} dx$$

$$\leq \fint_{Q_0} S_d(f\chi_{Q_0})(x)^q \, dx.$$

Since the dyadic square function is weak (1,1) (see [235]) by Kolmogorov's inequality,

$$\fint_{Q_0} S_d(f\chi_{Q_0})(x)^q \, dx \leq C \left(\fint_{Q_0} |f(x)| \, dx \right)^q \leq CM^d f(x)^q.$$

□

By combining Proposition 10.24 with the dyadic version of Theorem 9.9 we immediately get the following result.

Theorem 10.25. *Given p, $0 < p < 1$,*

$$\int_{\mathbb{R}^n} S_d f(x)^p u(x) \, dx \leq C \int_{\mathbb{R}^n} M^d f(x)^p M^d u(x) \, dx.$$

Remark 10.26. Theorem 10.25 can be improved. Using a careful decomposition argument, Wilson [234] (also see [235]) showed that you can replace $M^d f$ on the right-hand side with $M^d_* f$, where

$$M^d_* f(x) = \sup_{x\in Q\in\mathcal{D}} |f_Q|.$$

Given this result, it is tempting to conjecture that we can improve Proposition 10.24 in this way as well and prove

$$M_q^{d,\#}(S_d f)(x) \leq C_q M^d_* f(x).$$

Wilson [232] also proved the reverse of the inequality in Theorem 10.25: for $0 < p < 2$,

$$\int_{\mathbb{R}^n} M^d_* f(x)^p u(x) \, dx \leq C \int_{\mathbb{R}^n} S_d f(x)^p M^d u(x) \, dx. \tag{10.21}$$

As a corollary to (10.21) and Theorem 8.2 we get an inequality for $p \geq 2$.

Theorem 10.27. *Fix $p \geq 2$, and $r > p/2$. If (u, v) satisfy*

$$\|u^{1/r}\|_{A,Q} \|v^{-1/r}\|_{r',Q} \leq K < \infty,$$

where A is an r-Young function and $\bar{A} \in B_{r'}$, then

$$\int_{\mathbb{R}^n} M^d_* f(x)^p u(x) \, dx \leq C \int_{\mathbb{R}^n} S_d f(x)^p v(x) \, dx.$$

In particular, we can take the pair $(u, M^d_\Phi u)$, where $\Phi(t) = t\log(e+t)^{p/2-1+\delta}$, $\delta > 0$.

For A_1 pairs this is also due to Wilson [231, 230, 233] via a much more complicated proof. It was shown in [22] that in the scale of A_1 weights this result is sharp when $p = 2$, since the inequality is false for the pair $(u, M^d u)$.

As a corollary to the proof of (10.21), Wilson also proved the following more general two-weight result.

Corollary 10.28. *Given a pair of weights (u, v), suppose that for every cube $Q \in \mathcal{D}$,*

$$\int_Q M^d(u\chi_Q)(x)\,dx \le C\int_Q v(x)\,dx. \tag{10.22}$$

Then for $0 < p < 2$,

$$\int_{\mathbb{R}^n} M_*^d f(x)^p u(x)\,dx \le C\int_{\mathbb{R}^n} S_d f(x)^p v(x)\,dx.$$

In [235] he showed that the one-weight version of (10.22) holds if $w \in A_\infty^d$. Here we show that essentially the same is true in the two-weight case for factored weights.

Proposition 10.29. *For all $r > 1$, the pair of factored weights*

$$(\tilde{u}, \tilde{v}) = (w_1(M^d w_2)^{1-r}, (M^d w_1)w_2^{1-r})$$

in A_r^d satisfies (10.22).

Proof. Fix a dyadic cube Q; by homogeneity we may assume without loss of generality that $\|\tilde{u}\|_{1,Q} = 1$. Fix $a > 2^n$ and for each $k \ge 1$ let $\{Q_j^k\}$ be the Calderón-Zygmund cubes of $\tilde{u}\chi_Q$ at height a^k. (See Proposition A.1.) Since $a^k > 1$ and $Q \cap Q_j^k \neq \emptyset$, we must have that $Q_j^k \subsetneq Q$; if $Q \subseteq Q_j^k$ then we would have $\|\tilde{u}\|_{1,Q} > a^k$, a contradiction.

Recall that if we let $\Omega_k = \{x \in Q : M^d(\tilde{u}\chi_Q)(x) > a^k\}$, then the sets $E_j^k = Q_j^k \setminus \Omega_{k+1}$ are pairwise disjoint, and for some $\alpha > 0$, $|Q_j^k| < \alpha|E_j^k|$. Let $E_0 = Q \setminus \cup_{j,k} E_j^k$; then on E_0, $M^d(\tilde{u}\chi_Q) \le a$, and on E_j^k, $M^d(\tilde{u}\chi_Q) \le a^{k+1}$. By Proposition 6.2, $(M^d w_2)^{1-r} \in RH_\infty^d \subset A_\infty^d$, and so

$$\begin{aligned}
&\int_Q M^d(\tilde{u}\chi_Q)(x)\,dx \\
&\quad = \int_{E_0} M^d(\tilde{u}\chi_Q)(x)\,dx + \sum_{j,k}\int_{E_j^k} M^d(\tilde{u}\chi_Q)(x)\,dx \\
&\quad \le a|E_0| + \sum_{j,k} a^{k+1}|E_j^k| \\
&\quad \le a\int_Q \tilde{u}(x)\,dx + a\sum_{j,k} \fint_{Q_j^k} \tilde{u}(x)\,dx \cdot |E_j^k|
\end{aligned}$$

$$\leq C \int_Q \tilde{v}(x)\,dx + C \sum_{j,k} \fint_{Q_j^k} w_1(x)\,dx \cdot \int_{Q_j^k} M^d w_2(x)^{1-r}\,dx$$
$$\leq C \int_Q \tilde{v}(x)\,dx + C \sum_{j,k} \fint_{Q_j^k} w_1(x)\,dx \cdot \int_{E_j^k} M^d w_2(x)^{1-r}\,dx$$
$$\leq C \int_Q \tilde{v}(x)\,dx + C \sum_{j,k} \int_{E_j^k} M^d w_1(x) w_2(x)^{1-r}\,dx$$
$$\leq C \int_Q \tilde{v}(x)\,dx. \qquad \square$$

Remark 10.30. As was the case for Lemma 9.2, it follows from the proof that inequality (10.22) remains true if we replace the weight $\tilde{v}$ with the smaller weight $M^d w_1 (M^d w_2)^{1-r}$.

10.2 Vector-valued maximal operators

Given q, $1 < q < \infty$, and a vector-valued function $f = \{f_i\}$, we define the vector-valued maximal operator $Mf = \{Mf_i\}$ and we define the operator $\overline{M}_q$ by

$$\overline{M}_q f(x) = \|Mf(x)\|_{\ell^q}.$$

The vector-valued maximal operator was first introduced by Fefferman and Stein [75] as a generalization of both the Hardy-Littlewood maximal operator and the Marcinkiewicz integral.

To apply our extrapolation results to the vector-valued maximal operator, we define two families of pairs of functions. For strong type inequalities, we will hereafter assume that we are applying our extrapolation results to the family $\mathcal{F}$ consisting of pairs of the form $(\overline{M}_q f, \|f\|_{\ell^q})$, where each $f_i \in C_c^\infty$ with all the supports contained in a fixed compact set. Similarly, for weak type inequalities, let $\mathcal{F}$ be the family of pairs of the form $(\lambda \chi_{\{x : \overline{M}_q f(x) > \lambda\}}, \|f\|_{\ell^q})$, where $f \in C_c^\infty$ and $\lambda > 0$. (As before, we write $f \in \mathcal{F}$ to mean the corresponding pair is in $\mathcal{F}$.) By a standard density argument, if we can prove a weak or strong type two-weight norm inequality for $\overline{M}_q$ for $f \in \mathcal{F}$, then we can prove it for all f such that the right-hand side is finite.

The conjectures

As with the dyadic square function, there are two ways to view the vector-valued maximal operator. For all values of q the vector-valued maximal operator can be treated as a vector-valued singular integral (see, for instance, [88]), and we therefore again make conjectures analogous to Conjectures 9.18 and 9.20.

Conjecture 10.31. For each p, $1 < p < \infty$, and each q, $1 < q < \infty$, there exists a constant C such that for all $f = \{f_i\} \in L^p(v)$,

$$\int_{\mathbb{R}^n} \overline{M}_q f(x)^p u(x)\,dx \le C \int_{\mathbb{R}^n} \|f(x)\|_{\ell^q}^p v(x)\,dx \tag{10.23}$$

whenever the pair (u, v) satisfies

$$\|u^{1/p}\|_{A,Q} \|v^{-1/p}\|_{B,Q} \le K < \infty, \tag{10.24}$$

where $\bar{A} \in B_{p'}$ and $\bar{B} \in B_p$.

This conjecture is true when A is a log bump; it follows by modifying the proof of Theorem 9.21 in [47] as we did there for the dyadic square function to exploit the fact that the vector-valued maximal operator is more localized than a singular integral. We record this as Theorem 10.35 below. The general case when $\bar{A}$ is a $B_{p'}$ bump remains open.

In the case of weak type inequalities the corresponding conjecture is actually a theorem, as it was for the dyadic square function.

Theorem 10.32. *For each p, $1 < p < \infty$, and each q, $1 < q < \infty$, there exists a constant C such that for all $f = \{f_i\} \in L^p(v)$ and $\lambda > 0$,*

$$u(\{x \in \mathbb{R}^n : \overline{M}_q f(x) > \lambda\}) \le \frac{C}{\lambda^p} \int_{\mathbb{R}^n} \|f(x)\|_{\ell^q}^p v(x)\,dx$$

whenever the pair (u, v) satisfies

$$\|u^{1/p}\|_{A,Q} \|v^{-1/p}\|_{p',Q} \le K < \infty, \tag{10.25}$$

where $\bar{A} \in B_{p'}$.

The proof of this follows immediately from Corollary 8.4 and the weak $(1,1)$ inequality,

$$u(\{x \in \mathbb{R}^n : \overline{M}_q f(x) > \lambda\}) \le \frac{C}{\lambda} \int_{\mathbb{R}^n} \|f(x)\|_{\ell^q} Mu(x)\,dx, \tag{10.26}$$

that was proved in [176].

However, as with the dyadic square function, if we fix q, then $p = q$ becomes a critical index, with very different behavior depending on the size of p: if $1 < p \le q$, then the vector-valued maximal operator behaves like the Hardy-Littlewood maximal operator, and if $p > q$, then it is more singular, with its degree of singularity depending on the ratio p/q. From this perspective we make the following conjectures.

Conjecture 10.33. Given q, $1 < q < \infty$, for each p, $1 < p \le q$, there exists a constant C such that for all $f = \{f_i\} \in L^p(v)$,

$$\int_{\mathbb{R}^n} \overline{M}_q f(x)^p u(x)\,dx \le C \int_{\mathbb{R}^n} \|f(x)\|_{\ell^q}^p v(x)\,dx, \tag{10.27}$$

whenever the pair (u, v) satisfies

$$\|u^{1/p}\|_{p,Q}\|v^{-1/p}\|_{B,Q} \leq K < \infty, \tag{10.28}$$

where $\bar{B} \in B_p$.

For $q < p < \infty$, inequality (10.27) holds whenever the pair (u, v) satisfies

$$\|u^{1/r}\|_{A,Q}^{1/q}\|v^{-1/p}\|_{B,Q} \leq K < \infty, \tag{10.29}$$

where $r = p/q$, $\bar{A} \in B_{r'}$, $\bar{B} \in B_p$.

Conjecture 10.34. Given q, $1 < q < \infty$, for each p, $1 < p \leq q$, there exists a constant C such that for all $f = \{f_i\} \in L^p(v)$,

$$u(\{x \in \mathbb{R}^n : \overline{M}_q f(x) > \lambda\}) \leq \frac{C}{\lambda^p} \int_{\mathbb{R}^n} \|f(x)\|_{\ell^q}^p v(x)\, dx \tag{10.30}$$

whenever the pair (u, v) satisfies the two-weight A_p condition

$$\|u^{1/p}\|_{p,Q}\|v^{-1/p}\|_{p',Q} \leq K < \infty. \tag{10.31}$$

For $q < p < \infty$, inequality (10.30) holds whenever the pair (u, v) satisfies

$$\|u^{1/r}\|_{A,Q}^{1/q}\|v^{-1/p}\|_{p',Q} \leq K < \infty, \tag{10.32}$$

where $r = p/q$ and $\bar{A} \in B_{r'}$.

We can prove both of these conjectures in the case $p < q$; unlike the dyadic square function, we can adapt known results and techniques to prove our results for arbitrary pairs that satisfy bump conditions. In the case $p > q$ we have only partial results, primarily for factored weights. Below we will consider strong and weak type inequalities and then give results for factored weights.

Strong (p, p) inequalities

We first prove that Conjecture 10.31 is true when A is a log bump.

Theorem 10.35. *Given p, $1 < p < \infty$, suppose the pair of weights (u, v) satisfy*

$$\|u^{1/p}\|_{A,Q}\|v^{-1/p}\|_{B,Q} \leq K < \infty, \tag{10.33}$$

where $A(t) = t^p \log(e + t)^{p-1+\delta}$, $\delta > 0$, and $\bar{B} \in B_p$. Then for all $1 < q < \infty$,

$$\int_{\mathbb{R}^n} \overline{M}_q f(x)^p u(x)\, dx \leq \int_{\mathbb{R}^n} \|f(x)\|_{\ell^q}^p v(x)\, dx \tag{10.34}$$

Proof. We will first prove the desired strong (p,p) inequality for the dyadic operator $\overline{M}_q^d$ (i.e., the operator defined by replacing M with M^d). Our approach is nearly identical to the one taken in [47] where we proved Theorems 10.2 and 9.21 above: we use a decomposition argument as in the proof of Lemma 9.2. Therefore, here we will only sketch the outline of the proof, and refer the reader to [47] for complete details.

Without loss of generality we may assume that $f = \{f_i\}$ is non-negative: that is, $f_i \geq 0$ for all i. Fix r, $0 < r < 1$, and $\epsilon > 0$, such that $p/r - 1 + \epsilon = p - 1 + \delta$. By duality,

$$\|\overline{M}_q^d f\|_{L^p(u)}^r = \sup_h \int_{\mathbb{R}^n} \overline{M}_q^d f(x)^r u(x)^{r/p} h(x)\, dx,$$

where $\|h\|_{L^{(p/r)'}} = 1$ and h is bounded and has compact support. Fix h and let $\tilde{u} = u^{r/p}h$, and form the decomposition of $\tilde{u}$ as in the proof of Lemma 9.2. (See also the proof of Theorem 10.16.) This yields (with the same notation as in that proof)

$$\begin{aligned}\int_{\mathbb{R}^n} \overline{M}_q^d f(x)^r \tilde{u}(x)\, dx &= \sum_{k=-\infty}^{m-1} \int_{\mathbb{R}^n} \overline{M}_q^d f(x)^r (b_k(x) - b_{k+1}(x))\, dx \\ &\leq C \sum_{k,j} \tilde{u}_{Q_j^k} \int_{Q_j^k} |\overline{M}_q^d f(x) - c_j^k|^r\, dx.\end{aligned}$$

where each c_j^k is a constant to be chosen momentarily.

By Lemma 5.28 and Remark 5.29, for each index i and $x \in Q_j^k$, the term $M^d(f_i \chi_{\mathbb{R}^n \setminus Q_j^k})(x)$ is constant. Therefore, let

$$c_j^k = \overline{M}_q^d(f\chi_{\mathbb{R}^n \setminus Q_j^k})(x).$$

Hence, $\overline{M}_q^d f(x) \geq c_j^k$, and so

$$\begin{aligned}&\sum_{k,j} \tilde{u}_{Q_j^k} \int_{Q_j^k} |\overline{M}_q^d f(x) - c_j^k|^r\, dx \\ &\quad= \sum_{k,j} \tilde{u}_{Q_j^k} \int_{Q_j^k} (\overline{M}_q^d f(x) - c_j^k)^r\, dx \\ &\quad\leq \sum_{k,j} \tilde{u}_{Q_j^k} \int_{Q_j^k} (\overline{M}_q^d(f\chi_{Q_j^k})(x) + \overline{M}_q^d(f\chi_{\mathbb{R}\setminus Q_j^k})(x) - c_j^k)^r\, dx \\ &\quad= \sum_{k,j} \tilde{u}_{Q_j^k} \int_{Q_j^k} \overline{M}_q^d(f\chi_{Q_j^k})(x)^r\, dx, \\ &\quad\leq C \sum_{k,j} \tilde{u}_{Q_j^k} \left(\fint_{Q_j^k} \|f(x)\|_{\ell^q}\, dx \right)^r |E_j^k|.\end{aligned}$$

The last inequality follows from Kolmogorov's inequality and the weak (1, 1) inequality for $\overline{M}_q^d$; the sets E_j^k are the pairwise disjoint sets given in Proposition A.1. By the generalized Hölder's inequality and rescaling,

$$\tilde{u}_{Q_j^k}\left(\fint_{Q_j^k}\|f(x)\|_{\ell^q}\,dx\right)^r \leq C\|u^{1/p}\|_{A,Q_j^k}^r\|h\|_{\bar{C},Q_j^k}\|\|f\|_{\ell^q}v^{1/p}\|_{\bar{B},Q_j^k}^r\|v^{-1/p}\|_{B,Q_j^k},$$

where $C(t)=t^{p/r}\log(e+t)^{p/r-1+\epsilon}$. Hence, by (10.33),

$$\begin{aligned}
&\sum_{k,j}\tilde{u}_{Q_j^k}\left(\fint_{Q_j^k}\|f(x)\|_{\ell^q}\,dx\right)^r|E_j^k|\\
&\quad\leq C\sum_{k,j}\|h\|_{\bar{C},Q_j^k}\|\|f\|_{\ell^q}v^{1/p}\|_{\bar{B},Q_j^k}^r|E_j^k|\\
&\quad\leq C\sum_{k,j}\int_{E_j^k}M_{\bar{C}}h(x)M_{\bar{B}}(\|f\|_{\ell^q}v^{1/p})(x)^r\,dx\\
&\quad\leq\left(\int_{\mathbb{R}^n}M_{\bar{C}}h(x)^{(p/r)'}\,dx\right)^{1/(p/r)'}\left(\int_{\mathbb{R}^n}M_{\bar{B}}(\|f\|_{\ell^q}v^{1/p})(x)^p\,dx\right)^{r/p}.
\end{aligned}$$

Since $\bar{C}\in B_{(p/r)'}$ and $\bar{B}\in B_p$ by Theorem 5.13 we get the desired inequality for $\overline{M}_q^d$.

To complete the proof we need to pass from the dyadic maximal operator to the non-dyadic case. To do so, we use a lemma due to Sawyer [201] that is based on ideas due to Fefferman and Stein [75] (also see [88]).

Lemma 10.36. *Given $t\in\mathbb{R}^n$, let M_t^d denote the maximal operator defined with respect to translates of the dyadic grid by t. Let M_k denote the maximal operator gotten by taking the supremum of averages over cubes with side length at most 2^k. Then there exists a constant C independent of k such that*

$$M_kf(x)\leq\frac{C}{|Q_k|}\int_{Q_k}M_t^df(x)\,dt,$$

where Q_k is the cube centered at 0 with $\ell(Q_k)=2^{k+2}$.

For each x,

$$\overline{M}_qf(x)=\|Mf(x)\|_{\ell^q}=\lim_{k\to\infty}\|M_kf(x)\|_{\ell^q}.$$

The above argument shows that (10.34) holds with the dyadic maximal operator replaced by M_t^d with constant independent of t. Hence, by Fatou's lemma,

Minkowski's inequality, and Lemma 10.36,

$$
\begin{aligned}
&\int_{\mathbb{R}^n} \overline{M}_q f(x)^p u(x)\,dx \\
&\qquad \le \liminf_{k\to\infty} \int_{\mathbb{R}^n} \|M_k f(x)\|_{\ell^q}^p u(x)\,dx \\
&\qquad \le C \liminf_{k\to\infty} \int_{\mathbb{R}^n} \left(\sum_i \left(\frac{1}{|Q_k|}\int_{Q_k} M_t^d f_i(x)\,dt\right)^q\right)^{p/q} u(x)\,dx \\
&\qquad \le C \liminf_{k\to\infty} \int_{\mathbb{R}^n} \left(\frac{1}{|Q_k|}\int_{Q_k} \|M_t^d f(x)\|_{\ell^q}\,dt\right)^p u(x)\,dx \\
&\qquad \le C \liminf_{k\to\infty} \int_{\mathbb{R}^n} \frac{1}{|Q_k|}\int_{Q_k} \|M_t^d f(x)\|_{\ell^q}^p\,dt\,u(x)\,dx \\
&\qquad \le C \liminf_{k\to\infty} \frac{1}{|Q_k|}\int_{Q_k}\int_{\mathbb{R}^n} \|f(x)\|_{\ell^q}^p v(x)\,dx\,dt \\
&\qquad \le C \int_{\mathbb{R}^n} \|f(x)\|_{\ell^q}^p v(x)\,dx.
\end{aligned}
$$

This completes the proof. □

Remark 10.37. We can prove a weaker version of Theorem 10.35 analogous to Proposition 10.7 using extrapolation. By Theorem 10.38 below, for $1 < p < q$ the vector-valued maximal operator is strong (p, p) with respect to the weights (u, Mu). Therefore, by Corollary 8.7, we immediately get that (10.34) holds if the pair (u, v) satisfies (10.33) with A such that $\bar{A}$ is doubling and $\bar{A} \in B_{p'}$, and $B(t) = t^{sp'}$, $s > 1$.

If the vector-valued maximal operator were linear and self-adjoint, then by a duality argument we could switch the log bump and the power bump to get that (10.34) holds if we instead take $A(t) = t^{sp}$, $s > 1$, and $\bar{B}$ is doubling and in B_p. This is in fact true when B is a log bump, but the proof requires that we use the decomposition argument in [56] and the sharp maximal function inequality

$$
M_\delta^\#(\overline{N}_q f)(x) \le C_{q,\delta} M(\|f_i\|_{\ell^q})(x), \tag{10.35}
$$

$0 < \delta < 1$, $1 < q < \infty$, where $\overline{N}_q$ is the vector-valued operator associated with a smooth analog of the Hardy-Littlewood maximal operator. This inequality was proved in [57]. The smooth maximal function is pointwise comparable to the Hardy-Littlewood maximal function, so this approach actually yields the desired inequality for $\overline{M}_q$. Details of all these arguments are left to the reader. It would be interesting to have a proof of this result that used extrapolation.

We now consider results related to Conjecture 10.33. We first prove that it is true when $1 < p \le q$.

Theorem 10.38. *Given q, $1 < q < \infty$, suppose $1 < p \le q$. If the pair of weights (u, v) satisfies*

$$\|u^{1/p}\|_{p,Q}\|v^{-1/p}\|_{B,Q} \le K < \infty,$$

where $\bar{B} \in B_p$, then

$$\int_{\mathbb{R}^n} \overline{M}_q f(x)^p u(x)\, dx \le \int_{\mathbb{R}^n} \|f(x)\|_{\ell^q}^p v(x)\, dx.$$

In particular, this inequality holds for the A_1 pair (u, Mu).

Proof. When $p = q$ this follows immediately from Theorem 5.14:

$$\begin{aligned}\int_{\mathbb{R}^n} \overline{M}_q f(x)^p u(x)\, dx &= \sum_{i=1}^{\infty} \int_{\mathbb{R}^n} M f_i(x)^p u(x)\, dx \\ &\le C \sum_{i=1}^{\infty} \int_{\mathbb{R}^n} |f_i(x)|^p v(x)\, dx = C \int_{\mathbb{R}^n} \|f(x)\|_{\ell^q}^p v(x)\, dx.\end{aligned}$$

For $1 < p < q$, we get the desired inequality by extrapolation using Theorem 7.2 with $p_0 = q$, $A_0(t) = t^q$, and $\bar{B}_0 \in B_q$. □

Remark 10.39. Theorem 10.38 was proved in [176] by fixing p and using interpolation in the scale of ℓ^q spaces. Our proof has the (small) advantage that it treats q (and so the operator) as fixed, and extrapolates in the exponent p.

Remark 10.40. If we use Theorem 7.2 to extrapolate up to $p > q$, we get a weaker version of Theorem 10.35: in the scale of log-bumps we get the bump $B(t) = t^{p'} \log(e+t)^{q'-1+\delta}$ on the right instead of the smaller $B(t) = t^{p'} \log(e+t)^{p'-1+\delta}$.

In the case $p > q$ we have one result that is analogous to Theorem 10.9 for the dyadic square function.

Theorem 10.41. *Given q, $1 < q < \infty$, and p, $q < p < \infty$, suppose that the pair of weights (u, v) satisfies*

$$\|u^{1/p}\|_{A,Q}\|v^{-1/p}\|_{B,Q} \le K < \infty, \tag{10.36}$$

where $A(t) = C(t^q)$ with $\bar{C} \in B_{(p/q)'}$ and $B(t) = t^{q(p/q)'}$. Then

$$\int_{\mathbb{R}^n} \overline{M}_q f(x)^p u(x)\, dx \le \int_{\mathbb{R}^n} \|f(x)\|_q^p v(x)\, dx.$$

Proof. By Theorem 10.38, for $p = q$ we have the strong (p, p) inequality for $\overline{M}_q$ for the pair (u, Mu). The desired inequality now follows from Theorem 8.2 with $p_0 = q$ and a rescaling argument as in Remark 10.10. □

Remark 10.42. In Theorem 10.41 we can take $A(t) = t^p \log(e+t)^{p/q-1+\delta}$. In [176] this result was proved with a different bump on the right: $B(t) = t^{q'} \log(e+t)^{q'-1+\delta}$. Neither condition is universally better than the other. If $q \geq 2$ or if $1 < q < 2$ and $p < \frac{q}{2-q}$, then $q' < q(p/q)'$. On the other hand, if $1 < q < 2$ and $p > \frac{q}{2-q}$, then $q' > q(p/q)'$ Combining these two results, we see that for q fixed we always have some kind of power bump on the right, as p decreases to q we asymptotically get the log bump in Conjecture 10.33, and as p increases to infinity we approach the bump t^q. Compare these to the power bump conditions for the dyadic square function in Remark 10.10.

Weak (p,p) inequalities

We show that Conjecture 10.34 is true if $1 < p < q$.

Theorem 10.43. *Given q, $1 < q < \infty$, suppose $1 < p < q$. If the pair of weights (u,v) is in A_p, then*

$$u(\{x \in \mathbb{R}^n : \overline{M}_q f(x) > \lambda\}) \leq \frac{C}{\lambda^p} \int_{\mathbb{R}^n} \|f(x)\|_{\ell^q}^p v(x)\, dx. \tag{10.37}$$

Proof. The proof depends on a weak $(1,1)$ inequality for a dyadic vector-valued maximal operator defined with respect to an arbitrary (non-doubling) measure. The proof of this inequality is quite technical and is deferred to Appendix A below.

Fix a function $f = \{f_i\}$ and fix q, p, $1 < p < q$. Let $r = q/p$. Without loss of generality we may assume that each f_i is non-negative. We will first prove inequality (10.37) with the maximal operator replaced with the dyadic maximal operator.

Fix a cube $Q \in \mathcal{D}$; then for each i, since $(u,v) \in A_p$,

$$\begin{aligned}
\left(\frac{1}{|Q|}\int_Q f_i(x)\,dx\right)^p &\leq \left(\frac{1}{|Q|}\int_Q f_i(x)^p v\,dx\right)\left(\frac{1}{|Q|}\int_Q v(x)^{-p'/p}\right)^{p/p'} \\
&\leq \left(\frac{1}{u(Q)}\int_Q f_i(x)^p (v(x)/u(x))u(x)\,dx\right) \\
&\qquad \times \left(\frac{1}{|Q|}\int_Q u(x)\,dx\right)\left(\frac{1}{|Q|}\int_Q v(x)^{-p'/p}\,dx\right)^{p/p'} \\
&\leq \frac{C}{u(Q)}\int_Q f_i(x)^p (v(x)/u(x))u(x)\,dx.
\end{aligned}$$

If we fix x and take the supremum over all dyadic cubes containing x, we get that $M^d f_i(x) \leq C M_u^d(f_i^p(v/u))(x)^{1/p}$, where M_u^d is the dyadic maximal operator defined with respect to the measure $u\,dx$. (See (A.4).) Let $f^p = \{f_i^p\}$; then

$\overline{M}_q^d f(x) \le C\|M_u^d((v/u)f^p)(x)\|_{\ell^r}^{1/p}$. Therefore, by Theorem A.18,

$$\begin{aligned} u(\{x \in \mathbb{R}^n : \overline{M}_q^d f(x) > \lambda\}) &\le u(\{x \in \mathbb{R}^n : \|M_u^d((v/u)f^p)(x)\|_{\ell^r} > \lambda^p/C\}) \\ &\le \frac{C}{\lambda^p}\int_{\mathbb{R}^n} \|(v(x)/u(x))f(x)^p\|_{\ell^r} u(x)\,dx \\ &= \frac{C}{\lambda^p}\int_{\mathbb{R}^n} \|f(x)\|_{\ell^q}^p v(x)\,dx. \end{aligned}$$

To complete the proof, we must pass from the dyadic maximal operator to the Hardy-Littlewood maximal operator. To do so we argue exactly as in the proof of Theorem 10.35, noting only that for $p > 1$, $\|\cdot\|_{L^{p,\infty}(u)}$ is a norm, and so we can apply Minkowski's inequality to it. □

Remark 10.44. We have no new results for the case $p \ge q$ except for the factored weight results given below.

Inequalities for factored weights

For the dyadic square function, many of our results when $p < 2$ are only proved for factored weights. However, we have been able to prove that the analogous results (when $p < q$) for the vector-valued maximal operator hold for all weights. In this section we give three results for factored weights that are analogous to the results for $p > 2$ for the square function; their proofs are essentially the same.

The first two results are for strong type inequalities when $p > q$; they are the analogs of Theorems 10.14 and 10.16.

Theorem 10.45. *Given $p > q$, let $r = p/q$, $A(t) = t^p \log(e+t)^{r-1+\delta}$, and $B(t) = t^{p'} \log(e+t)^{q'-1+\delta}$, $\delta > 0$. Let $\Phi(t) = A(t^{1/p})$ and $\Psi(t) = B(t^{1/p'})$ and form the pair of factored weights*

$$(\tilde{u}, \tilde{v}) = (w_1(M_\Psi w_2)^{1-p}, (M_\Phi w_1)w_2^{1-p}).$$

Then the pair $(\tilde{u}, \tilde{v})$ satisfies

$$\|\tilde{u}^{1/p}\|_{A,Q}\|\tilde{v}^{-1/p}\|_{B,Q} \le K < \infty,$$

and the vector-valued maximal operator satisfies the strong (p,p) inequality

$$\int_{\mathbb{R}^n} \overline{M}_q f(x)^p \tilde{u}(x)\,dx \le C\int_{\mathbb{R}^n} \|f(x)\|_{\ell^q}^p \tilde{v}(x)\,dx. \tag{10.38}$$

The proof of Theorem 10.45 is essentially the same as the proof of Theorem 10.14. By Theorem 10.38 we have the strong (p,p) inequality when $p = q$; in particular it is true for factored weights. Therefore, we can apply Theorem 7.14 and extrapolate in the scale of factored weights exactly as before with 2 replaced by q in the argument.

Theorem 10.46. *Given $p > q$, let $r = p/q$, $A(t) = t^p \log(e+t)^{r+\delta}$, and $\bar{B} \in B_p$. Let $\Phi(t) = A(t^{1/p})$ and $\Psi(t) = B(t^{1/p'})$ and form the pair of factored weights*

$$(\tilde{u}, \tilde{v}) = (w_1(M_\Psi w_2)^{1-p}, (M_\Phi w_1)w_2^{1-p}).$$

Then $(\tilde{u}, \tilde{v})$ satisfies

$$\|\tilde{u}^{1/p}\|_{A,Q}\|\tilde{v}^{-1/p}\|_{B,Q} \leq K < \infty,$$

and the vector-valued maximal operator satisfies the strong (p,p) inequality (10.38).

Proof. The proof follows by combining arguments in the proofs of Theorems 10.16 and 10.35. First, exactly as in the proof of Theorem 10.35, it suffices to prove the corresponding inequality for the dyadic vector-valued maximal operator and for $f = \{f_i\}$ with each f_i non-negative. We then repeat the argument in the proof of Theorem 10.16, replacing $r = p/2$ with $r = p/q$. The first part, up to inequality (10.19) is primarily the decomposition of the weight, and only requires the fact that $\overline{M}_q$ is bounded on $L^p(\bar{w}_2)$ since $\bar{w}_2 \in A_p$; this was proved by Anderson and John [5].

To choose the constant $c_{j,k}$ in inequality (10.19) we need to argue as in the proof of Theorem 10.35, where we use the fact that $\overline{M}_q$ is a positive operator. We can then apply an inequality corresponding to Lemma 10.23 for the vector-valued maximal operator, with the dyadic maximal operator replaced by the Hardy-Littlewood maximal operator. The proof is exactly the same except that we use the sharp function estimate (10.35).

The remainder of the proof now continues exactly as in the proof of Theorem 10.16. □

Our only weak type result for factored weights is analogous to Theorem 10.19 for the dyadic square function, and the proof is exactly the same, replacing $p/2$ with p/q, and using Theorem 10.43 for factored pairs that satisfy the two-weight A_p condition.

Theorem 10.47. *Given q, $1 < q < \infty$, let $p \geq q$, $A(t) = t^p \log(e+t)^{p/q-1+\delta}$, $\delta > 0$, and $\Phi(t) = A(t^{1/p})$. Then the pair of factored weights*

$$(\tilde{u}, \tilde{v}) = (w_1(Mw_2)^{1-p}, (M_\Phi w_1)w_2^{1-p})$$

satisfies

$$\|\tilde{u}^{1/p}\|_{A,Q}\|\tilde{v}^{-1/p}\|_{p',Q} \leq K < \infty,$$

and the operator $\overline{M}_q$ satisfies the weak (p,p) inequality

$$\tilde{u}(\{x \in \mathbb{R}^n : \overline{M}_q(x) > \lambda\}) \leq \frac{C}{\lambda^p}\int_{\mathbb{R}^n} \|f(x)\|_{\ell^q}^p \tilde{v}(x)\,dx. \tag{10.39}$$

Remark 10.48. As was the case for the dyadic square function (see Remark 10.21), if Conjecture 10.34 is true for $p = q$ for factored weights, then by extrapolation it is true for $p > q$.

Appendix A

The Calderón-Zygmund Decomposition

In this appendix we state and prove some variations of the classical Calderón-Zygmund decomposition. (For this result we refer the reader to [68, 88, 91].) With the exception of the material in the last section none of these results are new, but the proofs are scattered across the literature. We include them here, with uniform notation, for the convenience of the reader.

Recall the following notation: $\mathcal{D}$ denotes the collection of all dyadic cubes; given $Q \in \mathcal{D}$, $\widetilde{Q}$ denotes the dyadic parent of Q: the unique dyadic cube containing Q such that $\ell(\widetilde{Q}) = 2\ell(Q)$.

A.1 The Calderón-Zygmund decomposition for M_Φ

The first result generalizes the Calderón-Zygmund decomposition by replacing the Hardy-Littlewood maximal operator by an Orlicz maximal operator. It originally appeared in [174] except for part (c) which is from [53].

Proposition A.1. *Given a Young function Φ, suppose that f is a measurable function such that $\|f\|_{\Phi,Q} \to 0$ as $|Q| \to \infty$. Then the following are true:*

(a) *For each $\lambda > 0$, there exists a disjoint collection of maximal dyadic cubes $\{Q_j\}$ such that*

$$E_\lambda^d = \{x \in \mathbb{R}^n : M_\Phi^d f(x) > \lambda\} = \bigcup_j Q_j,$$

and for every j,

$$\lambda < \|f\|_{\Phi,Q_j} \le 2^n \lambda.$$

(b) *For each* $\lambda > 0$, *let* $\{Q_j\}_j$ *be the collection of cubes in* (*a*) *corresponding to the level set of height* $\lambda/4^n$. *Then*

$$E_\lambda = \{x \in \mathbb{R}^n : M_\Phi f(x) > \lambda\} \subset \bigcup_j 3\,Q_j.$$

(c) *Given* $a > 2^n$, *for each* $k \in \mathbb{Z}$ *let* $\{Q_j^k\}_j$ *be the collection of maximal dyadic cubes in* (*a*) *with*

$$\Omega_k = \{x \in \mathbb{R}^n : M_\Phi^d f(x) > a^k\} = \bigcup_j Q_j^k.$$

Let $E_j^k = Q_j^k \setminus \Omega_{k+1}$. *Then* $E_j^k \subset Q_j^k$, $|E_j^k| \geq \frac{a-2^n}{a}\,|Q_j^k|$ *and the sets* E_j^k *are pairwise disjoint for all* j *and* k.

Remark A.2. When $\Phi(t) = t$, the cubes in (*a*) are called the Calderón-Zygmund cubes of f at height λ.

Proof. To prove (a) we may assume that $E_\lambda^d \neq \emptyset$ since otherwise there is nothing to prove. Let Λ_λ be the family of dyadic cubes such that $\lambda < \|f\|_{\Phi,Q}$; this is non-empty since $E_\lambda^d \neq \emptyset$. For each $Q \in \Lambda_\lambda$ there exists a maximal cube $Q' \in \Lambda_\lambda$ with $Q \subset Q'$, since $\|f\|_{\Phi,Q} \to 0$ as $|Q| \to \infty$. Let $\{Q_j\} \subset \Lambda_\lambda$ denote the family of such maximal cubes; clearly they are pairwise disjoint. Also if $\widetilde{Q}_j$ is the dyadic parent of Q_j, the maximality of Q_j and the convexity of Φ imply that

$$\lambda < \|f\|_{\Phi,Q_j} \leq 2^n\,\|f\|_{\Phi,\widetilde{Q}_j} \leq 2^n\,\lambda.$$

If $x \in E_\lambda^d$, there exists a dyadic cube $Q \ni x$ such that $\|f\|_{\Phi,Q} > \lambda$. Hence, $Q \subset Q_j$ for some j. Conversely, since $\|f\|_{\Phi,Q_j} > \lambda$, if $x \in Q_j$, then $M_\Phi^d f(x) > \lambda$, so $x \in E_\lambda^d$. Therefore, $E_\lambda^d = \cup_j Q_j$.

To prove (b), we may again assume that $E_\lambda \neq \emptyset$; otherwise the family of cubes is empty. Given $x \in E_\lambda$, by definition there exists a cube $Q \ni x$ such that $\|f\|_{\Phi,Q} > \lambda$. Let k be the unique integer so that $2^{-(k+1)} \leq \ell(Q) < 2^{-k}$. Then there exists N, $1 \leq N \leq 2^n$, and there exists a family of dyadic cubes $\{P_j\}_{j=1}^N$ such that $\ell(P_j) = 2^{-k}$, $Q \subset \cup_j P_j$, and $P_j \cap Q \neq \emptyset$. One of these cubes, say P_1, is such that $\|f\,\chi_{P_1}\|_{\Phi,Q} > \lambda/2^n$. For if not we get a contradiction:

$$\lambda < \|f\|_{\Phi,Q} = \left\| f \sum_{j=1}^N \chi_{P_j} \right\|_{\Phi,Q} \leq \sum_{j=1}^N \|f\,\chi_{P_j}\|_{\Phi,Q} \leq N\,\lambda/2^n \leq \lambda.$$

By the convexity of Φ and since $1 < \ell(P_1)/\ell(Q) \leq 2$,

$$\lambda/4^n < 2^{-n}\|f\,\chi_{P_1}\|_{\Phi,Q} \leq \|f\|_{\Phi,P_1}.$$

Since P_1 is dyadic, $P_1 \in \Lambda_{\lambda/4^n}$ and by maximality there exists $Q_j \supset P_1$. Furthermore, $Q \cap P_1 \neq \emptyset$ and $\ell(Q) < \ell(P_1)$, so $x \in Q \subset 3P_1 \subset 3Q_j$. Therefore, $E_\lambda \subset \cup_j 3Q_j$.

Finally, we prove (c). Since $\Omega_{k+1} \subset \Omega_k$ it is immediate that the sets E_j^k are pairwise disjoint. To estimate their measure, note that if $Q_j^k \cap Q_i^{k+1} \neq \emptyset$, then by maximality and the fact that $a > 2^n$, $Q_i^{k+1} \subsetneq Q_j^k$. Since $\|a^{-k-1}f\|_{\Phi,Q_i^{k+1}} > 1$, by the definition of the Luxemburg norm,

$$\fint_{Q_i^{k+1}} \Phi\left(\frac{|f(x)|}{a^{k+1}}\right) dx > 1.$$

Similarly, since $\|f\|_{\Phi,Q_j^k} \leq 2^n a^k$,

$$\fint_{Q_j^k} \Phi\left(\frac{|f(x)|}{2^n a^k}\right) dx \leq 1.$$

Therefore, by the convexity of Φ,

$$\begin{aligned} |Q_j^k \cap \Omega_{k+1}| = \sum_{i:Q_i^{k+1}\subsetneq Q_j^k} |Q_i^{k+1}| &\leq \sum_{i:Q_i^{k+1}\subsetneq Q_j^k} \int_{Q_i^{k+1}} \Phi\left(\frac{|f(x)|}{a^{k+1}}\right) dx \\ &\leq \int_{Q_j^k} \Phi\left(\frac{|f(x)|}{a^{k+1}}\right) dx \leq \frac{2^n}{a} |Q_j^k|. \end{aligned}$$

Thus, $|E_j^k| \geq \frac{a-2^n}{a} |Q_j^k|$ and the proof is complete. □

Remark A.3. As a corollary to Theorem A.1 we can immediately prove the weak modular inequality in Theorem 5.5:

$$|\{x \in \mathbb{R}^n : M_\Phi f(x) > \lambda\}| \leq 3^n \int_{\{x\in\mathbb{R}^n:|f(x)|>\lambda/2\}} \Phi\left(\frac{2\cdot 4^n|f(x)|}{\lambda}\right) dx. \tag{A.1}$$

To see this, write $f = f_\infty + f_0$ where $f_\infty = f\,\chi_{\{|f|>\lambda/2\}}$. Then $Mf(x) \leq Mf_\infty(x) + \lambda/2$. Further, $|f_\infty(x)| \leq |f(x)|$ so we can apply part (b) to f_∞ with $\lambda/2$ in place of λ. Let $\{Q_j\}$ be the resulting family of pairwise disjoint cubes. Then by the properties of these cubes and the definition of the Luxemburg norm,

$$\begin{aligned} &|\{x \in \mathbb{R}^n : M_\Phi f(x) > \lambda\}| < |\{x \in \mathbb{R}^n : M_\Phi f_\infty(x) > \lambda/2\}| \\ &\qquad \leq \Big|\bigcup_j 3Q_j\Big| \leq 3^n \sum_j |Q_j| \leq 3^n \sum_j \int_{Q_j} \Phi\left(\frac{2\cdot 4^n|f_\infty(x)|}{\lambda}\right) dx \\ &\qquad\qquad \leq 3^n \int_{\{x\in\mathbb{R}^n:|f(x)|>\lambda/2\}} \Phi\left(\frac{2\cdot 4^n|f(x)|}{\lambda}\right) dx. \end{aligned}$$

A.2 A weighted Calderón-Zygmund decomposition

In this section we give a Calderón-Zygmund decomposition defined with respect to the measure $w\,dx$, where $w \in A_\infty^d$. The first part of Proposition A.5 is usually given in the more general context of doubling measures. (See, for example, [88, 216].) The second half is part of the folklore of harmonic analysis but does not seem to have appeared explicitly in print.

In order to prove our results we need to make a definition and give a lemma with some basic properties of A_∞^d weights. (Recall that $A_\infty \subset A_\infty^d$, so these results hold for this class as well.)

A weight w is dyadic doubling with doubling constant C, if $w(\widetilde{Q}) \le C\,w(Q)$ for every $Q \in \mathcal{D}$. We denote the best constant C in this inequality by $[w]_D$.

Lemma A.4. *Given $w \in A_\infty^d$, there exist constants $r, C_w \ge 1$, such that for every $Q \in \mathcal{D}$ and every measurable set $E \subset Q$,*

$$\frac{|E|}{|Q|} \le C_w \left(\frac{w(E)}{w(Q)}\right)^r.$$

In particular, w is dyadic-doubling with $[w]_D \le 2^{n/r} C_w^{1/r}$. Conversely, there exist constants $t < 1$, $K_w \ge 1$ such that

$$\frac{w(E)}{w(Q)} \le K_w \left(\frac{|E|}{|Q|}\right)^t.$$

Proof. Both inequalities are well-known. Since $w \in A_\infty^d$, $w \in A_r^d$ for some $r > 1$. Then by Hölder's inequality and the definition of A_r^d,

$$\left(\frac{|E|}{|Q|}\right)^{\frac{1}{r}} \le \left(\fint_Q \chi_E(x)\,w(x)\,dx\right)\left(\fint_Q w(x)^{1-r'}\,dx\right)^{r-1} \le [w]_{A_r^d} \frac{w(E)}{w(Q)}.$$

Similarly, if $w \in A_\infty^d$, $w \in RH_s^d$ for some $s > 1$. Thus,

$$\begin{aligned} w(E) &= |Q| \fint_Q \chi_E(x) w(x)\,dx \\ &\le |Q| \left(\fint_Q w(x)^s\,dx\right)^{1/s} \left(\frac{|E|}{|Q|}\right)^{1/s'} \le [w]_{RH_s} w(Q) \left(\frac{|E|}{|Q|}\right)^{1/s'}. \end{aligned}$$ □

Given a weight w, and $f \in L^1_{\mathrm{loc}}(w)$, define the weighted dyadic maximal function

$$M_w^d f(x) = \sup_{x \in Q \in \mathcal{D}} \frac{1}{w(Q)} \int_Q |f(y)| w(y)\,dy.$$

(If $w(Q) = 0$ we define this average to be equal to 0.) When $w \equiv 1$ this becomes the standard dyadic maximal operator.

Proposition A.5. *Given $w \in A_\infty^d$ and $f \in L^1_{\mathrm{loc}}(w)$ such that $\frac{1}{w(Q)}\int_Q |f(x)|w(x)\,dx \to 0$ as $|Q| \to \infty$, fix $a > [w]_D$. Then for every $k \in \mathbb{Z}$ there exists a collection of maximal disjoint dyadic cubes $\{Q_j^k\}$ such that*

$$\Omega_k = \{x \in \mathbb{R}^n : M_w^d f(x) > a^k\} = \bigcup_j Q_j^k, \tag{A.2}$$

and for every j,

$$a^k < \frac{1}{w(Q_j^k)} \int_{Q_j^k} |f(x)|w(x)\,dx \le [w]_D a^k. \tag{A.3}$$

Furthermore, if we define $E_j^k = Q_j^k \setminus \Omega_{k+1}$, then the sets $E_j^k \subset Q_j^k$ are pairwise disjoint, and $w(E_j^k) \ge \frac{a-[w]_D}{a}\, w(Q_j^k)$.

Remark A.6. Proposition A.5 can be given in a more general form analogous to Proposition A.1 above. Here we prove the version needed in Chapter 10; we leave the details of the more general version to the interested reader.

Proof. The construction of the cubes $\{Q_j\}$ that satisfy (A.2) and the first inequality in (A.3) is identical to the proof of (a) in Proposition A.1 with $\lambda = a^k$ and $\Phi(t) = t$. The second inequality in (A.3) follows from the maximality of the cubes Q_j^k and Lemma A.4:

$$\frac{1}{w(Q_j^k)} \int_{Q_j^k} |f(x)|w(x)\,dx \le \frac{w(\widetilde{Q}_j^k)}{w(Q_j^k)} \frac{1}{w(\widetilde{Q}_j^k)} \int_{\widetilde{Q}_j^k} |f(x)|w(x)\,dx \le [w]_D\, a^k.$$

Finally, we prove the estimate for $w(E_j^k)$. Again, we proceed as in the proof of Proposition A.1:

$$\begin{aligned} w(Q_j^k \cap \Omega_{k+1}) = \sum_{i:Q_i^{k+1} \subsetneq Q_j^k} & w(Q_i^{k+1}) \\ \le a^{-(k+1)} & \sum_{i:Q_i^{k+1} \subsetneq Q_j^k} \int_{Q_i^{k+1}} |f(x)|\,w(x)\,dx \\ & \le a^{-(k+1)} \int_{Q_j^k} |f(x)|\,w(x)\,dx \le \frac{[w]_D}{a}\, w(Q_j^k). \end{aligned}$$

It follows that $w(E_j^k) \ge \frac{a-[w]_D}{a}\, w(Q_j^k)$ and the proof is complete. □

A.3 A fractional Calderón-Zygmund decomposition

Here we give a version of the Calderón-Zygmund decomposition that is adapted to the fractional maximal operators $M_{\Phi,\alpha}$, $0 < \alpha < n$. Proposition A.7 is taken from [32, 36] where it is stated without proof. (Note that there is an error in the statement of the constants in the second reference.)

Proposition A.7. *Given α, $0<\alpha<n$, and a Young function Φ, let f be a measurable function such that $|Q|^{\alpha/n}\,\|f\|_{\Phi,Q}\to 0$ as $|Q|\to\infty$. Then the following are true:*

(a) *For each $\lambda>0$, there exists a disjoint collection of maximal dyadic cubes $\{Q_j\}$ such that*

$$E_\lambda^d=\{x\in\mathbb{R}^n : M_{\Phi,\alpha}^d f(x)>\lambda\}=\bigcup_j Q_j,$$

and for every j,

$$\lambda<|Q_j|^{\alpha/n}\|f\|_{\Phi,Q_j}\leq 2^{n-\alpha}\,\lambda.$$

(b) *For each $\lambda>0$, let $\{Q_j\}$ be the collection of cubes in (a) corresponding to the level sets of height $\lambda/2^{2n-\alpha}$. Then*

$$E_\lambda=\{x\in\mathbb{R}^n : M_{\Phi,\alpha}f(x)>\lambda\}\subset\bigcup_j 3Q_j.$$

Remark A.8. For a fractional modular weak type inequality similar to (A.1) see Lemma 5.49 and the references given there. We can also construct pairwise disjoint sets analogous to the sets E_j^k in Proposition A.1; the argument is exactly the same. (See [36].)

Proof. The proof is adapted from that of Proposition A.1 and so we will omit some details that remain unchanged.

To prove (a), let Λ_λ be the family of dyadic cubes that satisfy $|Q|^{\alpha/n}\|f\|_{\Phi,Q}>\lambda$, and let $\{Q_j\}$ be the collection of maximal cubes in Λ_λ. This set exists since $|Q|^{\alpha/n}\,\|f\|_{\Phi,Q}\to 0$ as $|Q|\to\infty$. The maximality implies that the Q_j are pairwise disjoint. Further, by the maximality and the convexity of Φ,

$$\lambda<|Q_j|^{\alpha/n}\,\|f\|_{\Phi,Q_j}\leq 2^{n-\alpha}\,|\widetilde{Q}_j|^{\alpha/n}\,\|f\|_{\Phi,\widetilde{Q}_j}\leq 2^{n-\alpha}\,\lambda.$$

Finally, by the definition of $M_{\Phi,\alpha}^d$ we get $E_\lambda^d=\cup_j Q_j$.

To prove (b): given $x\in E_\lambda$, there exists a cube Q with $x\in Q$ such that $|Q|^{\alpha/n}\,\|f\|_{\Phi,Q}>\lambda$. Let k be the unique integer such that $2^{-(k+1)}\leq\ell(Q)<2^{-k}$. Then there exists N, $1\leq N\leq 2^n$, and a family of dyadic cubes $\{P_j\}_{j=1}^N$ such that $\ell(P_j)=2^{-k}$, $Q\subset\cup_j P_j$, and $P_j\cap Q\neq\emptyset$. Then one of these cubes, say P_1, satisfies $|Q|^{\alpha/n}\,\|f\,\chi_{P_1}\|_{\Phi,Q}>\lambda/2^n$. For if not we get a contradiction:

$$\begin{aligned}\lambda<|Q|^{\alpha/n}\,\|f\|_{\Phi,Q}=|Q|^{\alpha/n}\left\|f\sum_{j=1}^N\chi_{P_j}\right\|_{\Phi,Q}\\ \leq\sum_{j=1}^N|Q|^{\alpha/n}\,\|f\,\chi_{P_j}\|_{\Phi,Q}\leq N\lambda/2^n\leq\lambda.\end{aligned}$$

Therefore, by the convexity of Φ and since $1 < \ell(P_1)/\ell(Q) \le 2$,

$$\lambda/2^{2n-\alpha} < 2^{-n+\alpha}|Q|^{\alpha/n}\,\|f\,\chi_{P_1}\|_{\Phi,Q} \le 2^{-n+\alpha}\left(\frac{|P_1|}{|Q|}\right)^{1-\alpha/n}|P_1|^{\alpha/n}\,\|f\|_{\Phi,P_1} \le |P_1|^{\alpha/n}\,\|f\|_{\Phi,P_1}.$$

We can now argue exactly as before to get that $E_\lambda \subset \cup_j 3Q_j$. □

A.4 A Calderón-Zygmund decomposition for Borel measures

In this section we give a new Calderón-Zygmund decomposition valid for arbitrary Borel measures in $\mathbb{R}^n$ that shares the properties of the classical Calderón-Zygmund decomposition. A key feature is that it holds for non-doubling measures. This decomposition first appeared in [138]. Other versions of the Calderón-Zygmund decomposition for non-doubling measures have been considered by a number of authors; see, for instance, [143, 157, 166, 220, 221].

For clarity we introduce some notation and two classes of measures. Let $\mathbb{R}^n_j$, $1 \le j \le 2^n$, denote the n-dimensional quadrants in $\mathbb{R}^n$: that is, the sets $I^\pm \times I^\pm \times \cdots \times I^\pm$ where $I^+ = [0,\infty)$ and $I^- = (-\infty,0)$.

Let M be the class of all non-negative Borel measures μ such that for every $Q \in \mathcal{D}$, $\mu(Q) < \infty$. Let $M_\infty \subset M$ be the collection of measures such that $\mu(\mathbb{R}^n_j) = \infty$, $1 \le j \le 2^n$.

Remark A.9. While here and below we are restricting our attention to dyadic cubes, it will be clear that all of our results will hold for any translation of the dyadic grid. This is implicit in the proof of Theorem 10.43 above.

Given $\mu \in M$, we define the associated dyadic maximal operator by

$$M^d_\mu f(x) = \sup_{x\in Q\in\mathcal{D}} \frac{1}{\mu(Q)} \int_Q |f(y)|\,d\mu(y) = \sup_{x\in Q\in\mathcal{D}} \fint_Q |f(y)|\,d\mu(y); \qquad \text{(A.4)}$$

if $\mu(Q) = 0$, then we define the average to be equal to zero. Hereafter we will use the notation $\fint$ with the assumption that the underlying measure is μ and not Lebesgue measure as it was in the body of this monograph.

The decomposition for $\mu \in M_\infty$

The Calderón-Zygmund decomposition has a cleaner expression in the special case of measures $\mu \in M_\infty$. We will first consider this case, and then treat the general case below. We begin with the analog of Proposition A.1.

Proposition A.10. *Given $\mu \in M_\infty$, $f \in L^1(\mu)$, for every $\lambda > 0$, there exists a collection of disjoint maximal dyadic cubes $\{Q_j\}$ such that*

$$E_\lambda^d = \{x \in \mathbb{R}^n : M_\mu^d f(x) > \lambda\} = \bigcup_j Q_j, \tag{A.5}$$

where the Q_j are such that

$$\fint_{Q_j} |f(x)|\, d\mu(x) > \lambda \qquad \text{and} \qquad \fint_{\widetilde{Q}_j} |f(x)|\, d\mu(x) \leq \lambda. \tag{A.6}$$

Proof. The proof is identical to the proof of the classical Calderón-Zygmund decomposition (that is, the proof of Proposition A.1 when $\Phi(t) = t$), provided that for $Q \in \mathcal{D}$, $\fint_Q |f(x)|\, d\mu(x) \to 0$ as $\mu(Q) \to \infty$. (This allows us to find the maximal cube satisfying the first inequality in (A.6).) To see this, fix $Q_0 \in \mathcal{D}$ and let $Q_0 \subset Q_1 \subset Q_2 \subset \cdots$ be the sequence of dyadic parents—that is, $\ell(Q_j) = 2^j \ell(Q_0)$. Then, since $\mu \in M_\infty$, $\mu(Q_j) \to \infty$ as $j \to \infty$, and $\int_{Q_j} f(x)\, d\mu(x) \leq \int_{\mathbb{R}^n} |f(x)|\, d\mu(x) < \infty$, we get the desired limit. □

As an immediate consequence of Proposition A.10 we get that M_μ^d is weak $(1, 1)$.

Proposition A.11. *Given $\mu \in M_\infty$ and $f \in L^1(\mu)$, for every $\lambda > 0$,*

$$\mu(\{x \in \mathbb{R}^n : M_\mu^d f(x) > \lambda\}) \leq \frac{1}{\lambda} \int_{\mathbb{R}^n} |f(x)|\, d\mu(x). \tag{A.7}$$

Proof. Fix $\lambda > 0$ and let $\{Q_j\}$ be the cubes from Proposition A.10. Then

$$\mu(E_\lambda^d) = \sum_j \mu(Q_j) \leq \frac{1}{\lambda} \sum_j \int_{Q_j} |f(x)|\, d\mu(x) \leq \frac{1}{\lambda} \int_{\mathbb{R}^n} |f(x)|\, d\mu(x). \qquad \square$$

Remark A.12. Arguing as in the proof of Proposition A.1, we can prove the stronger inequality

$$\mu(\{x \in \mathbb{R}^n : M_\mu^d f(x) > \lambda\}) \leq \frac{2}{\lambda} \int_{\{x \in \mathbb{R}^n : |f(x)| > \lambda/2\}} |f(x)|\, d\mu(x).$$

Recall that in the classical case the Calderón-Zygmund cubes are used to decompose the function f as $g + b$, the so-called good and bad parts. We can generalize this decomposition to the case of Borel measures $\mu \in M_\infty$, but to do so we must add a third term.

Theorem A.13. ([138]) *Given $\mu \in M_\infty$ and $f \in L^1(\mu)$, for every $\lambda > 0$ let $\{Q_j\}$ be the collection of Calderón-Zygmund cubes in Proposition* A.10. *Then we can decompose f as $f = g + a + b$, where g, a and b are measurable functions such that:*

(a) *for every* p, $1 \le p < \infty$,

$$\|g\|_{L^p(\mu)}^p \le C_p\, \lambda^{p-1}\, \|f\|_{L^1(\mu)}.$$

(b) $a = \sum_j \alpha_j$, *where* $\operatorname{supp}(\alpha_j) \subset Q_j$, $\int_{Q_j} \alpha_j(x)\, d\mu(x) = 0$, *and*

$$\|a\|_{L^1(\mu)} = \sum_j \|\alpha_j\|_{L^1(\mu)} \le 2\, \|f\|_{L^1(\mu)}.$$

(c) $b = \sum_j \beta_j$, *where* $\operatorname{supp}(\beta_j) \subset \widetilde{Q}_j$, $\int_{\widetilde{Q}_j} \beta_j(x)\, d\mu(x) = 0$, *and*

$$\|b\|_{L^1(\mu)} \le \sum_j \|\beta_j\|_{L^1(\mu)} \le 4\, \|f\|_{L^1(\mu)}.$$

Proof. We sketch the proof, omitting some technical details. We refer the reader to [138] for the complete proof.

As before, let $E_\lambda^d = \cup_j Q_j$. For brevity, given a cube Q, let $f_Q = \fint_Q f\, d\mu$ and $|f|_Q = \fint_Q |f|\, d\mu$. We define g, a and b as follows:

$$\begin{aligned} g(x) &= g_1(x) + g_2(x) + g_3(x) \\ &= f(x)\, \chi_{\mathbb{R}^n \setminus E_\lambda^d}(x) + \sum_j f_{\widetilde{Q}_j}\, \chi_{Q_j}(x) + \sum_j \big(f_{Q_j} - f_{\widetilde{Q}_j}\big) \frac{\mu(Q_j)}{\mu(\widetilde{Q}_j)}\, \chi_{\widetilde{Q}_j}(x), \end{aligned}$$

$$a(x) = \sum_j \alpha_j(x) = \sum_j \big(f(x) - f_{Q_j}\big)\, \chi_{Q_j}(x),$$

$$b(x) = \sum_j \beta_j(x) = \sum_j \big(f_{Q_j} - f_{\widetilde{Q}_j}\big) \Big(\chi_{Q_j}(x) - \frac{\mu(Q_j)}{\mu(\widetilde{Q}_j)}\, \chi_{\widetilde{Q}_j}(x)\Big).$$

From the definition it is immediate that $f = g + a + b$, $\operatorname{supp}(\alpha_j) \subset Q_j$, $\operatorname{supp}(\beta_j) \subset \widetilde{Q}_j$, and

$$\int_{Q_j} \alpha_j(x)\, d\mu(x) = \int_{\widetilde{Q}_j} \beta_j(x)\, d\mu(x) = 0.$$

We first estimate the L^1 norm of a. Since the cubes Q_j are pairwise disjoint, we have that

$$\begin{aligned} \sum_j \|\alpha_j\|_{L^1(\mu)} &\le \sum_j \int_{Q_j} |f(x) - f_{Q_j}|\, d\mu(x) \\ &\le 2 \sum_j \int_{Q_j} |f(x)|\, d\mu(x) \le 2 \int_{\mathbb{R}^n} |f(x)|\, dx. \end{aligned}$$

To estimate the L^1 norm of b, note that since the Q_j are pairwise disjoint and $|f|_{\widetilde{Q}_j} \leq \lambda < |f|_{Q_j}$,

$$\begin{aligned}\sum_j \|\beta_j\|_{L^1(\mu)} &\leq \sum_j \left(|f|_{Q_j} + |f|_{\widetilde{Q}_j}\right) \int_{\widetilde{Q}_j} \left|\chi_{Q_j} - \frac{\mu(Q_j)}{\mu(\widetilde{Q}_j)} \chi_{\widetilde{Q}_j}\right| d\mu \\ &\leq 2 \sum_j \left(|f|_{Q_j} + \lambda\right) \mu(Q_j) \\ &\leq 4 \sum_j \int_{Q_j} |f(x)|\, d\mu(x) \\ &\leq 4 \int_{\mathbb{R}^n} |f(x)|\, dx.\end{aligned}$$

To complete the proof we must show that g satisfies the norm inequality in (a). We will show this for each g_i, $1 \leq i \leq 3$. For g_1, it is immediate that $\|g_1\|_{L^1(\mu)} \leq \|f\|_{L^1(\mu)}$. Further, since M_μ^d is weak type $(1,1)$, we can apply the Lebesgue differentiation theorem to get

$$\|g_1\|_{L^\infty(\mu)} \leq \|M_\mu^d f \cdot \chi_{\mathbb{R}^n \setminus E_\lambda^d}\|_{L^\infty(\mu)} \leq \lambda.$$

Therefore,

$$\int_{\mathbb{R}^n} |g_1(x)|^p\, d\mu(x) \leq \|g_1\|_{L^\infty(\mu)}^{p-1} \int_{\mathbb{R}^n} |g_1(x)|\, d\mu(x) \leq \lambda^{p-1} \|f\|_{L^1(\mu)}.$$

The estimate for g_2 is similar. Since $|f|_{\widetilde{Q}_j} \leq \lambda$, $|g_2(x)| \leq \lambda\, \chi_{E_\lambda^d}(x)$ and so

$$\|g_2\|_{L^1(\mu)} \leq \lambda\, \mu(E_\lambda^d) \leq \|f\|_{L^1(\mu)}, \qquad \|g_2\|_{L^\infty(\mu)} \leq \lambda.$$

The estimate for g_3 is the most difficult. The terms in its definition are supported in $\widetilde{Q}_j$ which are not disjoint, so an $L^\infty(\mu)$ estimate is not expected. However, we do have that

$$\begin{aligned}|g_3(x)| &\leq \sum_j \left(|f|_{Q_j} + |f|_{\widetilde{Q}_j}\right) \frac{\mu(Q_j)}{\mu(\widetilde{Q}_j)} \chi_{\widetilde{Q}_j}(x) \\ &\leq \sum_j \left(|f|_{Q_j} + \lambda\right) \frac{\mu(Q_j)}{\mu(\widetilde{Q}_j)} \chi_{\widetilde{Q}_j}(x) \\ &\leq 2 \sum_j \int_{Q_j} |f(x)|\, d\mu(x) \frac{\chi_{\widetilde{Q}_j}(x)}{\mu(\widetilde{Q}_j)} \\ &= 2Tf(x).\end{aligned}$$

Since the Q_j are disjoint,

$$\|Tf\|_{L^1(\mu)} = \int_{E_\lambda^d} |f(x)|\, d\mu(x) \leq \|f\|_{L^1(\mu)},$$

and so $\|g_3\|_{L^1(\mu)} \le 2\,\|f\|_{L^1(\mu)}$. By a careful induction argument (see [138]) we in fact have that for every $m \in \mathbb{N}$,

$$\|Tf\|^m_{L^m(\mu)} \le m!\Big(\sup_j \fint_{\widetilde{Q}_j} |f(x)|\,d\mu(x)\Big)^{m-1} \int_{E^d_\lambda} |f(x)|\,d\mu(x) \le m!\,\lambda^{m-1}\|f\|_{L^1(\mu)};$$

hence,

$$\|g_3\|^m_{L^m(\mu)} \le 2^m\,\|Tf\|^m_{L^m(\mu)} \le 2^m\,m!\,\lambda^{m-1}\,\|f\|_{L^1(\mu)}.$$

Finally, by Hölder's inequality we get the desired estimate for every p, $1 \le p < \infty$. This completes the proof. □

Remark A.14. Theorem A.13 can be extended to vector-valued functions $f = \{f_i\}$. Fix q, $1 < q < \infty$, and $f = \{f_i\}$ such that $\|f\|_{\ell^q} \in L^1(\mu)$. For a given $\lambda > 0$, take the cubes Q_j to be the Calderón-Zygmund decomposition of the level set of $M^d_\mu(\|f\|_{\ell^q})$. Now define $g = \{g_i\}$, $a = \{a_i\}$ and $b = \{b_i\}$ as vector-valued functions with the averages taken component-wise. Then these functions satisfy the same properties. The proof is essentially the same if absolutes values are replaced by ℓ^q-norms and Minkowski's inequality is used to show that $\|f_Q\|_{\ell^q} \le (\|f\|_{\ell^q})_Q$.

For $1 < q < \infty$, define the vector-valued dyadic maximal operator by applying the dyadic Hardy-Littlewood maximal operator to each component:

$$M^d_\mu f(x) = \{M^d_\mu f_i(x)\}, \qquad \overline{M}^d_{\mu,q} f(x) = \|M^d_\mu f(x)\|_{\ell^q}.$$

Our principal application of Theorem A.13 is to show that the operator $\overline{M}^d_{\mu,q}$ is weak type $(1,1)$ even though μ may be non-doubling.

Theorem A.15. *Given $\mu \in M_\infty$, then for all q, $1 < q < \infty$, and for every $f = \{f_i\}$,*

$$\mu(\{x \in \mathbb{R}^n : \overline{M}^d_{\mu,q} f(x) > \lambda\}) \le \frac{C}{\lambda} \int_{\mathbb{R}^n} \|f(x)\|_{\ell^q}\,d\mu(x). \tag{A.8}$$

Proof. We introduce a new maximal operator

$$\widetilde{M}^d_\mu h(x) = \sup_{x \in Q \in \mathcal{D}} \Big|\fint_Q h(y)\,d\mu(y)\Big|;$$

clearly, $\widetilde{M}^d_\mu h(x) \le M^d_\mu h(x)$, and $\widetilde{M}^d_\mu h(x) = M^d_\mu h(x)$ if h is non-negative.

If we split each component f_i into its positive and negative parts, then, since $\overline{M}^d_{\mu,q}$ is sublinear, we may assume without loss of generality that $f_i \ge 0$ for every i. Therefore, it will suffice to prove (A.8) for non-negative functions with $\widetilde{M}^d_\mu$ replacing M^d_μ.

Fix $\lambda > 0$ and $f = \{f_i\} \in L^1$ with $f_i \geq 0$ a.e.. By Theorem A.13 and Remark A.14 we can write $f = g + a + b$. Then

$$\|\widetilde{M}_\mu^d f(x)\|_{\ell^q} \leq \|\widetilde{M}_\mu^d g(x) + \widetilde{M}_\mu^d a(x) + \widetilde{M}_\mu^d b(x)\|_{\ell^q} \\ \leq \|\widetilde{M}_\mu^d g(x)\|_{\ell^q} + \|\widetilde{M}_\mu^d a(x) + \widetilde{M}_\mu^d b(x)\|_{\ell^q},$$

and so

$$\begin{aligned} &\mu(\{x \in \mathbb{R}^n : \|\widetilde{M}_\mu^d f(x)\|_{\ell^q} > \lambda\}) \\ &\quad \leq \mu(\{x \in \mathbb{R}^n : \|\widetilde{M}_\mu^d g(x)\|_{\ell^q} > \lambda/2\}) \\ &\qquad + \mu(E_\lambda^d) \\ &\qquad + \mu(\{x \in \mathbb{R}^n \setminus E_\lambda^d : \|\widetilde{M}_\mu^d a(x) + \widetilde{M}_\mu^d b(x)\|_{\ell^q} > \lambda/2\}) \\ &\quad = I_1 + I_2 + I_3. \end{aligned}$$

We estimate each term separately. By (A.7) we immediately get

$$I_2 \leq \frac{1}{\lambda} \int_{\mathbb{R}^n} \|f(x)\|_{\ell^q} \, d\mu(x).$$

To estimate I_1, note first that M_μ^d is bounded on $L^q(\mu)$: this follows by interpolation from the weak $(1,1)$ inequality in Proposition A.11 and the fact that it is bounded on $L^\infty(\mu)$. Then if we apply Chebyshev's inequality and the vector-valued version of (a) in Theorem A.13, we get

$$\begin{aligned} I_1 &\leq \frac{2^q}{\lambda^q} \int_{\mathbb{R}^n} \|\widetilde{M}_\mu^d g(x)\|_{\ell^q}^q \, d\mu(x) \\ &\leq \frac{2^q}{\lambda^q} \sum_i \int_{\mathbb{R}^n} M_\mu^d g_i(x)^q \, d\mu(x) \\ &\leq \frac{C}{\lambda^q} \sum_i \int_{\mathbb{R}^n} |g_i(x)|^q \, d\mu(x) \\ &= \frac{C}{\lambda^q} \big\| \|g\|_{\ell^q} \big\|_{L^q(\mu)}^q \\ &\leq \frac{C}{\lambda} \big\| \|f\|_{\ell^q} \big\|_{L^1(\mu)}. \end{aligned}$$

To estimate I_3 we will first show that if $x \in \mathbb{R}^n \setminus E_\lambda^d$, then $\widetilde{M}_\mu^d a(x) = 0$. Recall that $a = \sum_j \alpha_j$, where $\alpha_j = \{\alpha_{i,j}\}$. Fix $x \in \mathbb{R}^n \setminus Q_j$. If Q is a dyadic cube such that $Q \ni x$ and $Q \cap Q_j \neq \emptyset$, then $Q \supsetneq Q_j$. Therefore, since $\operatorname{supp}(\alpha_{i,j}) \subset Q_j$ and $\int_{Q_j} \alpha_{i,j}(x) \, d\mu(x) = 0$, we have that for all i,

$$\widetilde{M}_\mu^d \alpha_{i,j}(x) = \sup_{\substack{x \in Q \in \mathcal{D} \\ Q \supsetneq Q_j}} \left| \fint_Q \alpha_{i,j}(y) \, d\mu(y) \right| = \sup_{\substack{x \in Q \in \mathcal{D} \\ Q \supsetneq Q_j}} \left| \frac{1}{\mu(Q)} \int_{Q_j} \alpha_{i,j}(y) \, d\mu(y) \right| = 0.$$

It follows immediately that for every $x \in \mathbb{R}^n \setminus E_\lambda^d$,

$$0 \le \|\widetilde{M}_\mu^d a(x)\|_{\ell^q} \le \sum_j \|\widetilde{M}_\mu^d \alpha_j(x)\|_{\ell^q} = 0.$$

Therefore,

$$\begin{aligned}
I_3 &= \mu(\{x \in \mathbb{R}^n \setminus E_\lambda^d : \|\widetilde{M}_\mu^d b(x)\|_{\ell^q} > \lambda/2\}) \\
&\le \frac{2}{\lambda} \int_{\mathbb{R}^n \setminus E_\lambda^d} \|\widetilde{M}_\mu^d b(x)\|_{\ell^q}\, d\mu(x) \\
&\le \frac{2}{\lambda} \sum_j \int_{\mathbb{R}^n \setminus E_\lambda^d} \|\widetilde{M}_\mu^d \beta_j(x)\|_{\ell^q}\, d\mu(x) \\
&\le \frac{2}{\lambda} \sum_j \int_{\mathbb{R}^n \setminus \widetilde{Q}_j} \|\widetilde{M}_\mu^d \beta_j(x)\|_{\ell^q}\, d\mu(x) + \int_{\widetilde{Q}_j \setminus Q_j} \|\widetilde{M}_\mu^d \beta_j(x)\|_{\ell^q}\, d\mu \\
&= \frac{2}{\lambda} \sum_j J_j + K_j.
\end{aligned}$$

We estimate the summands in the last term. Arguing exactly as we did before, if we fix $x \notin \widetilde{Q}_j$, and if Q is a dyadic cube such that $Q \ni x$ and $Q \cap \widetilde{Q}_j \neq \emptyset$, then $Q \supsetneq \widetilde{Q}_j$. Then, since $\operatorname{supp} \beta_{i,j} \subset \widetilde{Q}_j$ and $\int_{\widetilde{Q}_j} \beta_{i,j}\, d\mu = 0$,

$$\widetilde{M}_\mu^d \beta_{i,j}(x) = \sup_{\substack{x \in Q \in \mathcal{D} \\ Q \supsetneq \widetilde{Q}_j}} \left| \fint_Q \beta_{i,j}(x)\, d\mu(x) \right| = \sup_{\substack{x \in Q \in \mathcal{D} \\ Q \supsetneq \widetilde{Q}_j}} \left| \frac{1}{\mu(Q)} \int_{\widetilde{Q}_j} \beta_{i,j}(x)\, d\mu(x) \right| = 0.$$

Therefore, for each j, $J_j = 0$.

To estimate K_j, let $x \in \widetilde{Q}_j \setminus Q_j$ and take any dyadic cube Q containing x such that $Q \cap \widetilde{Q}_j \neq \emptyset$. If $\widetilde{Q}_j \subseteq Q$, then we again have $\fint_Q \beta_{i,j}\, d\mu = 0$. Therefore, we may assume that $Q \subsetneq \widetilde{Q}_j$. In this case we must have that $Q \cap Q_j = \emptyset$: we cannot have $Q \subset Q_j$ since $x \in Q$ and $x \notin Q_j$, and we cannot have $Q_j \subsetneq Q$ since this contradicts $Q \subsetneq \widetilde{Q}_j$. Thus,

$$\begin{aligned}
\left| \fint_Q \beta_{i,j}(x)\, d\mu(x) \right| &= |(f_i)_{Q_j} - (f_i)_{\widetilde{Q}_j}| \fint_Q \left| \chi_{Q_j}(x) - \frac{\mu(Q_j)}{\mu(\widetilde{Q}_j)} \chi_{\widetilde{Q}_j}(x) \right| d\mu(x) \\
&= |(f_i)_{Q_j} - (f_i)_{\widetilde{Q}_j}| \fint_Q \frac{\mu(Q_j)}{\mu(\widetilde{Q}_j)} \chi_{\widetilde{Q}_j}(x)\, d\mu(x) \\
&= |(f_i)_{Q_j} - (f_i)_{\widetilde{Q}_j}| \frac{\mu(Q_j)}{\mu(\widetilde{Q}_j)}.
\end{aligned}$$

This implies that $\widetilde{M}_\mu^d \beta_{i,j}(x) = |(f_i)_{Q_j} - (f_i)_{\widetilde{Q}_j}|\, \mu(Q_j)/\mu(\widetilde{Q}_j)$, and so

$$\begin{aligned}
\|\widetilde{M}_\mu^d \beta_j(x)\|_{\ell^q} &= \|f_{Q_j} - f_{\widetilde{Q}_j}\|_{\ell^q} \frac{\mu(Q_j)}{\mu(\widetilde{Q}_j)} \\
&\le \left(\|f_{Q_j}\|_{\ell^q} + \|f_{\widetilde{Q}_j}\|_{\ell^q}\right) \frac{\mu(Q_j)}{\mu(\widetilde{Q}_j)} \le \left((\|f\|_{\ell^q})_{Q_j} + (\|f\|_{\ell^q})_{\widetilde{Q}_j}\right) \frac{\mu(Q_j)}{\mu(\widetilde{Q}_j)} \\
&\le \left((\|f\|_{\ell^q})_{Q_j} + \lambda\right) \frac{\mu(Q_j)}{\mu(\widetilde{Q}_j)} \le 2\,(\|f\|_{\ell^q})_{Q_j} \frac{\mu(Q_j)}{\mu(\widetilde{Q}_j)}.
\end{aligned}$$

Therefore,

$$K_j \le 2\,(\|f\|_{\ell^q})_{Q_j} \frac{\mu(Q_j)}{\mu(\widetilde{Q}_j)} \mu(\widetilde{Q}_j \setminus Q_j) \le 2 \int_{Q_j} \|f(x)\|_{\ell^q}\, d\mu(x),$$

and so

$$I_3 \le \frac{2}{\lambda} \sum_j J_j + K_j \le \frac{4}{\lambda} \sum_j \int_{Q_j} \|f(x)\|_{\ell^q}\, d\mu(x) \le \frac{4}{\lambda} \int_{\mathbb{R}^n} \|f(x)\|_{\ell^q}\, d\mu(x).$$

This completes the proof. □

The decomposition for $\mu \in M$

Here we generalize the results in the previous section by considering the case of arbitrary $\mu \in M$—that is, measures such that $\mu(\mathbb{R}^n_j) < \infty$ for some (or all) j. Let $\mathcal{D}_j$ be the collection of dyadic cubes contained in $\mathbb{R}^n_j$, and define

$$M_\mu^{d,j} f(x) = \sup_{x \in Q \in \mathcal{D}_j} \fint_Q |f(y)|\, d\mu(y) = M_\mu^d (f\, \chi_{\mathbb{R}^n_j})(x)\, \chi_{\mathbb{R}^n_j}(x).$$

Note that $\operatorname{supp}(M_\mu^{d,j} f) \subset \mathbb{R}^n_j$. Hence, given a function f we have that

$$f(x) = \sum_{j=1}^{2^n} f(x)\, \chi_{\mathbb{R}^n_j}(x), \qquad M_\mu^d f(x) = \sum_{j=1}^{2^n} M_\mu^{d,j} f(x)\, \chi_{\mathbb{R}^n_j}(x),$$

and in each sum there is at most only one non-zero term. Because of this decomposition, to extend our results it will suffice to assume that f is supported in $\mathbb{R}^n_j$ for some j (say $j = 1$) and obtain the corresponding decompositions and estimates in $\mathbb{R}^n_j$.

Hereafter we will assume that f is supported in $\mathbb{R}^n_1$. In particular, $M_\mu^d f = M_\mu^{d,1} f$ and this function is supported in $\mathbb{R}^n_1$. In particular, for any $\lambda > 0$,

$$E_\lambda^d = \{x \in \mathbb{R}^n : M_\mu^d f(x) > \lambda\} = \{x \in \mathbb{R}^n_1 : M_\mu^{d,1} f(x) > \lambda\},$$

and so any decomposition of this set will consist of cubes in $\mathcal{D}_1$. We modify our notation and define $\fint_{\mathbb{R}^n_1} f(x)\,d\mu(x) = \frac{1}{\mu(\mathbb{R}^n_1)} \int_{\mathbb{R}^n_1} f(x)\,d\mu(x)$ if $\mu(\mathbb{R}^n_1) < \infty$ and $\fint_{\mathbb{R}^n_1} f(x)\,d\mu(x) = 0$ if $\mu(\mathbb{R}^n_1) = \infty$.

The following result is the analog of Propositions A.10 and A.11.

Proposition A.16. *Given $\mu \in M$ and $f \in L^1(\mathbb{R}^n_1)$ with $\operatorname{supp}(f) \in \mathbb{R}^n_1$, for every $\lambda > \fint_{\mathbb{R}^n_1} |f(x)|\,d\mu(x)$ there exists a collection of disjoint maximal dyadic cubes $\{Q_j\} \subset \mathcal{D}_1$ such that* (A.5) *and* (A.6) *hold. Further,* (A.7) *holds for every $\lambda > 0$.*

Proof. If $\mu(\mathbb{R}^n_1) = \infty$, then the proofs given above go through without change. If $\mu(\mathbb{R}^n_1) < \infty$, then in the notation used above, $\fint_{Q_j} |f(x)|\,d\mu(x) \to \fint_{\mathbb{R}^n_1} |f(x)|\,d\mu(x) < \lambda$. Hence, if $Q \in \mathcal{D}_1$ is such that $\fint_Q |f(x)|\,d\mu(x) > \lambda$, then Q must be contained in a maximal cube with the same property. But then we can argue exactly as before to form the cubes $\{Q_j\}$.

Finally, to prove the weak $(1,1)$ inequality, we note that the proof goes through exactly as before if $\lambda > \fint_{\mathbb{R}^n_1} |f(x)|\,d\mu(x)$. On the other hand, if $0 < \lambda \leq \fint_{\mathbb{R}^n_1} |f(x)|\,d\mu(x)$, then we immediately have

$$\mu(E^d_\lambda) \leq \mu(\mathbb{R}^n_1) \leq \frac{1}{\lambda} \int_{\mathbb{R}^n_1} |f(x)|\,d\mu(x). \qquad \square$$

Using Proposition A.16 we can immediately repeat the arguments in the proof of Theorem A.13 to get the following result.

Theorem A.17. *Given $\mu \in M$ and $f \in L^1(\mathbb{R}^n_1)$ with $\operatorname{supp}(f) \in \mathbb{R}^n_1$, for every $\lambda > \fint_{\mathbb{R}^n_1} |f(x)|\,d\mu(x)$ we can write $f = g + a + b$ with g, a and b as in Theorem* A.13 *with the additional property that all the cubes involved are contained in $\mathbb{R}^n_1$.*

With this Calderón-Zygmund decomposition we can now prove the weak $(1,1)$ inequality for the vector-valued maximal operator defined for any $\mu \in M$.

Theorem A.18. *Given $\mu \in M$, for all q, $1 < q < \infty$, the vector-valued dyadic maximal operator $\overline{M}^d_{\mu,q}$ satisfies the weak $(1,1)$ inequality*

$$\mu(\{x \in \mathbb{R}^n : \|M^d_\mu f(x)\|_{\ell^q} > \lambda\}) \leq \frac{C}{\lambda} \int_{\mathbb{R}^n} \|f(x)\|_{\ell^q}\,d\mu(x). \tag{A.9}$$

Proof. We will first prove this in the special case that $\operatorname{supp}(f) \subset \mathbb{R}^n_1$. In this case, we can repeat the proof of Theorem A.15 using Proposition A.16 and Theorem A.17. If $\mu(\mathbb{R}^n_1) = \infty$, then the argument goes through without change. If $\mu(\mathbb{R}^n_1) < \infty$, then we only need to prove the weak $(1,1)$ inequality for $0 < \lambda \leq \fint_{\mathbb{R}^n_1} \|f(x)\|_{\ell^q}\,d\mu(x)$ since in this range we cannot form the Calderón-Zygmund decomposition. But this is immediate: since $\operatorname{supp}(M^d_\mu f_i) \subset \mathbb{R}^n_1$ for every i,

$$\mu(\{x \in \mathbb{R}^n : \|M^d_\mu f(x)\|_{\ell^q} > \lambda\}) \leq \mu(\mathbb{R}^n_1) \leq \frac{1}{\lambda} \int_{\mathbb{R}^n_1} \|f(x)\|_{\ell^q}\,d\mu(x).$$

To prove the weak type inequality in the general case, fix f and write $f = \sum_{j=1}^{2^n} f\,\chi_{\mathbb{R}^n_j}$. Then

$$\|M^d_\mu f(x)\|_{\ell^q} = \sum_{j=1}^{2^n} \|M^d_\mu(f\,\chi_{\mathbb{R}^n_j})(x)\|_{\ell^q}\ \chi_{\mathbb{R}^n_j}(x),$$

and so by the above argument (and its analogs for each $\mathbb{R}^n_j$),

$$\begin{aligned}
\mu(\{x \in \mathbb{R}^n : \|M^d_\mu f(x)\|_{\ell^q} > \lambda\}) \\
&= \sum_{j=1}^{2^n} \mu(\{x \in \mathbb{R}^n_j : \|M^d_\mu(f\,\chi_{\mathbb{R}^n_j})\|_{\ell^q} > \lambda\}) \\
&\le \sum_{j=1}^{2^n} \frac{C}{\lambda} \int_{\mathbb{R}^n_j} \|f(x)\,\chi_{\mathbb{R}^n_j}(x)\|_{\ell^q}\, d\mu(x) \\
&= \frac{C}{\lambda} \int_{\mathbb{R}^n} \|f(x)\|_{\ell^q}\, d\mu(x).
\end{aligned}$$

□

Bibliography

[1] D. R. Adams. A note on Riesz potentials. *Duke Math. J.*, 42(4):765–778, 1975.

[2] H. A. Aimar, A. L. Bernardis, and F. J. Martín-Reyes. Multiresolution approximations and wavelet bases of weighted L^p spaces. *J. Fourier Anal. Appl.*, 9(5):497–510, 2003.

[3] J. Álvarez, J. Hounie, and C. Pérez. A pointwise estimate for the kernel of a pseudo-differential operator, with applications. *Rev. Un. Mat. Argentina*, 37(3-4):184–199 (1992), 1991. X Latin American School of Mathematics (Spanish) (Tanti, 1991).

[4] J. Alvarez and C. Pérez. Estimates with A_∞ weights for various singular integral operators. *Boll. Un. Mat. Ital. A (7)*, 8(1):123–133, 1994.

[5] K. F. Andersen and R. T. John. Weighted inequalities for vector-valued maximal functions and singular integrals. *Studia Math.*, 69(1):19–31, 1980/81.

[6] K. Astala, T. Iwaniec, and E. Saksman. Beltrami operators in the plane. *Duke Math. J.*, 107(1):27–56, 2001.

[7] P. Auscher. On necessary and sufficient conditions for L^p-estimates of Riesz transforms associated to elliptic operators on $\mathbb{R}^n$ and related estimates. *Mem. Amer. Math. Soc.*, 186(871):xviii+75, 2007.

[8] P. Auscher and J. M. Martell. Weighted norm inequalities, off-diagonal estimates and elliptic operators. Part III. Harmonic analysis of elliptic operators. *J. Funct. Anal.*, 241(2):703–746, 2006.

[9] P. Auscher and J. M. Martell. Weighted norm inequalities, off-diagonal estimates and elliptic operators. Part I. General operator theory and weights. *Adv. Math.*, 212(1):225–276, 2007.

[10] R. J. Bagby and J. D. Parsons. Orlicz spaces and rearranged maximal functions. *Math. Nachr.*, 132:15–27, 1987.

[11] A. Benedek, A.-P. Calderón, and R. Panzone. Convolution operators on Banach space valued functions. *Proc. Nat. Acad. Sci. U.S.A.*, 48:356–365, 1962.

[12] C. Bennett and R. Sharpley. *Interpolation of operators*, volume 129 of *Pure and Applied Mathematics*. Academic Press Inc., Boston, MA, 1988.

[13] S. Bloom. Solving weighted norm inequalities using the Rubio de Francia algorithm. *Proc. Amer. Math. Soc.*, 101(2):306–312, 1987.

[14] R. P. Boas and S. Bochner. On a theorem of M. Riesz for Fourier series. *J. Lond. Math. Soc.*, 14:62–73, 1939.

[15] S. M. Buckley. Estimates for operator norms on weighted spaces and reverse Jensen inequalities. *Trans. Amer. Math. Soc.*, 340(1):253–272, 1993.

[16] S. M. Buckley. Summation conditions on weights. *Michigan Math. J.*, 40(1):153–170, 1993.

[17] D. L. Burkholder and R. F. Gundy. Extrapolation and interpolation of quasi-linear operators on martingales. *Acta Math.*, 124:249–304, 1970.

[18] C. Capone, D. Cruz-Uribe, and A. Fiorenza. The fractional maximal operator and fractional integrals on variable L^p spaces. *Rev. Mat. Iberoam.*, 23(3):743–770, 2007.

[19] M. Carozza and A. Passarelli Di Napoli. Composition of maximal operators. *Publ. Mat.*, 40(2):397–409, 1996.

[20] M. J. Carro and H. Heinig. Modular inequalities for the Calderón operator. *Tohoku Math. J. (2)*, 52(1):31–46, 2000.

[21] M. J. Carro, C. Pérez, F. Soria, and J. Soria. Maximal functions and the control of weighted inequalities for the fractional integral operator. *Indiana Univ. Math. J.*, 54(3):627–644, 2005.

[22] S.-Y. A. Chang, J. M. Wilson, and T. H. Wolff. Some weighted norm inequalities concerning the Schrödinger operators. *Comment. Math. Helv.*, 60(2):217–246, 1985.

[23] S. Chanillo and R. L. Wheeden. Some weighted norm inequalities for the area integral. *Indiana Univ. Math. J.*, 36(2):277–294, 1987.

[24] R. R. Coifman. Distribution function inequalities for singular integrals. *Proc. Nat. Acad. Sci. U.S.A.*, 69:2838–2839, 1972.

[25] R. R. Coifman and C. Fefferman. Weighted norm inequalities for maximal functions and singular integrals. *Studia Math.*, 51:241–250, 1974.

[26] R. R. Coifman, P. W. Jones, and J. L. Rubio de Francia. Constructive decomposition of BMO functions and factorization of A_p weights. *Proc. Amer. Math. Soc.*, 87(4):675–676, 1983.

[27] R. R. Coifman and R. Rochberg. Another characterization of BMO. *Proc. Amer. Math. Soc.*, 79(2):249–254, 1980.

[28] A. Cordoba and C. Fefferman. A weighted norm inequality for singular integrals. *Studia Math.*, 57(1):97–101, 1976.

[29] M. Cotlar and C. Sadosky. On the Helson-Szegő theorem and a related class of modified Toeplitz kernels. In *Harmonic analysis in Euclidean spaces (Proc. Sympos. Pure Math., Williams Coll., Williamstown, Mass., 1978), Part 1*, Proc. Sympos. Pure Math., XXXV, pages 383–407. Amer. Math. Soc., Providence, R.I., 1979.

[30] M. Cotlar and C. Sadosky. On some L^p versions of the Helson-Szegő theorem. In *Conference on harmonic analysis in honor of Antoni Zygmund, Vol. I, II (Chicago, Ill., 1981)*, Wadsworth Math. Ser., pages 306–317. Wadsworth, Belmont, CA, 1983.

[31] D. Cruz-Uribe. The class $A_\infty^+(g)$ and the one-sided reverse Hölder inequality. *Canad. Math. Bull.*, 40(2):169–173, 1997.

[32] D. Cruz-Uribe. New proofs of two-weight norm inequalities for the maximal operator. *Georgian Math. J.*, 7(1):33–42, 2000.

[33] D. Cruz-Uribe. A new proof of weighted weak-type inequalities for fractional integrals. *Comment. Math. Univ. Carolin.*, 42(3):481–485, 2001.

[34] D. Cruz-Uribe, L. Diening, and A. Fiorenza. A new proof of the boundedness of maximal operators on variable Lebesgue spaces. *Boll. Unione Mat. Ital. (9)*, 2(1):151–173, 2009.

[35] D. Cruz-Uribe and A. Fiorenza. The A_∞ property for Young functions and weighted norm inequalities. *Houston J. Math.*, 28(1):169–182, 2002.

[36] D. Cruz-Uribe and A. Fiorenza. Endpoint estimates and weighted norm inequalities for commutators of fractional integrals. *Publ. Mat.*, 47(1):103–131, 2003.

[37] D. Cruz-Uribe and A. Fiorenza. Approximate identities in variable L^p spaces. *Math. Nachr.*, 280(3):256–270, 2007.

[38] D. Cruz-Uribe and A. Fiorenza. Weighted endpoint estimates for commutators of fractional integrals. *Czechoslovak Math. J.*, 57(132)(1):153–160, 2007.

[39] D. Cruz-Uribe and A. Fiorenza. Convergence in variable Lebesgue spaces. *Publ. Mat.*, 54(2):441–459, 2010.

[40] D. Cruz-Uribe, A. Fiorenza, J. M. Martell, and C. Pérez. The boundedness of classical operators on variable L^p spaces. *Ann. Acad. Sci. Fenn. Math.*, 31(1):239–264, 2006.

[41] D. Cruz-Uribe, A. Fiorenza, and C. J. Neugebauer. The maximal function on variable L^p spaces. *Ann. Acad. Sci. Fenn. Math.*, 28(1):223–238, 2003.

[42] D. Cruz-Uribe, A. Fiorenza, and C. J. Neugebauer. Corrections to: "The maximal function on variable L^p spaces" [Ann. Acad. Sci. Fenn. Math. **28** (2003), no. 1, 223–238;]. *Ann. Acad. Sci. Fenn. Math.*, 29(1):247–249, 2004.

[43] D. Cruz-Uribe, L. Forzani, and D. Maldonado. The structure of increasing weights on the real line. *Houston J. Math.*, 34(3):951–983, 2008.

[44] D. Cruz-Uribe, J. M. Martell, and C. Pérez. Extrapolation from A_∞ weights and applications. *J. Funct. Anal.*, 213(2):412–439, 2004.

[45] D. Cruz-Uribe, J. M. Martell, and C. Pérez. Weighted weak-type inequalities and a conjecture of Sawyer. *Int. Math. Res. Not.*, (30):1849–1871, 2005.

[46] D. Cruz-Uribe, J. M. Martell, and C. Pérez. Extensions of Rubio de Francia's extrapolation theorem. *Collect. Math.*, (Vol. Extra):195–231, 2006.

[47] D. Cruz-Uribe, J. M. Martell, and C. Pérez. Sharp two-weight inequalities for singular integrals, with applications to the Hilbert transform and the Sarason conjecture. *Adv. Math.*, 216(2):647–676, 2007.

[48] D. Cruz-Uribe, J. M. Martell, and C. Pérez. Sharp weighted estimates for approximating dyadic operators. *Electron. Res. Announc. Math. Sci.*, 17, 2010.

[49] D. Cruz-Uribe, J. M. Martell, and C. Pérez. Sharp weighted estimates for classical operators. *Preprint*, 2010.

[50] D. Cruz-Uribe and C. J. Neugebauer. The structure of the reverse Hölder classes. *Trans. Amer. Math. Soc.*, 347(8):2941–2960, 1995.

[51] D. Cruz-Uribe and C. J. Neugebauer. Weighted norm inequalities for the centered maximal operator on $\mathbf{R}^+$. *Ricerche Mat.*, 48(2):225–241 (2000), 1999.

[52] D. Cruz-Uribe, C. J. Neugebauer, and V. Olesen. The one-sided minimal operator and the one-sided reverse Hölder inequality. *Studia Math.*, 116(3):255–270, 1995.

[53] D. Cruz-Uribe and C. Pérez. Sharp two-weight, weak-type norm inequalities for singular integral operators. *Math. Res. Lett.*, 6(3-4):417–427, 1999.

[54] D. Cruz-Uribe and C. Pérez. Two weight extrapolation via the maximal operator. *J. Funct. Anal.*, 174(1):1–17, 2000.

[55] D. Cruz-Uribe and C. Pérez. Two-weight, weak-type norm inequalities for fractional integrals, Calderón-Zygmund operators and commutators. *Indiana Univ. Math. J.*, 49(2):697–721, 2000.

[56] D. Cruz-Uribe and C. Pérez. On the two-weight problem for singular integral operators. *Ann. Sc. Norm. Super. Pisa Cl. Sci. (5)*, 1(4):821–849, 2002.

[57] G. Curbera, J. García-Cuerva, J. M. Martell, and C. Pérez. Extrapolation with weights, rearrangement-invariant function spaces, modular inequalities and applications to singular integrals. *Adv. Math.*, 203(1):256–318, 2006.

[58] G. David and J.-L. Journé. A boundedness criterion for generalized Calderón-Zygmund operators. *Ann. of Math. (2)*, 120(2):371–397, 1984.

[59] L. Diening. Maximal function on generalized Lebesgue spaces $L^{p(\cdot)}$. *Math. Inequal. Appl.*, 7(2):245–253, 2004.

[60] L. Diening. Maximal function on Musielak-Orlicz spaces and generalized Lebesgue spaces. *Bull. Sci. Math.*, 129(8):657–700, 2005.

[61] L. Diening. *Lebesgue and Sobolev spaces with variable exponent.* Habilitation, Universität Freiburg, 2007.

[62] L. Diening, P. Harjulehto, P. Hästö, Y. Mizuta, and T. Shimomura. Maximal functions in variable exponent spaces: limiting cases of the exponent. *Ann. Acad. Sci. Fenn. Math.*, 34(2):503–522, 2009.

[63] L. Diening, P. Hästö, and A. Nekvinda. Open problems in variable exponent Lebesgue and Sobolev spaces. In *FSDONA04 Proceedings (Drabek and Rakosnik (eds.); Milovy, Czech Republic*, pages 38–58. Academy of Sciences of the Czech Republic, Prague, 2005.

[64] L. Diening and M. Růžička. Calderón-Zygmund operators on generalized Lebesgue spaces $L^{p(\cdot)}$ and problems related to fluid dynamics. *J. Reine Angew. Math.*, 563:197–220, 2003.

[65] O. Dragičević, L. Grafakos, M. C Pereyra, and S. Petermichl. Extrapolation and sharp norm estimates for classical operators on weighted Lebesgue spaces. *Publ. Mat.*, 49(1):73–91, 2005.

[66] J. Duoandikoetxea. *Análisis de Fourier*, volume 29 of *Colección de Estudios.* Ediciones de la Universidad Autónoma de Madrid, Spain, 1991.

[67] J. Duoandikoetxea. Weighted norm inequalities for homogeneous singular integrals. *Trans. Amer. Math. Soc.*, 336(2):869–880, 1993.

[68] J. Duoandikoetxea. *Fourier analysis*, volume 29 of *Graduate Studies in Mathematics.* American Mathematical Society, Providence, RI, 2001.

[69] J. Duoandikoetxea. Extrapolation of weights revisited: new proofs and sharp bounds, Preprint, 2010.

[70] J. Duoandikoetxea, A. Moyua, O. Oruetxebarria, and E. Siejo. Radial A_p weights with applications to the disc multiplier and the Bochner-Riesz operators. *Indiana Univ. Math. J.*, 57:1239–1258, 2008.

[71] J. Duoandikoetxea and J. L. Rubio de Francia. Maximal and singular integral operators via Fourier transform estimates. *Invent. Math.*, 84(3):541–561, 1986.

[72] E. M. Dyn′kin and B. P. Osilenker. Weighted estimates for singular integrals and their applications. In *Mathematical analysis, Vol. 21*, Itogi Nauki i Tekhniki, pages 42–129. Akad. Nauk SSSR Vsesoyuz. Inst. Nauchn. i Tekhn. Inform., Moscow, 1983.

[73] X. Fan and D. Zhao. On the spaces $L^{p(x)}(\Omega)$ and $W^{m,p(x)}(\Omega)$. *J. Math. Anal. Appl.*, 263(2):424–446, 2001.

[74] C. Fefferman. The uncertainty principle. *Bull. Amer. Math. Soc. (N.S.)*, 9(2):129–206, 1983.

[75] C. Fefferman and E. M. Stein. Some maximal inequalities. *Amer. J. Math.*, 93:107–115, 1971.

[76] C. Fefferman and E. M. Stein. H^p spaces of several variables. *Acta Math.*, 129(3-4):137–193, 1972.

[77] R. Fefferman. A note on singular integrals. *Proc. Amer. Math. Soc.*, 74(2):266–270, 1979.

[78] R. Fefferman and J. Pipher. Multiparameter operators and sharp weighted inequalities. *Amer. J. Math.*, 119(2):337–369, 1997.

[79] A. Fiorenza and M. Krbec. Indices of Orlicz spaces and some applications. *Comment. Math. Univ. Carolin.*, 38(3):433–451, 1997.

[80] B. Franchi, C. Pérez, and R. L. Wheeden. Sharp geometric Poincaré inequalities for vector fields and non-doubling measures. *Proc. London Math. Soc. (3)*, 80(3):665–689, 2000.

[81] N. Fujii. A proof of the Fefferman-Stein-Strömberg inequality for the sharp maximal functions. *Proc. Amer. Math. Soc.*, 106(2):371–377, 1989.

[82] N. Fujii. A condition for a two-weight norm inequality for singular integral operators. *Studia Math.*, 98(3):175–190, 1991.

[83] J. García-Cuerva. An extrapolation theorem in the theory of A_p weights. *Proc. Amer. Math. Soc.*, 87(3):422–426, 1983.

[84] J. García-Cuerva. José Luis Rubio de Francia (1949–1988). *Collect. Math.*, 38(1):3–15, 1987.

[85] J. García-Cuerva and K. S. Kazarian. Spline wavelet bases of weighted L^p spaces, $1 \leq p < \infty$. *Proc. Amer. Math. Soc.*, 123(2):433–439, 1995.

[86] J. García-Cuerva and J. M Martell. Two-weight norm inequalities for maximal operators and fractional integrals on non-homogeneous spaces. *Indiana Univ. Math. J.*, 50(3):1241–1280, 2001.

[87] J. García-Cuerva and J. M. Martell. Wavelet characterization of weighted spaces. *J. Geom. Anal.*, 11(2):241–264, 2001.

[88] J. García-Cuerva and J. L. Rubio de Francia. *Weighted norm inequalities and related topics*, volume 116 of *North-Holland Mathematics Studies*. North-Holland Publishing Co., Amsterdam, 1985.

[89] F. W. Gehring. The L^p-integrability of the partial derivatives of a quasiconformal mapping. *Acta Math.*, 130:265–277, 1973.

[90] I. Genebashvili, A. Gogatishvili, V. Kokilashvili, and M. Krbec. *Weight theory for integral transforms on spaces of homogeneous type*, volume 92 of *Pitman Monographs and Surveys in Pure and Applied Mathematics*. Longman, Harlow, 1998.

[91] L. Grafakos. *Classical Fourier Analysis*, volume 249 of *Graduate Texts in Mathematics*. Springer, New York, 2nd edition, 2008.

[92] L. Grafakos. *Modern Fourier Analysis*, volume 250 of *Graduate Texts in Mathematics*. Springer, New York, 2nd edition, 2008.

[93] L. Grafakos and J. M. Martell. Extrapolation of weighted norm inequalities for multivariable operators and applications. *J. Geom. Anal.*, 14(1):19–46, 2004.

[94] R. F. Gundy and R. L. Wheeden. Weighted integral inequalities for the nontangential maximal function, Lusin area integral, and Walsh Paley series. *Studia Math.*, 49:107–124, 1973/74.

[95] J. Gustavsson and J. Peetre. Interpolation of Orlicz spaces. *Studia Math.*, 60(1):33–59, 1977.

[96] E. Harboure, R. A. Macías, and C. Segovia. Extrapolation results for classes of weights. *Amer. J. Math.*, 110(3):383–397, 1988.

[97] E. Harboure, R. A. Macías, and C. Segovia. An extrapolation theorem for pairs of weights. *Rev. Un. Mat. Argentina*, 40(3-4):37–48, 1997.

[98] G. H. Hardy and J. E. Littlewood. A maximal theorem with function-theoretic applications. *Acta Math.*, 54(1):81–116, 1930.

[99] P. Harjulehto and P. Hästö. Sobolev inequalities with variable exponent attaining the values 1 and n. *Publ. Mat.*, 52(2):347–363, 2008.

[100] P. Hästö. Local-to-global results in variable exponent spaces. *Math. Res. Letters*, 16(2):263–278, 2009.

[101] H. Helson and G. Szegö. A problem in prediction theory. *Ann. Mat. Pura Appl. (4)*, 51:107–138, 1960.

[102] E. Hernández. Factorization and extrapolation of pairs of weights. *Studia Math.*, 95(2):179–193, 1989.

[103] E. Hernández and G. Weiss. *A first course on wavelets*. Studies in Advanced Mathematics. CRC Press, Boca Raton, FL, 1996. With a foreword by Yves Meyer.

[104] R. Hunt, B. Muckenhoupt, and R. L. Wheeden. Weighted norm inequalities for the conjugate function and Hilbert transform. *Trans. Amer. Math. Soc.*, 176:227–251, 1973.

[105] T. Hytönen. On Petermichl's dyadic shift and the Hilbert transform. *C. R. Math. Acad. Sci. Paris*, 346(21-22):1133–1136, 2008.

[106] M. Izuki. Wavelets and modular inequalities in variable L^p spaces. *Georgian Math. J.*, 15(2):281–293, 2008.

[107] S. Janson. Mean oscillation and commutators of singular integral operators. *Ark. Mat.*, 16(2):263–270, 1978.

[108] B. Jawerth. Weighted inequalities for maximal operators: linearization, localization and factorization. *Amer. J. Math.*, 108(2):361–414, 1986.

[109] R. Johnson and C. J. Neugebauer. Change of variable results for A_p- and reverse Hölder RH_r-classes. *Trans. Amer. Math. Soc.*, 328(2):639–666, 1991.

[110] W. B. Johnson and G. Schechtman. Sums of independent random variables in rearrangement invariant function spaces. *Ann. Probab.*, 17(2):789–808, 1989.

[111] P. W. Jones. Factorization of A_p weights. *Ann. of Math. (2)*, 111(3):511–530, 1980.

[112] L. Kahanpää and L. Mejlbro. Some new results on the Muckenhoupt conjecture concerning weighted norm inequalities connecting the Hilbert transform with the maximal function. In *Proceedings of the second Finnish-Polish summer school in complex analysis (Jyväskylä, 1983)*, volume 28 of *Bericht*, pages 53–72, Jyväskylä, 1984. Univ. Jyväskylä.

[113] K. S. Kazarian. On bases and unconditional bases in the spaces $L^p(\mu)$, $1 \le p < \infty$. *Studia Math.*, 71(3):227–249, 1981/82.

[114] R. A. Kerman and A. Torchinsky. Integral inequalities with weights for the Hardy maximal function. *Studia Math.*, 71(3):277–284, 1981/82.

[115] V. Kokilashvili and M. Krbec. *Weighted inequalities in Lorentz and Orlicz spaces.* World Scientific Publishing Co. Inc., River Edge, NJ, 1991.

[116] O. Kováčik and J. Rákosník. On spaces $L^{p(x)}$ and $W^{k,p(x)}$. *Czechoslovak Math. J.*, 41(116)(4):592–618, 1991.

[117] M. A. Krasnosel′skiĭ and Ja. B. Rutickiĭ. *Convex functions and Orlicz spaces.* Translated from the first Russian edition by Leo F. Boron. P. Noordhoff Ltd., Groningen, 1961.

[118] M. Krbec. Modular interpolation spaces. I. *Z. Anal. Anwendungen*, 1(1):25–40, 1982.

[119] D. S. Kurtz and R. L. Wheeden. Results on weighted norm inequalities for multipliers. *Trans. Amer. Math. Soc.*, 255:343–362, 1979.

[120] M. Lacey, K. Moen, C. Pérez, and R. Torres. Sharp weighted bounds for fractional integral operators. *J. Funct. Anal.*, 259(5):1073–1097, 2010.

[121] M. Lacey, S. Petermichl, and M. Reguera. Sharp A_2 inequality for Haar shift operators. *Math. Annalen*, 348(1):127–141, 2010.

[122] M. A. Leckband. Structure results on the maximal Hilbert transform and two-weight norm inequalities. *Indiana Univ. Math. J.*, 34(2):259–275, 1985.

[123] M. A. Leckband and C. J. Neugebauer. Weighted iterates and variants of the Hardy-Littlewood maximal operator. *Trans. Amer. Math. Soc.*, 279(1):51–61, 1983.

[124] P. G. Lemarié-Rieusset. Ondelettes et poids de Muckenhoupt. *Studia Math.*, 108(2):127–147, 1994.

[125] A. K. Lerner. Weighted norm inequalities for the local sharp maximal function. *J. Fourier Anal. Appl.*, 10(5):465–474, 2004.

[126] A. K. Lerner. Some remarks on the Hardy-Littlewood maximal function on variable L^p spaces. *Math. Z.*, 251(3):509–521, 2005.

[127] A. K. Lerner. An elementary approach to several results on the Hardy-Littlewood maximal operator. *Proc. Amer. Math. Soc.*, 136(8):2829–2833, 2008.

[128] A. K. Lerner and S. Ombrosi. A boundedness criterion for general maximal operators. *Publ. Mat.*, 54(1):53–71, 2010.

[129] A. K. Lerner, S. Ombrosi, and C. Pérez. A_1 bounds for Calderón-Zygmund operators related to a problem of Muckenhoupt and Wheeden. *Math. Res. Lett.*, 16(1):149–156, 2009.

[130] A. K. Lerner, S. Ombrosi, and C. Pérez. Weak type estimates for singular integrals related to a dual problem of Muckenhoupt-Wheeden. *J. Fourier Anal. Appl.*, 15(3):394–403, 2009.

[131] A. K. Lerner and C. Pérez. A new characterization of the Muckenhoupt A_p weights through an extension of the Lorentz-Shimogaki theorem. *Indiana Univ. Math. J.*, 56(6):2697–2722, 2007.

[132] J. Lindenstrauss and L. Tzafriri. *Classical Banach spaces. II*, volume 97 of *Ergebnisse der Mathematik und ihrer Grenzgebiete [Results in Mathematics and Related Areas]*. Springer-Verlag, Berlin, 1979. Function spaces.

[133] R. L. Long and F. S. Nie. Weighted Sobolev inequality and eigenvalue estimates of Schrödinger operators. In *Harmonic analysis (Tianjin, 1988)*, volume 1494 of *Lecture Notes in Math.*, pages 131–141. Springer, Berlin, 1991.

[134] R. A. Macías and M. S. Riveros. One-sided extrapolation at infinity and singular integrals. *Proc. Roy. Soc. Edinburgh Sect. A*, 130(5):1081–1102, 2000.

[135] L. Maligranda. *Orlicz spaces and interpolation*, volume 5 of *Seminars in Mathematics*. IMECC, Universidad Estadual de Campinas, Campinas, Brazil, 1989.

[136] J. Marcinkiewicz and A. Zygmund. Quelques inégalités pour les opérations linéaires. *Fundam. Math.*, 32:115–121, 1939.

[137] J. M. Martell. Fractional integrals, potential operators and two-weight, weak type norm inequalities on spaces of homogeneous type. *J. Math. Anal. Appl.*, 294(1):223–236, 2004.

[138] J. M. Martell and J. Parcet. Calderón-Zygmund decomposition for arbitrary measures on $\mathbb{R}^n$ and applications. *Preprint*, 2007.

[139] F. J. Martín-Reyes. New proofs of weighted inequalities for the one-sided Hardy-Littlewood maximal functions. *Proc. Amer. Math. Soc.*, 117(3):691–698, 1993.

[140] F. J. Martín-Reyes and A. de la Torre. One-sided BMO spaces. *J. London Math. Soc. (2)*, 49(3):529–542, 1994.

[141] F. J. Martín-Reyes, P. Ortega Salvador, and A. de la Torre. Weighted inequalities for one-sided maximal functions. *Trans. Amer. Math. Soc.*, 319(2):517–534, 1990.

[142] F. J. Martín-Reyes, L. Pick, and A. de la Torre. A_∞^+ condition. *Canad. J. Math.*, 45(6):1231–1244, 1993.

[143] J. Mateu, P. Mattila, A. Nicolau, and J. Orobitg. BMO for nondoubling measures. *Duke Math. J.*, 102(3):533–565, 2000.

[144] V. G. Maz'ja. *Sobolev spaces.* Springer Series in Soviet Mathematics. Springer-Verlag, Berlin, 1985. Translated from the Russian by T. O. Shaposhnikova.

[145] Y. Meyer. *Wavelets and operators*, volume 37 of *Cambridge Studies in Advanced Mathematics*. Cambridge University Press, Cambridge, 1992. Translated from the 1990 French original by D. H. Salinger.

[146] T. Miyamoto. On some interpolation theorems of quasi-linear operators. *Math. Japon.*, 42(3):545–556, 1995.

[147] B. Muckenhoupt. Private communication.

[148] B. Muckenhoupt. Weighted norm inequalities for the Hardy maximal function. *Trans. Amer. Math. Soc.*, 165:207–226, 1972.

[149] B. Muckenhoupt. The equivalence of two conditions for weight functions. *Studia Math.*, 49:101–106, 1973/74.

[150] B. Muckenhoupt. Weighted norm inequalities for classical operators. In *Harmonic analysis in Euclidean spaces (Proc. Sympos. Pure Math., Williams Coll., Williamstown, Mass., 1978), Part 1*, Proc. Sympos. Pure Math., XXXV, pages 69–83. Amer. Math. Soc., Providence, R.I., 1979.

[151] B. Muckenhoupt. Norm inequalities relating the Hilbert transform to the Hardy-Littlewood maximal function. In *Functional analysis and approximation (Oberwolfach, 1980)*, volume 60 of *Internat. Ser. Numer. Math.*, pages 219–231. Birkhäuser, Basel, 1981.

[152] B. Muckenhoupt. Problems. In J. Marshall Ash and R. L. Jones, editors, *Harmonic Analysis: Calderón-Zygmund and Beyond*, volume 411 of *Contemp. Math.*, pages 131–135. Amer. Math. Soc., Providence, RI, 2006.

[153] B. Muckenhoupt and R. L. Wheeden. Norm inequalities for the Littlewood-Paley function g_λ^*. *Trans. Amer. Math. Soc.*, 191:95–111, 1974.

[154] B. Muckenhoupt and R. L. Wheeden. Weighted norm inequalities for fractional integrals. *Trans. Amer. Math. Soc.*, 192:261–274, 1974.

[155] B. Muckenhoupt and R. L. Wheeden. Two weight function norm inequalities for the Hardy-Littlewood maximal function and the Hilbert transform. *Studia Math.*, 55(3):279–294, 1976.

[156] J. Musielak. *Orlicz spaces and modular spaces*, volume 1034 of *Lecture Notes in Mathematics*. Springer-Verlag, Berlin, 1983.

[157] F. Nazarov, S. Treil, and A. Volberg. Weak type estimates and Cotlar inequalities for Calderón-Zygmund operators on nonhomogeneous spaces. *Internat. Math. Res. Notices*, (9):463–487, 1998.

[158] F. Nazarov, S. Treil, and A. Volberg. The Bellman functions and two-weight inequalities for Haar multipliers. *J. Amer. Math. Soc.*, 12(4):909–928, 1999.

[159] F. Nazarov, S. Treil, and A. Volberg. Accretive system Tb-theorems on nonhomogeneous spaces. *Duke Math. J.*, 113(2):259–312, 2002.

[160] F. Nazarov, S. Treil, and A. Volberg. The Tb-theorem on non-homogeneous spaces. *Acta Math.*, 190(2):151–239, 2003.

[161] F. Nazarov, S. Treil, and A. Volberg. Two weight inequalities for individual Haar multipliers and other well localized operators. *Math. Res. Lett.*, 15(3):583–597, 2008.

[162] A. Nekvinda. Hardy-Littlewood maximal operator on $L^{p(x)}(\mathbb{R})$. *Math. Inequal. Appl.*, 7(2):255–265, 2004.

[163] C. J. Neugebauer. Inserting A_p-weights. *Proc. Amer. Math. Soc.*, 87(4):644–648, 1983.

[164] C. J. Neugebauer. A double weight extrapolation theorem. *Proc. Amer. Math. Soc.*, 93(3):451–455, 1985.

[165] R. O'Neil. Fractional integration in Orlicz spaces. I. *Trans. Amer. Math. Soc.*, 115:300–328, 1965.

[166] J. Orobitg and C. Pérez. A_p weights for nondoubling measures in $\mathbf{R}^n$ and applications. *Trans. Amer. Math. Soc.*, 354(5):2013–2033 (electronic), 2002.

[167] R.E.A.C. Paley. A remarkable series of orthogonal functions. I. *Proc. Lond. Math. Soc., II. Ser.*, 34:241–264, 1932.

[168] A. Passarelli Di Napoli. A local version of Rubio de Francia's extrapolation theorem. *Ricerche Mat.*, 44(2):291–301, 1995.

[169] M. C. Pereyra. Lecture notes on dyadic harmonic analysis. In *Second Summer School in Analysis and Mathematical Physics (Cuernavaca, 2000)*, volume 289 of *Contemp. Math.*, pages 1–60. Amer. Math. Soc., Providence, RI, 2001.

[170] C. Pérez. Weighted norm inequalities for general maximal operators. *Publ. Mat.*, 35(1):169–186, 1991.

[171] C. Pérez. Two weighted inequalities for potential and fractional type maximal operators. *Indiana Univ. Math. J.*, 43(2):663–683, 1994.

[172] C. Pérez. Weighted norm inequalities for singular integral operators. *J. London Math. Soc. (2)*, 49(2):296–308, 1994.

[173] C. Pérez. Endpoint estimates for commutators of singular integral operators. *J. Funct. Anal.*, 128(1):163–185, 1995.

[174] C. Pérez. On sufficient conditions for the boundedness of the Hardy-Littlewood maximal operator between weighted L^p-spaces with different weights. *Proc. London Math. Soc. (3)*, 71(1):135–157, 1995.

[175] C. Pérez. Sharp estimates for commutators of singular integrals via iterations of the Hardy-Littlewood maximal function. *J. Fourier Anal. Appl.*, 3(6):743–756, 1997.

[176] C. Pérez. Sharp weighted inequalities for the vector-valued maximal function. *Trans. Amer. Math. Soc.*, 352(7):3265–3288, 2000.

[177] C. Pérez. Calderón–Zygmund theory related to Poincaré-Sobolev inequalities, fractional integrals and singular integral operators. Lecture Notes, Neuquén, Unión Matemática de Argentina, 2004. Available at `http://grupo.us.es/anaresba/miembros/papers-carlos.htm`.

[178] C. Pérez and R. Trujillo-González. Sharp weighted estimates for vector-valued singular integral operators and commutators. *Tohoku Math. J. (2)*, 55(1):109–129, 2003.

[179] Carlos Pérez. *A Course on Singular Integrals and Weights.* Advanced Courses in Mathematics. CRM Barcelona. Birkhäuser Verlag, Basel, 2011.

[180] S. Petermichl. Dyadic shifts and a logarithmic estimate for Hankel operators with matrix symbol. *C. R. Acad. Sci. Paris Sér. I Math.*, 330(6):455–460, 2000.

[181] S. Petermichl. The sharp bound for the Hilbert transform on weighted Lebesgue spaces in terms of the classical A_p characteristic. *Amer. J. Math.*, 129(5):1355–1375, 2007.

[182] S. Petermichl. The sharp weighted bound for the Riesz transforms. *Proc. Amer. Math. Soc.*, 136(4):1237–1249, 2008.

[183] S. Petermichl, S. Treil, and A. Volberg. Why the Riesz transforms are averages of the dyadic shifts? In *Proceedings of the 6th International Conference on Harmonic Analysis and Partial Differential Equations (El Escorial, 2000)*, number Vol. Extra:209–228, 2002.

[184] S. Petermichl and A. Volberg. Heating of the Ahlfors-Beurling operator: weakly quasiregular maps on the plane are quasiregular. *Duke Math. J.*, 112(2):281–305, 2002.

[185] L. Pick and M. Růžička. An example of a space $L^{p(x)}$ on which the Hardy-Littlewood maximal operator is not bounded. *Expo. Math.*, 19(4):369–371, 2001.

[186] Y. Rakotondratsimba. Two-weight norm inequality for Calderón-Zygmund operators. *Acta Math. Hungar.*, 80(1-2):39–54, 1998.

[187] Y. Rakotondratsimba. Weighted norm inequalities for the Hardy-Littlewood maximal operator on radial and nonincreasing functions. *Rend. Mat. Appl. (7)*, 18(3):487–496, 1998.

[188] Y. Rakotondratsimba. Two-weight inequality for commutators of singular integral operators. *Kobe J. Math.*, 16(1):1–20, 1999.

[189] M. M. Rao and Z. D. Ren. *Theory of Orlicz spaces*, volume 146 of *Monographs and Textbooks in Pure and Applied Mathematics.* Marcel Dekker Inc., New York, 1991.

[190] F. Ricci and E. M. Stein. Multiparameter singular integrals and maximal functions. *Ann. Inst. Fourier (Grenoble)*, 42(3):637–670, 1992.

[191] M. Rosenblum. Summability of Fourier series in $L^p(d\mu)$. *Trans. Amer. Math. Soc.*, 105:32–42, 1962.

[192] H. L. Royden. *Real analysis.* Macmillan Publishing Company, New York, third edition, 1988.

[193] J. L. Rubio de Francia. Factorization and extrapolation of weights. *Bull. Amer. Math. Soc. (N.S.)*, 7(2):393–395, 1982.

[194] J. L. Rubio de Francia. A new technique in the theory of A_p weights. In *Topics in modern harmonic analysis, Vol. I, II (Turin/Milan, 1982)*, pages 571–579. Ist. Naz. Alta Mat. Francesco Severi, Rome, 1983.

[195] J. L. Rubio de Francia. Factorization theory and A_p weights. *Amer. J. Math.*, 106(3):533–547, 1984.

[196] J. L. Rubio de Francia, F. J. Ruiz, and J. L. Torrea. Calderón-Zygmund theory for operator-valued kernels. *Adv. in Math.*, 62(1):7–48, 1986.

[197] W. Rudin. *Real and complex analysis.* McGraw-Hill Book Co., New York, third edition, 1987.

[198] F. J. Ruiz and J. L. Torrea. Factorization and extrapolation of pairs of weights in two different measure spaces. *Math. Nachr.*, 150:175–183, 1991.

[199] S. Samko. On a progress in the theory of Lebesgue spaces with variable exponent: maximal and singular operators. *Integral Transforms Spec. Funct.*, 16(5-6):461–482, 2005.

[200] E. T. Sawyer. Weighted norm inequalities for fractional maximal operators. In *1980 Seminar on Harmonic Analysis (Montreal, Que., 1980)*, volume 1 of *CMS Conf. Proc.*, pages 283–309. Amer. Math. Soc., Providence, R.I., 1981.

[201] E. T. Sawyer. A characterization of a two-weight norm inequality for maximal operators. *Studia Math.*, 75(1):1–11, 1982.

[202] E. T. Sawyer. Two weight norm inequalities for certain maximal and integral operators. In *Harmonic analysis (Minneapolis, Minn., 1981)*, volume 908 of *Lecture Notes in Math.*, pages 102–127. Springer, Berlin, 1982.

[203] E. T. Sawyer. Norm inequalities relating singular integrals and the maximal function. *Studia Math.*, 75(3):253–263, 1983.

[204] E. T. Sawyer. A two weight weak type inequality for fractional integrals. *Trans. Amer. Math. Soc.*, 281(1):339–345, 1984.

[205] E. T. Sawyer. A weighted weak type inequality for the maximal function. *Proc. Amer. Math. Soc.*, 93(4):610–614, 1985.

[206] E. T. Sawyer. Weighted inequalities for the one-sided Hardy-Littlewood maximal functions. *Trans. Amer. Math. Soc.*, 297(1):53–61, 1986.

[207] E. T. Sawyer. A characterization of two weight norm inequalities for fractional and Poisson integrals. *Trans. Amer. Math. Soc.*, 308(2):533–545, 1988.

[208] E. T. Sawyer and R. L. Wheeden. Weighted inequalities for fractional integrals on Euclidean and homogeneous spaces. *Amer. J. Math.*, 114(4):813–874, 1992.

[209] C. Segovia and J. L. Torrea. Extrapolation for pairs of related weights. In *Analysis and partial differential equations*, volume 122 of *Lecture Notes in Pure and Appl. Math.*, pages 331–345. Dekker, New York, 1990.

[210] C. Segovia and J. L. Torrea. Weighted inequalities for commutators of fractional and singular integrals. *Publ. Mat.*, 35(1):209–235, 1991. Conference on Mathematical Analysis (El Escorial, 1989).

[211] P. M. Soardi. Wavelet bases in rearrangement invariant function spaces. *Proc. Amer. Math. Soc.*, 125(12):3669–3673, 1997.

[212] F. Soria and G. Weiss. A remark on singular integrals and power weights. *Indiana Univ. Math. J.*, 43(1):187–204, 1994.

[213] E. M. Stein. *On certain operators on L_p spaces.* PhD thesis, University of Chicago, Chicago IL, 1955.

[214] E. M. Stein. Note on singular integrals. *Proc. Amer. Math. Soc.*, 8:250–254, 1957.

[215] E. M. Stein. Note on the class $L \log L$. *Studia Math.*, 32:305–310, 1969.

[216] E. M. Stein. *Harmonic analysis: real-variable methods, orthogonality, and oscillatory integrals*, volume 43 of *Princeton Mathematical Series.* Princeton University Press, Princeton, NJ, 1993. With the assistance of Timothy S. Murphy, Monographs in Harmonic Analysis, III.

[217] E. M. Stein and G. Weiss. Interpolation of operators with change of measures. *Trans. Amer. Math. Soc.*, 87:159–172, 1958.

[218] E. M. Stein and G. Weiss. *Introduction to Fourier analysis on Euclidean spaces.* Princeton University Press, Princeton, N.J., 1971. Princeton Mathematical Series, No. 32.

[219] J. O. Strömberg and A. Torchinsky. Weights, sharp maximal functions and Hardy spaces. *Bull. Amer. Math. Soc. (N.S.)*, 3(3):1053–1056, 1980.

[220] X. Tolsa. BMO, H^1, and Calderón-Zygmund operators for non doubling measures. *Math. Ann.*, 319(1):89–149, 2001.

[221] X. Tolsa. A proof of the weak $(1, 1)$ inequality for singular integrals with non doubling measures based on a Calderón-Zygmund decomposition. *Publ. Mat.*, 45(1):163–174, 2001.

[222] A. Torchinsky. *Real-variable methods in harmonic analysis*, volume 123 of *Pure and Applied Mathematics.* Academic Press Inc., Orlando, FL, 1986.

[223] J. L. Torrea, J. García-Cuerva, J. Duoandikoetxea, and A. Carbery. The work of José Luis Rubio de Francia. I, II, III, IV. *Publ. Mat.*, 35(1):9–25, 27–63, 65–80, 81–93, 1991. Conference on Mathematical Analysis (El Escorial, 1989).

[224] S. Treil, A. Volberg, and D. Zheng. Hilbert transform, Toeplitz operators and Hankel operators, and invariant A_∞ weights. *Rev. Mat. Iberoamericana*, 13(2):319–360, 1997.

[225] B. O. Turesson. *Nonlinear potential theory and weighted Sobolev spaces*, volume 1736 of *Lecture Notes in Mathematics*. Springer-Verlag, Berlin, 2000.

[226] A. Uchiyama. L^p weighted inequalities for the dyadic square function. *Studia Math.*, 115(2):135–149, 1995.

[227] A. Volberg. *Calderón-Zygmund capacities and operators on nonhomogeneous spaces*, volume 100 of *CBMS Regional Conference Series in Mathematics*. Published for the Conference Board of the Mathematical Sciences, Washington, DC, 2003.

[228] D. K. Watson. Weighted estimates for singular integrals via Fourier transform estimates. *Duke Math. J.*, 60(2):389–399, 1990.

[229] D. K. Watson. Vector-valued inequalities, factorization, and extrapolation for a family of rough operators. *J. Funct. Anal.*, 121(2):389–415, 1994.

[230] J. M. Wilson. A sharp inequality for the square function. *Duke Math. J.*, 55(4):879–887, 1987.

[231] J. M. Wilson. Weighted inequalities for the dyadic square function without dyadic A_∞. *Duke Math. J.*, 55(1):19–50, 1987.

[232] J. M. Wilson. L^p weighted norm inequalities for the square function, $0 < p < 2$. *Illinois J. Math.*, 33(3):361–366, 1989.

[233] J. M. Wilson. Weighted norm inequalities for the continuous square function. *Trans. Amer. Math. Soc.*, 314(2):661–692, 1989.

[234] J. M. Wilson. Chanillo-Wheeden inequalities for $0 < p \leq 1$. *J. London Math. Soc. (2)*, 41(2):283–294, 1990.

[235] J. M. Wilson. *Weighted Littlewood-Paley theory and exponential-square integrability*, volume 1924 of *Lecture Notes in Math.* Springer, Berlin, 2007.

[236] S. J. Wu. A wavelet characterization for weighted Hardy spaces. *Rev. Mat. Iberoamericana*, 8(3):329–349, 1992.

[237] K. Yabuta. Sharp maximal function and C_p condition. *Arch. Math. (Basel)*, 55(2):151–155, 1990.

Index of Symbols

Bases, extrapolation and measures

Weight classes

Operators

Modular spaces and Banach function spaces

Young functions and bump conditions

Author Index

Subject Index

MIX
Papier aus verantwortungsvollen Quellen
Paper from responsible sources
FSC® C105338

If you have any concerns about our products,
you can contact us on
ProductSafety@springernature.com

In case Publisher is established outside the EU,
the EU authorized representative is:
Springer Nature Customer Service Center GmbH
Europaplatz 3, 69115 Heidelberg, Germany

Printed by Libri Plureos GmbH
in Hamburg, Germany